Feature Review in Agricultural Soils—Intensification of Soil Health

Feature Review in Agricultural Soils—Intensification of Soil Health

Guest Editors

Ryusuke Hatano
Rosa Francaviglia

MDPI

Basel • Beijing • Wuhan • Barcelona • Belgrade • Novi Sad • Cluj • Manchester

Guest Editors
Ryusuke Hatano
Research Faculty of Agriculture
Hokkaido University
Sapporo
Japan

Rosa Francaviglia
Council for Agricultural Research and Economics
Research Centre for Agriculture and Environment
Rome
Italy

Editorial Office
MDPI AG
Grosspeteranlage 5
4052 Basel, Switzerland

This is a reprint of the Special Issue, published open access by the journal *Agriculture* (ISSN 2077-0472), freely accessible at: https://www.mdpi.com/journal/agriculture/special_issues/7446W86ZWX.

For citation purposes, cite each article independently as indicated on the article page online and as indicated below:

Lastname, A.A.; Lastname, B.B. Article Title. *Journal Name* **Year**, *Volume Number*, Page Range.

ISBN 978-3-7258-6179-8 (Hbk)
ISBN 978-3-7258-6180-4 (PDF)
https://doi.org/10.3390/books978-3-7258-6180-4

Contents

About the Editors

Ryusuke Hatano

Ryusuke Hatano is an emeritus professor of Hokkaido University. He has worked for various international projects related to nutrient cycling in terrestrial ecosystems including nutrient discharge to the aquasphere and greenhouse gas emissions to the atmosphere in East Siberia, South and West China, Japan and tropical peatlands in Indonesia and Malaysia. He is a past president of the Japanese Society of Soil Physics (2010–2012), a past president of the Japanese Society of Soil Science and Plant Nutrition (2019–2021), a past chair of Commission 4.3 of the International Union of Soil Sciences (2011–2018), and a past chair of Division 2 of the International Union of Soil Sciences (2018–2022). He has published more than 240 peer-reviewed papers.

Rosa Francaviglia

Rosa Francaviglia, with expertise in agronomy, is a senior researcher at the Council for Agricultural Research and Economics, Research Centre for Agriculture and Environment, in Rome, Italy, since 1981 (now retired). Her main research topics include the effects of climate change on agriculture, carbon sinks and agricultural soils, soil organic carbon simulation models, soil fertility, conservation agriculture, crop diversification, agro-environmental evaluations, soil quality indicators, and good agro-environmental conditions (GAEC) under the EU Common Agricultural Policy.

Preface

A healthy planet is impossible without healthy soil. Due to declining productivity caused by population growth and climate change, agricultural land reclamation is increasing, raising concerns about increased soil degradation. Improving agricultural soil health is a critical issue. This Special Issue "Feature Review in Agricultural Soils—Intensification of Soil Health" examines issues related to soil conservation and fertility management for enhancing agricultural soil health.

Soil conservation techniques are important for reducing soil degradation, increasing soil resilience, and promoting soil health. Published papers on this topic discussed the remediation of saline-alkaline soils, the sustainability of rotational shifting cultivation, and the development of soil contamination countermeasures. As for soil fertility management techniques, the following were discussed: effective utilization of available nutrients accumulated in the soil to reduce the use of chemical fertilizers; a fertility management system that evaluates soil nitrogen fertility; making organic phosphorus available by enhancing phosphatase activity; making inorganic phosphorus available using biochar; prevention and recovery of the deterioration of soil physicochemical properties due to cultivation; the effects of organic soil amendments; and improving soil carbon sequestration through organic matter management.

Ryusuke Hatano and Rosa Francaviglia
Guest Editors

agriculture

Editorial

Research Advances in Intensification of Soil Health in Agriculture

Rosa Francaviglia [1] and Ryusuke Hatano [2,*]

1 Council for Agricultural Research and Economics, Research Centre for Agriculture and Environment, 00184 Rome, Italy; r.francaviglia@gmail.com
2 Research Faculty of Agriculture, Hokkaido University, Sapporo 060-8589, Japan
* Correspondence: hatano@agr.hokudai.ac.jp

Soil supports the primary production of plants, which is the basis of the food chain, and nourishes all living things on land. Furthermore, soil provides ecosystem-specific services, such as biological production, atmospheric gas composition, water quality, and biodiversity. In other words, without healthy soil, the planet cannot be healthy. Soil health is defined as "the ability of the soil to sustain the productivity, diversity, and environmental services of terrestrial ecosystems" [1]. However, of the nine key indicators of planetary health that can sustain human society, seven—climate change (CO_2 emissions), alteration of biosphere integrity, (biodiversity loss), variation in freshwater (rivers flow and soil moisture content), land system change (forest cover), alteration of biogeochemical fluxes (nitrogen and phosphorus loadings), introduction of new pollutants (anthropogenic pollution such as microplastics and radioactive wastes), and ocean acidification—are estimated to have exceeded the "planetary boundary" of the safe operating zone, with only two remaining within it: aerosol loading and stratospheric ozone depletion [2].

Agriculture contributes significantly to CO_2 emissions through deforestation and the destruction of natural grasslands, and is the main cause of biodiversity loss and increased nitrogen and phosphorus loads. It is directly and indirectly related to the seven global issues. While 30% of the Earth's surface is used for agriculture and 10% is used for cropland, 23% of total land is degraded due to inappropriate soil management [3]. Furthermore, the conversion of natural ecosystems into cropland has recently increased, particularly in South America, Asia, and Africa [4–6]. One factor contributing to such expansion is the decline in crop yields due to decreased rainfall [6]. However, in South America, the biological productivity of cropland is lower than that of natural land, which has been pointed out as a cause of land degradation. On the other hand, in Asia and North America, biological productivity has increased on expanded cropland, but this has led to greater consumption of water resources [4].

There is a significant correlation between the proportion of agricultural land in a watershed and the nitrogen load discharged into rivers. Excessive application of nitrogen chemical fertilizers increases N_2O emissions. In particular, peatland drainage significantly increases CO_2 and N_2O emissions. In other words, the expansion of agricultural land increases the nutrient load in the hydrosphere and greenhouse gas emissions in the atmosphere [7].

Sustainable intensification is needed, meaning producing more food from the same agricultural area while reducing environmental impact. To achieve this, it is important to improve soil health on agricultural land. Soil health is essential for improving crop productivity, strengthening ecosystem resilience, and reducing environmental impact, as well as for achieving a balanced agricultural and natural ecosystem. The importance of conservation agriculture, which maximizes natural soil functions and restores ecosystem services, is emphasized [8]. In recent years, it has been recognized that the carbon sequestration potential of soils has sufficient scope to prevent global warming [9], so conservation agriculture approaches that sequester carbon in soils are gaining importance.

Therefore, this Special Issue, "Feature Review in Agricultural Soils—Intensification of Soil Health", examines issues related to soil conservation and fertility management to improve soil health in agricultural soils. The collection includes three articles on soil conservation (remediation of saline–alkaline soils [10], sustainability of rotational shifting cultivation [11], and development of technologies of countermeasures to soil contamination [12]), and five articles on fertility management (fertility management system for soil nitrogen fertility assessment [13]; basic research on soil phosphorus fertility assessment: availability of organic phosphorus by enhancing phosphatase activity [14]; availability of inorganic phosphorus by biochar [15]; effects of organic amendments: effects on soil physicochemical properties [16]; and effects on soil carbon sequestration and soil structure [17]).

Intensification of soil health for soil conservation and restration: Soil conservation and restoration are key techniques for reducing soil degradation, increasing soil resilience and promoting soil health.

(1) For saline–alkaline soils, soil improvement measures closely related to soil health have been developed and implemented, based on physical, chemical, and biological indicators. The specific measures vary depending on regional characteristics, such as salinity, alkalinity, soil ionic composition, pH, hydrological processes, groundwater levels, soil properties, and regional policies. The three major Chinese regions with saline–alkaline soils use different techniques depending on the region. However, with the maturation of technology and the dissemination of information, comprehensive measures for various saline–alkaline soils have been widely promoted [10].

(2) Fires associated with rotational shifting cultivation tend to reduce soil porosity, clay content, cohesion, and cation-exchange capacity; increase sand content, pH, available phosphorus, and organic nitrogen; and increase the diversity of some bacteria while reducing fungal communities. Soil erosion is also a major concern. Effective management strategies, including controlled burning, appropriate land zoning, and sustainable agricultural practices such as agroforestry, cover crops, and crop rotation, are essential to mitigate the negative impacts on soil health and microbial communities [11].

(3) The number of articles on soil health related to agricultural soil contámination increased from 27 in 2020 to 54 in 2024 (Web of Science search), confirming the growing interest in soil health. Furthermore, the article emphasizes the need to integrate the latest AI-based technologies into soil health research, such as machine learning, natural language processing, and computer visualization for monitoring soil contaminants. These tools can be used as preventive measures to minimize the adverse effects of contaminants on soil [12].

Improving soil health through soil fertility management: Soil fertility management techniques require the restoration of soil physicochemical and biological properties, which are depleted by cultivation, and appropriate fertilizer application. To reduce the use of chemical fertilizers, it is important to effectively utilize the available nutrients accumulated in the soil as well as ensure the appropriate use of organic resources.

(1) Improving soil health in agricultural fields requires a fertilization management system that can adapt to variations in soil fertility within a region and within a field. As part of the Japanese Hokkaido Clean Agriculture Project, the use of compost as a partial substitute for chemical fertilizer reduces the amount of chemical fertilizer applied. Remote sensing technology has also been integrated to address spatial variations in soil properties between fields with different levels of nitrogen fertility. Specifically, for topdressing of winter wheat, satellite imagery is used to estimate the nitrogen uptake in each field based

on the NDVI, while simultaneously measuring nitrogen fertility through soil analysis. This allows for optimal and variable nitrogen application levels within the field and enables the development of waste-efficient nitrogen application techniques using variable-rate fertilizer broadcasters [13].

(2) In soil, the main source of phosphorus is found in organic matter. Therefore, it is important to enhance the activity of acid and alkaline phosphatase (APase), which is essential for the release of phosphorus from organic matter through hydrolysis. Previous research has demonstrated a strong correlation between APase and soil pH, with positive effects of clay content, organic matter, microbial biomass carbon, and nitrogen. Proper soil management practices, such as balanced use of organic fertilizer, optimal soil moisture levels, reduced tillage, crop rotation, and the use of beneficial plant microorganisms, help increase both APase activity and soil pH. Further research on the contribution of APase activity to crop productivity is needed to apply phosphorus to fertilizer management [14].

(3) Inorganic phosphorus is also an important component of the soil phosphorus pool. The effects of biochar application on the availability and mobility of inorganic phosphorus depend on the biochar properties (raw material, pyrolysis temperature and time, C:N ratio, pH, ash content, and phosphorus content) and soil properties (pH, soil texture, and phosphorus content). Biochar application significantly increased various inorganic phosphorus fractions and soil available phosphorus. Except for biochar derived from wood residues with a high C/N ratio (>200), biochar significantly increased available phosphorus (water-extractable soil phosphorus, Olsen phosphorus and phosphorus bound to soil calcium compounds). Furthermore, application of biochar derived from crop residues significantly increased soil phosphorus associated with iron and aluminum oxides [15].

(4) The application of organic soil amendments (compost, vermicompost, biochar, olive pomace, etc.) plays a crucial role in improving soil physicochemical quality by increasing soil organic matter content, promoting aggregate formation, and improving soil structure in the short term. They have been shown to have positive effects on water retention capacity, pH level, nutrient availability, and carbon sequestration. However, some studies have not found any change in soil quality due to organic soil amendments. This suggests that the effects of organic soil amendments vary depending on the initial level of soil quality, the application rate of organic soil amendments, and the cropping system, and that strengthening soil management at the local level is important [16].

(5) Improving soil health not only improves crop yields and reduces environmental impacts but also achieves a positive carbon balance in the field (increasing soil carbon). Fertilization management has been examined for this purpose. In four managed grasslands across Japan, compost carbon application at a rate greater than 2.5 t C ha^{-1} y^{-1} was necessary to maintain standard yields and achieve a net positive carbon balance. Although compost application can reduce chemical fertilizer use, fertilization generates N_2O emissions, and an additional 1 t C ha^{-1} y^{-1} of compost carbon is estimated to be needed to offset the greenhouse effect of N_2O emissions. Without compost application, soil organic carbon was expected to decrease. In these fields, the ratio of soil organic carbon to clay content (SOC/Clay) dropped to less than 1/13 after 39 to 68 years, indicating significant soil structural degradation [17].

In conclusion, proper organic matter management is essential to improving agricultural soil health. Soils degraded by salinization, alkalinization, slash-and-burn farming, and heavy metal contamination are prone to soil erosion due to sparse vegetation and poor microbial activity. Protecting sparsely vegetated soils is a key challenge, and implementing vegetation restoration technologies is a priority. Since vegetation recovery in degraded soils takes a long time, it is crucial to identify the areas to be protected and introduce technologies to accelerate restoration.

Furthermore, improved agricultural fertility management is essential for sustainable intensification. A sufficient supply of organic carbon and an adequate nutrient supply are fundamental to improving soil health. Reducing the use of chemical fertilizers requires developing fertility management systems that respond to the variability in soil fertility across fields based on basic research on improving soil phosphorus and nitrogen fertility.

Author Contributions: R.F. and R.H. contributed equally to this article. All authors have read and agreed to the published version of the manuscript.

Acknowledgments: The authors would like to thank all manuscript contributors and peer reviewers of this Special Issue of *Agriculture*.

Conflicts of Interest: The authors declare no conflicts of interest.

References

1. FAO-ITPS. Towards a Definition of Soil Health. ITPS Soil Letters #1. 2020. Available online: https://openknowledge.fao.org/handle/20.500.14283/cb1110en (accessed on 17 October 2025).
2. EAT-Lancet Global Food Transformation Needed to Ease Pressure on the Planet and Save Millions of Lives. 2025. Available online: https://www.planetaryhealthcheck.org/wp-content/uploads/PlanetaryHealthCheck2025_ExecutiveSummary.pdf (accessed on 17 October 2025).
3. UNEP. Assessing Global Land Use: Balancing Consumption with Sustainable Supply. A Report of the Working Group on Land and Soils of the International Resource Panel. 2014. Available online: https://www.unep.org/resources/report/assessing-global-land-use-balancing-consumption-sustainable-supply-0 (accessed on 17 October 2025).
4. Liu, Z.; Liu, Y.; Wang, J. A global analysis of agricultural productivity and water resource consumption changes over cropland expansion regions. *Agric. Ecosyst. Environ.* **2021**, *321*, 107630. [CrossRef]
5. Tan, M.; Li, Y. Spatial and temporal variation of cropland at the global level from 1992 to 2015. *J. Resour. Ecol.* **2019**, *10*, 235–245. [CrossRef]
6. Zaveri, E.; Russ, J.; Damania, R. Rainfall anomalies are a significant driver of cropland expansion. *Proc. Natl. Acad. Sci. USA* **2020**, *117*, 10225–10233. [CrossRef] [PubMed]
7. Hatano, R. Study on the occurrence, impact, and mitigation of soil-derived environmental loads in agricultural ecosystems. *Soil Sci. Plant Nutr.* **2025**, *71*, 371–384. [CrossRef]
8. Lal, R. Soil health and carbon management. *Food Energy Secur.* **2016**, *5*, 212–222. [CrossRef]
9. Georgiou, K.; Ahlstrom, A.; Polley, H.W.; Jackson, R.B.; Vindu, O.; Abramoff, R.Z.; Feng, W.; Harden, J.W.; Pellegrini, A.F.A. Global stocks and capacity of mineral-associated soil organic carbon. *Nat. Commun.* **2022**, *13*, 3797. [CrossRef] [PubMed]
10. Zhu, W.; Gu, S.; Jiang, R.; Zhang, X.; Hatano, R. Saline–Alkali Soil Reclamation Contributes to Soil Health Improvement in China. *Agriculture* **2024**, *14*, 1210. [CrossRef]
11. Arunrat, N.; Kongsurakan, P.; Solomon, L.W.; Sereenonchai, S. Fire Impacts on Soil Properties and Implications for Sustainability in Rotational Shifting Cultivation: A Review. *Agriculture* **2024**, *14*, 1660. [CrossRef]
12. Rosca, C.-M.; Stancu, A. Emerging Trends in AI-Based Soil Contamination Monitoring and Prevention. *Agriculture* **2025**, *15*, 1280. [CrossRef]
13. Ishikura, K.; Fueki, N.; Niwa, K. Progress in "Clean Agriculture" for Nitrogen Management to Enhance the Soil Health of Arable Fields and Its Application by Remote Sensing in Hokkaido, Japan. *Agriculture* **2025**, *15*, 1192. [CrossRef]
14. Campdelacreu Rocabruna, P.; Domene, X.; Preece, C.; Peñuelas, J. Relationship among Soil Biophysicochemical Properties, Agricultural Practices and Climate Factors Influencing Soil Phosphatase Activity in Agricultural Land. *Agriculture* **2024**, *14*, 288. [CrossRef]
15. Lin, L.; Peng, Y.; Zhou, L.; Zhang, B.; Chen, Q.; Chen, H. Impacts of Biochar Application on Inorganic Phosphorus Fractions in Agricultural Soils. *Agriculture* **2025**, *15*, 103. [CrossRef]
16. Matisic, M.; Dugan, I.; Bogunovic, I. Challenges in Sustainable Agriculture—The Role of Organic Amendments. *Agriculture* **2024**, *14*, 643. [CrossRef]
17. Hatano, R.; Mukumbuta, I.; Shimizu, M. Soil Health Intensification through Strengthening Soil Structure Improves Soil Carbon Sequestration. *Agriculture* **2024**, *14*, 1290. [CrossRef]

agriculture

MDPI

Review

Emerging Trends in AI-Based Soil Contamination Monitoring and Prevention

Cosmina-Mihaela Rosca [1] and Adrian Stancu [2,*]

1 Department of Automatic Control, Computers, and Electronics, Faculty of Mechanical and Electrical Engineering, Petroleum-Gas University of Ploiesti, 39 Bucharest Avenue, 100680 Ploiesti, Romania; cosmina.rosca@upg-ploiesti.ro

2 Department of Business Administration, Faculty of Economic Sciences, Petroleum-Gas University of Ploiesti, 39 Bucharest Avenue, 100680 Ploiesti, Romania

* Correspondence: astancu@upg-ploiesti.ro

Abstract: Soil health directly impacts food security, so investigating contaminants is a topic of interest for the anticipatory study of the action–effect correlation. This paper conducts a systematic literature review through seven analyses, identifying researchers' interest in soil health using artificial intelligence tools. The first study examines the distribution of articles over the years to assess researchers' interest in soil health, and subsequently, the same analysis is conducted regarding artificial intelligence (AI) methods. Additionally, the productivity of authors, the distribution of articles by country, relevant publications, and the frequency of keywords are analyzed to identify areas of interest associated with soil health. Subsequently, the branches of AI and examples of applications that have already been investigated in the specialized literature are identified, allowing areas that are currently underexplored to be pinpointed. This paper also proposes a specialized analysis using an algorithm specifically developed by the author for this investigation, which evaluates the interdisciplinary potential of the articles analyzed in the literature. In this way, the authors of the present research will propose new research directions that include machine learning, natural language processing, computer visualization, and other artificial intelligence techniques for monitoring soil contaminants. They will also suggest using these tools as preventive measures to minimize the negative impact of contaminants on the soil. The direct consequence is the protection of soil health and its effects on human health.

Keywords: AI in agriculture; AI for soil; soil contamination; ML for soil; ML for soil monitoring; ML for soil prediction; ML for soil contaminants

1. Introduction

The agricultural sector, alongside the medical industry, is one of humanity's most important fields of activity. This is because the agricultural sector provides a vital component of human existence [1]. For these reasons, current technologies are being explored intensively to streamline crop production. One of the most intensively studied components in the literature is the machine learning (ML) component of the artificial intelligence (AI) field [2]. This field is actively integrated into many human activity sectors, and the agricultural industry has also been included in the list of AI priorities. The ML component contributes to the progress of the farm field by anticipating outcomes through the analysis of different scenarios. This behavior allows farmers to identify the best scenario to achieve the best harvest. One of the directions in which ML operates as a component of anticipating crop evolution is related to soil health, including fertility, erosion, and the parameters that influence agricultural productivity. Soil health is a topic intensely debated by researchers because it influences crop quality and consumer health.

This paper aims to be a review article, presenting a bibliometric and systematic analysis of the scientific literature regarding the use of AI technologies in the field of soil health, especially in the context of contamination and contamination prevention in agriculture. The purpose of this paper is to map the trends and research directions related to AI technologies used in soil health studies. In this way, gaps in the literature are identified and future directions are proposed. This paper contains thematic reviews and literature syntheses, as is typical of review-type works. Thus, concrete applications of AI in contamination monitoring, organic carbon prediction, etc., are analyzed. Additionally, this paper includes lists of references selected according to a WOS search protocol. The bibliometric nature is also supported by the inclusion of figures with visual representations of academic collaborations, co-keyword maps, annual publication distributions, etc. Additionally, the use of the VOSViewer 1.6.20 tool for mapping research trends demonstrates the bibliometric nature of the review.

The central objective of this study is to demonstrate the interdisciplinary nature of AI tools in the agricultural field by identifying how ML models contribute to the assessment, monitoring, prevention, and action regarding soil contamination. Soil health encompasses a multitude of factors, including erosion, the decline of organic carbon, the use of chemical fertilizers, and microbiological biodiversity. These are not the central factors of the present work. These factors constitute contextual elements for the specialized literature. This study focuses on the intersection between AI technologies and soil contamination, aiming to highlight that the field has not been sufficiently explored to date and that researchers should delve more into this dual approach of AI and soil health.

The authors emphasize, through the approaches in this work, the necessity of intensifying interdisciplinary research between AI and soil health. Although this paper analyzes all AI tools, the emphasis is on ML models, as they are primarily used in the assessment, monitoring, prevention, and intervention of soil contamination. Therefore, the main focus is on ML due to its predictive capabilities in agricultural and ecological fields. These explanations justify the central purpose of this study, which is to demonstrate the necessity of an interdisciplinary approach between ML and soil contamination.

This study will identify the factors influencing soil health and the ML elements that anticipate each type of scenario. This paper aims to identify future research directions using ML to find less-researched subdomains. This paper focuses particularly on soil pollution from plastic, which directly affects consumer health in the long term.

This research paper has the following contributions:

1. This paper identifies the major factors influencing soil health. This research identifies the soil contaminants and discusses their health implications for consumers.
2. This paper establishes correlations between the types of soil contamination that can be addressed proactively through the AI component. Additionally, this paper evaluates researchers' interest in exploring AI technologies for preventive assessment and proactive action for soil health.
3. This paper identifies the correlations between the problem typologies modeled through AI and the problem typologies with direct implications on soil health. This paper discusses the measures through which soil contamination level forecasts can generate proactive measures to reduce the effects of pollution.
4. This paper outlines the less explored elements in the literature regarding soil health treated in advance through AI.

By achieving these four significant contributions, this paper will indicate future research directions that can, in the long term, improve the techniques used in soil health monitoring to increase quality and productivity.

The primary objective of this study is to analyze soil contamination, considering preventive measures informed by AI technologies. Since the number of scientific papers directly addressing this topic is limited, it has been observed that the current state of research in which AI tools contribute to monitoring soil contaminants is insufficiently explored at this time. The bibliographic selection includes studies directly related to soil contamination and remediation, as well as works that address soil health, fertility, the impact on the food chain, and preventive measures. In this way, this work provides the reader with the necessary foundations for the further development of research in the preventive direction. Identifying current gaps highlights the potential of underutilized AI technologies concerning soil contamination. This study encourages future interdisciplinary research on AI tools in predicting, monitoring, preventing, and intervening in soil contamination.

2. Literature Review

Soil health refers to how the ecosystem associated with the soil is managed. The composition and functionality of the microbial communities in the soil influence the soil's health. The nutrient cycle, the decomposition of organic matter, and the overall fertility of the soil constitute the microbial diversity in soil ecosystems [3]. They directly influence healthy plant growth as well as soil productivity. Griffiths and Philippot [4] mention that microbial diversity leads to the increased resilience and resistance of soil systems to various disturbances [5]. In this way, healthy soil management is ensured. Specialized works mention the importance of agricultural practices in soil health. The research by Lense et al. [6] specifies crop rotation and intercropping practices to reduce soil erosion. In this way, microbial diversity and activity are stimulated.

On the other hand, Raliya and Cappellari et al. [7,8] mention the excessive use of chemical fertilizers, which tend to diminish microbial diversity. This leads to a decline in soil health in the long term. Organic fertilizers, such as urine, have demonstrated benefits for soil microbial health and promote a diverse microbial community to support nutrient cycling and soil fertility [9]. The soil's physicochemical properties influence microbial communities' structure and functionality. Imarhiagbe and Onwudiwe [10] present the variations in soil properties at different depths. These variations have directly impacted microbial diversity and activity through the interconnection between soil parameters and microbial health. Abu-Qaoud et al. [11] explore the integration of microorganisms into agricultural practices to improve soil quality by creating a favorable environment for microbial populations that enhance plant growth and yield under stress conditions. Urban soils are also explored in the specialized literature due to the unique challenges related to contamination and degradation that negatively affect the microbial community and soil health. The researches [12–14] study the efforts of urban greening through diverse vegetation on microbial health, with positive implications for ecological stability and human health. The incorporation of different types of vegetation in urban spaces to effectively restore soil microbiota promotes a healthy soil system in urban landscapes.

2.1. Soil Contaminants

Soil pollution is an extensively studied phenomenon in the specialized literature. It is interesting from the perspective of identifying sources, the direct impact on plants, the overall impact on agriculture, and human health. Factors that pollute the soil include heavy metals, pollution with polycyclic aromatic hydrocarbons (PAHs), pollution with phthalate esters (PAEs), pollution with microplastics (MPs) and macroplastics, and other less-studied factors.

Heavy metals are the most common cause of soil pollution. They affect ecosystems through accumulation and transfer in the food chain. The main identified toxic metals include the following:

- Cadmium (Cd), which appears in high concentrations in agricultural soils in China [15–17], Peru [18], and other regions;
- Lead (Pb), which is present in industrial and mining areas, such as Chile [19] and China [15];
- Arsenic (As), which is identified in contaminated soils in China [20,21] and Serbia [22];
- Chromium (Cr), Copper (Cu), Zinc (Zn), and Nickel (Ni), which represent metals associated with geogenic and farming chemical fertilizers [23], mining activities [16,19], and industrial processes [17];
- Mercury (Hg), which is present in mining areas [16,24] and intensive agriculture [25].

Health risks include carcinogenic and non-carcinogenic exposure. Children represent the most vulnerable group. The primary sources of contamination are anthropogenic activities, including intensive agriculture, mining, and traffic [26,27].

PAHs are present in agricultural and urban soils in Bangladesh [28] and China [29]. PAHs originate from pyrogenic sources (biomass burning, coal) and petrogenic sources (industrial activities). These persistent compounds accumulate in the soil and enter the food chain. The direct consequences are visible at the level of plant and human health. The carcinogenic risks are considered moderate, and higher concentrations appear in industrial and coastal areas.

PAE compounds, such as di(2-ethylhexyl) phthalate (DEHP) and di-n-butyl phthalate (DnBP), are common pollutants in agricultural soils in Xinjiang [30] and the coastal areas of South China [31]. These endocrine-disrupting compounds affect human health through bioaccumulation in plants and soil. Higher concentrations are attributed to rapid urbanization and intensive agricultural activities [32].

Microplastics affect agricultural areas through irrigation systems and organic manure. Sharmin et al. [33] highlights the presence of various types of MPs (HDPE, LDPE, PP, PET, PVC) and forms (fiber, film, fragment) in agricultural soils.

Mercury pollution is a specific topic analyzed regarding artisanal and small-scale gold mining (ASGM) operations in Sudan [34]. The waste from amalgamation leads to extreme pollution in mining areas and moderate levels in agricultural and residential areas. The inhalation of mercury vapors is the main route of exposure and poses risks for children [35].

Zoghlami et al. [36] use deinking sludge (DS) and combined deinking sludge (DSC) as a soil amendment and offers benefits such as improving soil fertility and waste management. However, these materials may contain litter contaminants and can reduce soil porosity, affecting soil health. The study by Moriarity et al. [37] evaluates the risks introduced by the use of pesticides to the health of Indigenous communities in Australia and Canada. Children are at increased risk of exposure to Pb and As. This paper mentions that pesticides affect the health of agricultural soils. Ma et al. [38] assess exposure to herbicides (HBCs) in black soil regions, with maximum detected values of 6288 ng/L. Acetochlor, fomesafen, bentazon, and atrazine (ATZ) exhibit high detection rates. Non-carcinogenic risks are below the acceptable limit (<1), but the carcinogenic risk associated with ATZ is 10^{-5} for children and 10^{-6} for adults. Monitoring organic contaminants in the soil is a measure to support soil health and human health [39].

Residual nanomaterials [40] and zeolite [41] are being investigated as solutions for remediating contaminated soil. Nanobiochar (nB) and nano-water treatment residues (nWTRs) reduce heavy metal concentrations and improve soil fertility. Zeolite immobilizes potential toxic elements (PTEs), reducing their availability for plants.

Microcystins (MCs) affect the health of the soil and fava bean plants by inhibiting plant growth and contaminating the food chain [42]. The excess of fertilizers with NO_3 increases the concentrations of nitrates and nitrites in soil, water, and food. Secondary compounds, such as nitrosamines, are toxic. Kundu et al. [43] analyze the sources of contamination that affect soil health. The soil contaminants, such as Pb, regulate their mobility and bioavailability. In urban agriculture, the association of Pb with Al oxide phases and phosphates suggests an increased immobility through natural "soil aging" processes. Entwistle et al. [44] show that urban gardens are safe even at Pb levels 10 times above the recommended limits. Furthermore, the issue of these contaminants in agricultural soils arises.

Pollution with hydrophobic organic compounds, such as petroleum hydrocarbons, affects soil, water, and air and has toxic and carcinogenic effects on biota. These compounds, resistant to degradation, persist in the environment, generating risks for human health and ecosystems and affecting soil health [45]. Bioremediation methods, including solubilizing agents and surfactants, offer ecological solutions for environmental decontamination that should be intensively investigated in the specialized literature.

Air, water, and soil pollution cause diseases and premature deaths. Heavy metals, pesticides, plastics, and over-fertilization affect the soil by reducing beneficial microorganisms and contaminating groundwater. Volatile organic compounds (VOCs) emit greenhouse gases and particles that exacerbate climate change and cause respiratory, cardiovascular, and cancer-related diseases [46].

Figure 1 illustrates the main soil contaminant types, including traditional and emerging pollutants. Heavy metals (Cd, Pb, As, Cr, Cu, Zn, Ni, and Hg) represent inorganic soil pollutants due to their high toxicity. PAHs, PAE, and VOCs represent organic contaminants impacting human health and the environment. Additionally, Figure 1 includes petroleum residues, pesticides, and herbicides, all common in agricultural and industrial areas.

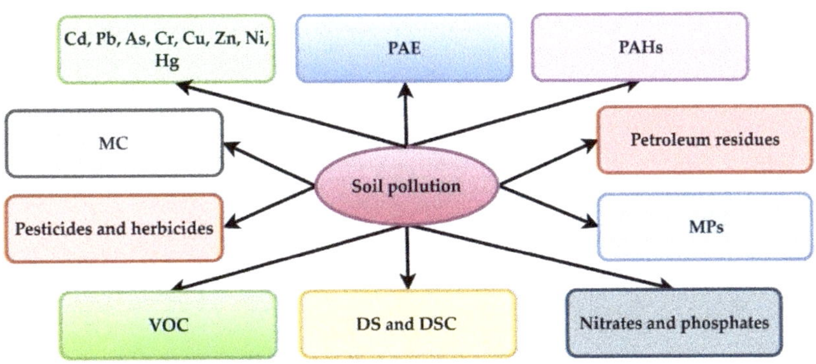

Figure 1. Main source of soil pollutants and emerging contaminants. Note: PAE—phthalate ester, PAH—polycyclic aromatic hydrocarbon, MC—Microcystins, MP—microplastic, VOC—Volatile organic compound, DS—deinking sludge, DSC—deinking sludge combined.

MPs, nitrates, phosphates from fertilizers, and pharmaceutical residues including DS and DSC highlight the emerging contaminants in Figure 1. MCs are a general factor that designates biologically active substances in very low concentrations. Overall, Figure 1 brings together various sources of soil pollution, highlighting the complexity of this phenomenon.

2.2. Soil Health and AI

AI is a discipline subdivided into six categories, each dedicated to different aspects of imitating human cognitive functions. These categories are represented by ML, natural language processing (NLP), computer vision (CV), robotics, expert systems (ESs), and evolutionary computation (EC). Figure 2 presents a synthesis of the main component elements in AI classification.

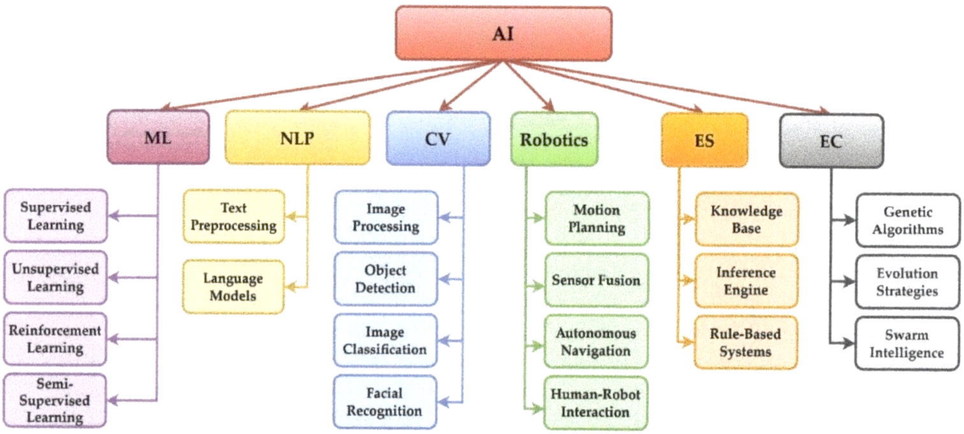

Figure 2. AI conceptual subfields.

ML is one of the most studied subfields of AI [47]. This involves a series of algorithms that allow machines to learn from large datasets and also enable them, through these algorithms, to develop performance over time without being explicitly programmed [48–50]. In the category of ML algorithms, there are supervised learning, unsupervised learning, reinforcement learning, and last but not least, semi-supervised learning [51,52].

NLP is an AI component that uses natural language to interact with computers and humans. It enables speech recognition, sentiment analysis, and machine translation [53]. This field also relies on ML techniques to interpret and generate human language.

In addition to ML and NLP, another subdomain of AI is CV. This refers to the ability of machines to interpret and make decisions based on visual data. This component overlaps with deep learning (DL) through convolutional neural networks (CNNs). These represent a specific type of DL model and are used in image classification, object detection, and facial recognition [54–56].

Robotics is a component of AI in which intelligent agents are designed to perform a series of physical tasks. Robots are used in sectors such as healthcare, manufacturing, and logistics, where they can assist in performing complex tasks [57].

Additionally, the field of AI includes more specialized areas, such as expert systems based on knowledge bases, inference engines, and rule-based systems [58]. There are also fields such as evolutionary computation, which include genetic algorithms, evolutionary strategies, and swarm intelligence [59–61].

The interdisciplinary approach between soil health and artificial intelligence techniques is expanding within agricultural sciences. The application of advanced computational techniques is still in its early stages when analyzing and simulating soil properties. ML evaluates soil properties and predicts future crops based on soil conditions. Imam et al. [62] demonstrate that ML models predict potato yields by analyzing the structures of microbial communities in the soil. The correlation between prediction and crop yields was demonstrated in this study. In addition, specialized ML algorithms such as the random forest (RF) algorithm have shown the ability to quantitatively evaluate the influence of certain soil indicators on agricultural productivity in the study by Su et al. [63]. Other algorithms, such as support vector machines (SVMs), have demonstrated soil fertility evaluation through ML. Thus, the paper by Shevchenko et al. [64] highlights complex relationships between soil characteristics and crop health. ML algorithms are integrated into real-time applications for predicting and monitoring soil fertility in the context of [65]. This predicts crops based on soil fertility and weather conditions. Rosca et al. [66] integrates the Internet of Things (IoT) and the RF algorithm to analyze environmental data in real time. The practical implications of ML techniques are demonstrated in the paper by Condran et al. [67], which monitors soil moisture and nutrient levels in irrigation and fertilization optimization strategies. Additionally, Hengl et al. [68] show the importance of RF algorithms in mapping soil properties with direct implications for agricultural practices. In this way, large volumes of data from soil-associated sensors, satellite images, and environmental variables are exploited, which help in making informed decisions based on real-time agricultural data. Moreover, the paper by Tripathi et al. [69] integrates DL methodologies in assessing soil health by exploring large volumes of data. In this way, researchers have discovered soil characteristics that contribute to plant health and productivity. The dynamic soil models presented by Yang et al. [70] ensure ecological and agricultural practices concerning soil health.

The use of NLP in soil health refers to agricultural productivity and ecosystem sustainability. Singh et al. [71] demonstrated that the unbalanced use of fertilizers directly affects soil health and, implicitly, human health. Textual data from farmers' reports are analyzed by integrating NLP technology into agriculture [72]. The article by Metwally et al. [73] ecologically evaluates, through specific techniques, how to discern patterns related to soil management practices that positively influence soil health. For example, NLP groups soil property data to identify site-specific management practices. In this way, Peter-Jerome et al. [74] align agronomic interventions with the varied state of the soil in different regions.

The chemical, physical, and biological properties of soil are studied in relation to agricultural productivity and ecosystem health in [75,76]. CV studies soil images to extract, through remote sensing techniques, values associated with moisture properties and nutrient levels over extensive areas. This approach enables decision-making in farm management [77,78]. Using CV techniques can also measure soil organic matter through visual and spectral analyses. The works by Zhou et al. and Jordán Vidal [79,80] mention the implications of CV in improving soil structure through the analysis of moisture retention and nutrient availability. CV algorithms trained on large datasets target soil characteristics through automated analyses on specific soil health indicators. Ghazal et al. [81] describe the application of fertilizers, irrigation needs, and crop rotation strategies as elements studied through CV algorithms. The analysis of these algorithms can also extend to the analysis of the soil microbiome. Venturini et al. and Babin et al. [82,83] mention the study of soil health through these algorithms. Visualizing soil microorganism structures and distributions through CV allows for understanding the relationships between soil microbial communities and their health indicators. The growth and health of plants at the soil level are closely correlated with the nutrient cycle and the decomposition of organic matter. Understanding the complexity of these subsystems of soil ecology through CV algorithms requires the analysis of chemical fertilizers that contribute to improving environmental quality [84].

The distribution and sources of heavy metals in the soils of the European Union (EU) are investigated in many studies, which illustrate the concerns of researchers in the field. Based on the LUCAS survey conducted by Ballabio et al. [85], it is shown that 5.5% of soil samples exceed the critical threshold of 1 mg Cd/kg, with an average level of 0.20 mg/kg, which is influenced by pH, texture, and the use of phosphate fertilizers. The paper by

Fendrich et al. [86] presents research on arsenic concentrations in different countries in the EU. This study identifies the RF-based model as the most accurate in mapping associated risks. In the paper by Ballabio et al. [87], the concentrations of Cu are noted with values of 49.3 mg/kg due to the use of Cu-based fungicides. Ballabio et al. [88] analyze Hg concentrations in the EU and observes correlations with emissions from coal-fired thermal power plants and mining areas.

Wang el al. [89] combine statistical methods and knowledge-based ML methods to identify the main factors of soil contamination with HMs and PAHs in agricultural areas of China. Additionally, for predicting the distribution of contaminants, the paper by Salgado et al. [90] reported an R^2 value ranging from 50.1% to 63.0% using the RF algorithm. These results encourage the potential of these technologies in mapping soil contamination.

Remote sensing is used in assessing the condition of oil-contaminated soils in [91,92]. They use images to investigate the increase in these values using a dataset of images from the center of the Tamsag–Bulag oil field in Mongolia [91]. These investigations show the use of the satellite monitoring of areas affected by extractive activities. Dean et al. [92] investigate the impact of oil refining and storage on soil. This is based on images analyzed using ML models. Monitoring soil quality in areas with limited data to manage soil pollution can be achieved using ML techniques, as demonstrated in [93,94].

3. Methodology

The methodology for evaluating the impact of research on soil health will include seven types of analysis, focusing on the papers published between 1 January 2020 and 30 April 2025. Documentation will be carried out through the Web of Science (WOS) platform. This research aims to highlight directions that have not yet been investigated or promise to significantly improve soil health but have not yet been extensively studied. Through this approach, experts in the field can identify the directions to investigate in future research.

The articles included in this analysis are cited in the reference section when they are examined in detail in the literature review or detailed analysis sections (Sections 4.6 and 4.7). Additionally, synthesis articles were included in the research to highlight the researchers' interest in the respective topic. The inclusion of review articles is justified by the fact that they provide an overview of the current state of research, the interest of researchers, research trends, existing gaps, as well as future directions of investigation that need to be addressed in subsequent contribution works.

The evaluation of researchers' interest in soil health regarding contaminants will be studied through seven analyses. Each analysis will also conduct a comparative analysis of researchers' interest in soil health regarding AI technologies. The seven analyses include the following:

1. The distribution of articles by year is investigated. This analysis aims to identify whether interest has been ascending or descending in soil health related to the contaminants and, further, in the combination of soil health investigated through specific AI methods.
2. Author productivity is investigated concerning soil health. This analysis is conducted using the VOSviewer 1.6.20 tool through the co-authorship map, for which the minimum number of documents per author will be set to 2 and the minimum number of citations per author to 1. This reflects the author's interest in exploring the topic of soil health.
3. The distribution of articles by country is investigated to determine if there is a correlation between countries with a high level of agriculture and the number of research

articles conducted by those countries. Subsequently, research in the field is investigated using AI tools to determine these countries' interest in the agricultural sector's technological advancement while adhering to soil health standards. This analysis highlights whether countries with a high level of agriculture have invested in soil health research.

4. Publications that include the most research in soil health are investigated to provide authors with an overview of publishers supporting efforts to improve soil health through AI technologies and other auxiliary technologies. Thus, all searches in WOS will be narrowed to the publishers that have published these studies.
5. The fifth analysis targets the frequency of keywords in research addressing soil health. This search identifies the fields that relate to this type of issue. The investigation is conducted through the co-occurrence map of keywords associated with each article identified in soil health. This analysis highlights the co-domains of analysis, meaning those related fields that address the issue of soil health. This study is conducted through the clusters created by the VOSviewer 1.6.20 tool.

Figure 3 presents the methodology related to this first set of five analyses. This figure shows that all these analyses are based on a WOS search that includes soil health elements concerning contaminants and the agricultural field for the five years analyzed. The use of AI technologies in addressing soil contamination is still in development at the time of writing this material, as AI technologies themselves are rapidly advancing. In the WOS searches, the authors of this material opted for an extended formulation to identify all relevant reference works in the field, even if they do not use the exact terms "preventive measures" or "contaminants". This approach avoids excluding important studies solely due to differences in terminology. This study is bibliometric in nature and maps the global research trends of the AI versus soil health relationship. In addition to the bibliometric approach, the authors contributed with a customized analysis of the current state of research in the field and added their personal expertise to the area of soil health.

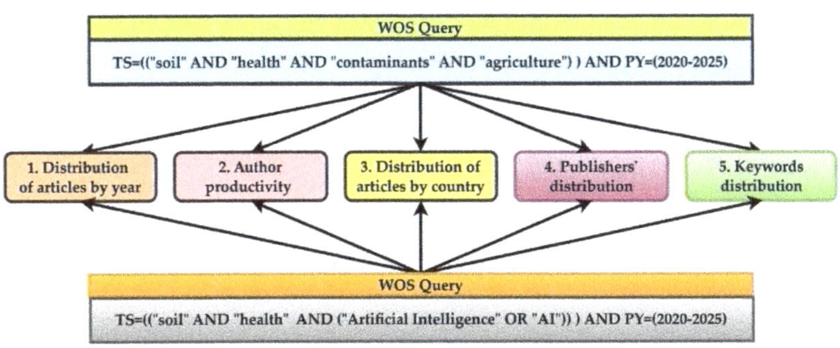

Figure 3. The methodology for the first set of five analyses.

6. Analysis number six presents the WOS branches of AI and examples of applications identified in the literature that address soil health issues related to each branch. It will highlight the branches that have not yet been explored.
7. The Cross AI Components Innovation Potential (CAI-CIP) analysis is a customized analysis that has not been conducted in any other review-type article. The CAI-CIP analysis identifies the interdisciplinary, innovative potential of the evaluated articles. To implement this analysis, the authors developed a program in C# that assesses the potential for multidisciplinary innovation by semantically analyzing articles between soil health and AI components. Thus, the key elements of the analysis and an interdisciplinary semantic similarity are identified. This similarity

is achieved using a predefined list of 200 AI techniques. This way, the relationships between transversal concepts are identified according to the logical scheme in Figure 4. Through this analysis, new research opportunities are discovered by highlighting unexplored connections and prioritizing those studies that impact soil quality.

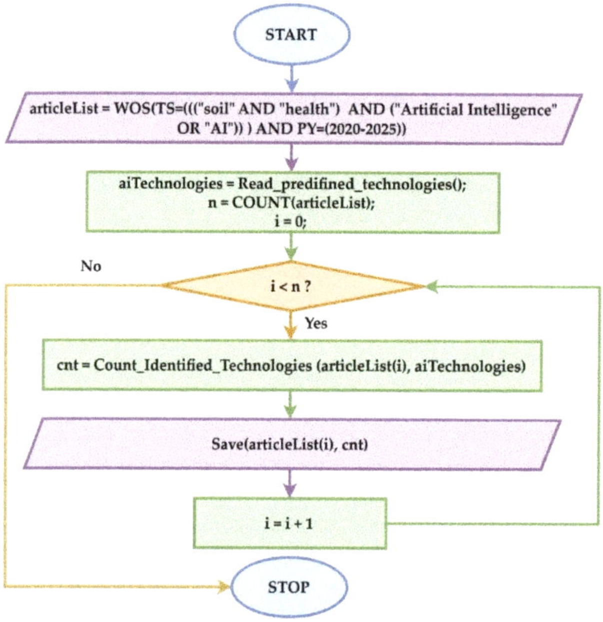

Figure 4. Logical flow of the CAI-CIP algorithm for analyzing soil health and contaminant research.

The algorithm presented in Figure 4 analyzes the abstracts provided by the general search in WOS. It identifies the predefined AI technologies as components of all existing technologies within these abstracts. Subsequently, the number of identified technologies at each abstract level is calculated, and the correlation between the analyzed paper and the number of identified technologies is saved. This algorithm allows for the analysis of AI trends concerning soil health. The algorithm identifies AI technologies in research related to soil health, contaminants, and agriculture. Thus, current research trends are explored through AI technologies, allowing for further innovations in the field. Additionally, the algorithm presented in Figure 4 analyzes the impact of AI technologies by the frequency of their mention at the abstract level. In this way, the obtained results will allow for the prioritization of research in specific directions, including AI technologies not mentioned in the papers identified in the WOS search.

The initial search strategy was based on general terms, such as artificial intelligence and AI. This initial approach acknowledged that this study's purpose was not to conduct a comprehensive analysis of every type of AI algorithm, but rather to identify general trends in the application of AI in the field of soil health and its contamination. Moreover, most scientific journals recommend using more general terms in the initial selection stage to avoid the excessive fragmentation of the literature. In this way, the entirety of reference works for the respective field is identified. The methodology initially aimed to highlight the global level of interest in the integration of AI in agriculture and soil monitoring, regardless of the specific type of technology used. Subsequently, the methodology included conducting a detailed search regarding the specific technologies presented in Figure 2. For this purpose, the WOS search expression is as follows:

TS=(

("soil health" OR "soil contamination")

AND

(

"artificial intelligence" OR "AI" OR "machine learning" OR "deep learning" OR "neural network" OR "convolutional neural network" OR "recurrent neural network" OR "support vector machine" OR "decision tree" OR "random forest" OR "gradient boosting" OR "natural language processing" OR "text preprocessing" OR "language models" OR

"computer vision" OR "image processing" OR "object detection" OR "image classification" OR "facial recognition" OR

"robotics" OR "motion planning" OR "sensor fusion" OR "autonomous navigation" OR "human-robot interaction" OR

"expert systems" OR "knowledge base" OR "inference engine" OR "rule-based systems" OR

"evolutionary computation" OR "genetic algorithms" OR "evolution strategies" OR "swarm intelligence"

)

) AND PY=(2020–2025)

This expression refers to AI technologies in studies related to health and soil contamination. The inclusion of all subfields and the most important algorithms identified in the literature highlights the breadth of existing research and, furthermore, illustrates the variety of AI applications in this interdisciplinary field.

To demonstrate the degree of fragmentation in the research, the search was reperformed to evaluate the impact of each AI component from Figure 2 on soil health through searches in WOS. These searches were structured to include in the keywords ML, NLP, computer vision, robots with AI, expert systems, and evolutionary computation. All searches were limited to the period 2020–2025.

4. Results

Each type of analysis proposed in the methodology section will generate results whose comments will highlight future research directions that specialists in the field should address in an interdisciplinary manner. This interdisciplinarity refers to integrating AI techniques in the preventive study of soil contamination. Alongside this idea related to soil contamination, researchers should expand their searches for preventive simulations in the actions they propose or undertake in the agricultural field concerning soil health.

4.1. Distribution of Articles by Year

The first search, "TS=(("soil" AND "health" AND "contaminants" AND "agriculture")) AND PY=(2020–2025)", generated a result of 217 articles on the WOS platform. The results of the soil health query related to agricultural contaminants highlighted in Figure 5 show the increased interest of researchers in this field. Thus, in 2020, there were only 27 studies, while in 2024, the total reached 54 studies, with the trend continuing to rise.

The annual distribution of research primarily aims to illustrate the dynamics of researchers' interest in soil health, both in the general context of contaminants and agriculture, as well as in the context of integrating AI technologies. Figure 5 compares the evolution of the number of published papers between 2020 and 2025 based on the two initially proposed distinctive search strategies. These analyses regarding the number of publications are standard for review articles and represent a way in which researchers are encouraged to explore certain technologies.

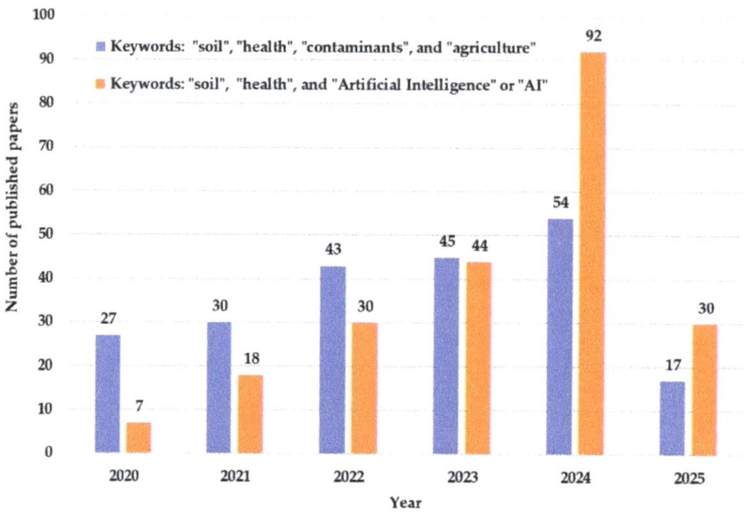

Figure 5. Number of published papers by searched keywords in WOS (2020–2025).

The second search, "TS=(("soil" AND "health" AND ("Artificial Intelligence" OR "AI"))) AND PY=(2020–2025)", generated 222 articles in WOS. The second search targeting soil health concerning AI techniques showed upward results regarding researchers' interest, increasing from seven articles in 2020 to ninety-two in 2024. The first quarter of 2025 has already recorded 30 articles in the field of artificial intelligence on soil health.

These results reflect the researchers' interest in soil health over the past 5 years, but the results are low considering the volume of work in the specialized literature. Moreover, the search that includes soil health concerning contaminants in the agricultural field and AI "TS=(("soil" AND "health" AND "contaminants" AND "agriculture") AND ("artificial intelligence" OR "AI")) AND PY=(2020–2025)" yielded only one article in WOS. This result signifies the need to intensify research in the agricultural field regarding soil health and contaminants through specific AI techniques.

4.2. Author Productivity and Co-Authorship Analysis

The analysis of author productivity started from the constraint of a minimum threshold of two articles per author and at least one citation per author, so the work would be considered a reference in soil health. The two initial constraints were established to limit the authors' scientific output and provide a clearer picture of the academic community's interest in soil health. The co-authorship network analysis in soil health studies is presented in Figure 6. The VOSViewer 1.6.20 analysis divided the WOS results into two clusters.

The first cluster contains seven elements, while the second contains five. The first cluster indicates the existence of a scientific community that is constantly active in soil health. The second cluster indicates an emerging area of research where collaborations are rarer or more recent. The ideas that emerge from this division into the two clusters highlight the need to broaden the scientific community involved in soil health studies. Attracting new specialists from various interdisciplinary fields, such as artificial intelligence, soil biology, agricultural engineering, and meteorological engineering, along with fostering international collaborations, could significantly expand research in soil health.

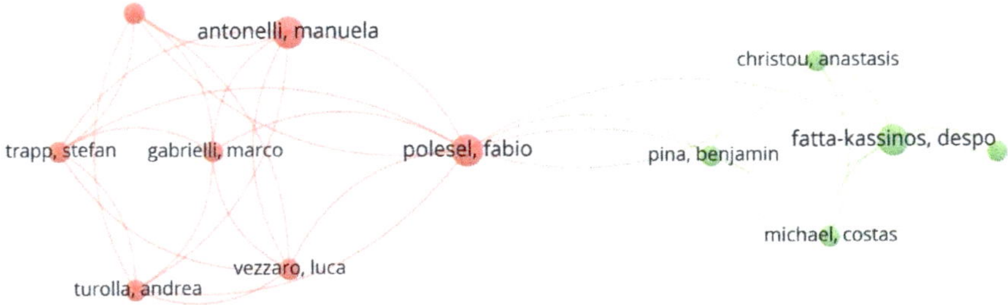

Figure 6. Author collaboration map in soil health research (2020–2025).

The second approach interprets results regarding the productivity of authors in soil health investigated through the lens of AI. The results provided a single cluster of six authors, as seen in Figure 7.

Figure 7. Author collaboration network in AI-integrated soil health research (2020–2025).

Compared to the first approach, this limited result indicates a minimal base of researchers who systematically publish in soil health within an AI approach. The fact that there are no more clusters indicates no broad collaboration networks centered around AI concerning soil health. Comparing these results with the approach that included terms such as contaminants and agriculture, it can be observed that the community is much more restricted, leading to the observation that the topic related to AI is much less explored. This finding demonstrates that the potential for developing interdisciplinary research in soil health using AI technologies is currently unexplored.

4.3. Geographical Distribution of Research Output

The analysis of the distribution of articles by country provided 24 items grouped into four clusters. The analysis investigates the existence of a correlation between the level of agricultural activities in a given country and the number of studies published by that country in the field of soil health. The results are presented in Figure 8, where the countries China, India, the USA, Pakistan, and Australia can be identified with a dominant number of studies in the field.

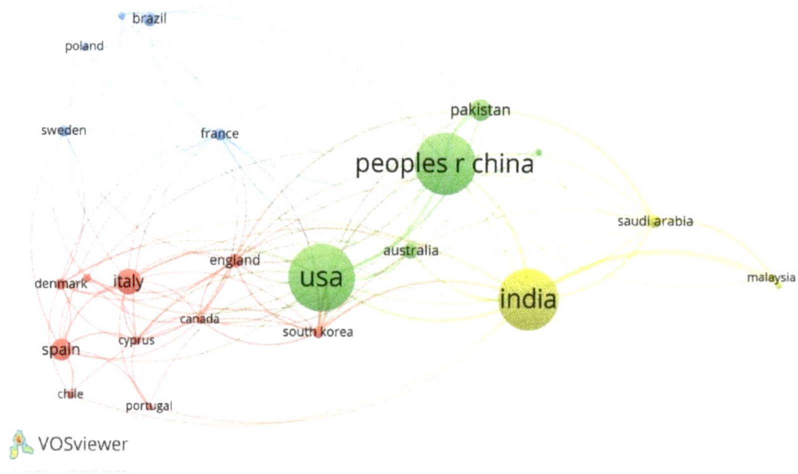

Figure 8. Geographical distribution of research on soil health and agriculture (2020–2025).

Countries with high agricultural production levels essentially correspond with the results identified in this map. For example, China (the world's largest agricultural country) and the USA (the third-largest agricultural country in the world) are in the same green cluster from Figure 8, with the highest number of research studies on soil health and agriculture. Another noteworthy aspect is that India (the second-largest agricultural country in the world) belongs to the second most important academic research cluster (yellow marked), along with Saudi Arabia and Malaysia. The third cluster, colored in red, includes countries such as Italy, Spain, England, South Korea, Denmark, Portugal, Chile, Cyprus, and Canada, whereas the fourth cluster (blue marked) comprises Brazil, France, Sweden, and Poland. Thus, the countries with high agricultural production are simultaneously investing in soil health research.

Additionally, the map shows many connections among countries, suggesting collaboration among authors in the global academic community. These results highlight the importance of international cooperation in addressing soil health issues. These observations serve as benchmarks for subsequent research that should explore the implications of modern technology in optimizing agricultural systems, especially at the soil health level.

India, China, and the USA are the only countries that have provided results regarding academic research related to the integration of AI techniques in soil health. These are the only results grouped into a single cluster regarding the second approach. These results reinforce the idea that other states should invest equally in soil health because this field is a fundamental source of human life.

4.4. Identification of Key Publishers

Regarding identifying the leading publishers disseminating research in soil health, it can be noted that in the first approach, where the theme is traditional, ecological, chemical, and agricultural, publishers dominate (e.g., Amer Chemical Soc). In the second approach, where the central element is AI, a series of differences emerge due to the transition to technical publishers, such as IEEE or Taylor & Francis. Figure 9 presents the top publishers in soil health research, while Figure 10 shows the prominent publishers in AI-augmented soil health investigations.

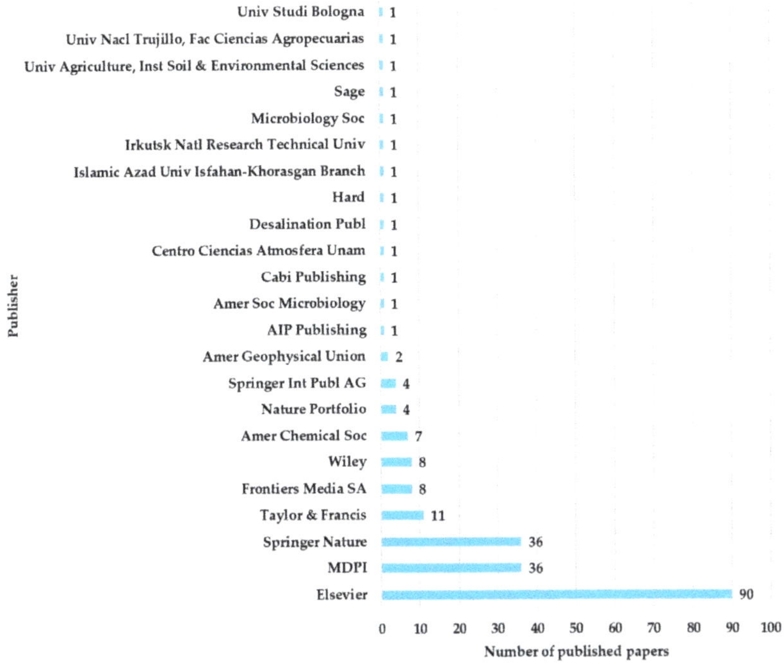

Figure 9. Publishers in soil health research focused on contaminants and agriculture (2020–2025).

Beyond these delimitations, the top publishers remain the same in both cases: Elsevier, MDPI, and Springer Nature. These results show a noticeable increased interest from technical publishers in a multidisciplinary approach, the dominance of top publishers who stand out through the many articles disseminating soil health issues, and opportunities for new publishers who should accept more and more of such interdisciplinary research.

4.5. Keyword Frequency and Thematic Mapping

This analysis focuses on the frequency of authors' keywords in research dedicated to soil health. This study is reported on the co-occurrence map of terms associated with each article. Figure 11 presents the results obtained from the WOS query, processed with the VosViewer 1.6.20 tool. In this figure, the 29 elements grouped into seven distinct clusters can be observed.

The central cluster is "agriculture" and it is marked in green. This includes approaches targeting microplastics, groundwater, environment, and degradation. Directly connected to this cluster is the blue cluster, with "heavy metals" as its central element. This is associated with the medical area, and in relation to this cluster is the purple one which targets the soil and its derivatives. Connected with the "soil" cluster are the red clusters associated with "soil health" and the yellow cluster, which has existing soil and water contamination as its central element. This interconnection of the clusters demonstrates the need for a simultaneous approach to agriculture in relation to soil health, contamination, water pollution, and heavy metals. The results presented in Figure 11 illustrate the diverse research themes related to soil health. These observations provide a global picture of the dominant themes in the current literature. The observations that emerge from this analysis focus on prioritizing interdisciplinary research, especially that which includes elements of AI.

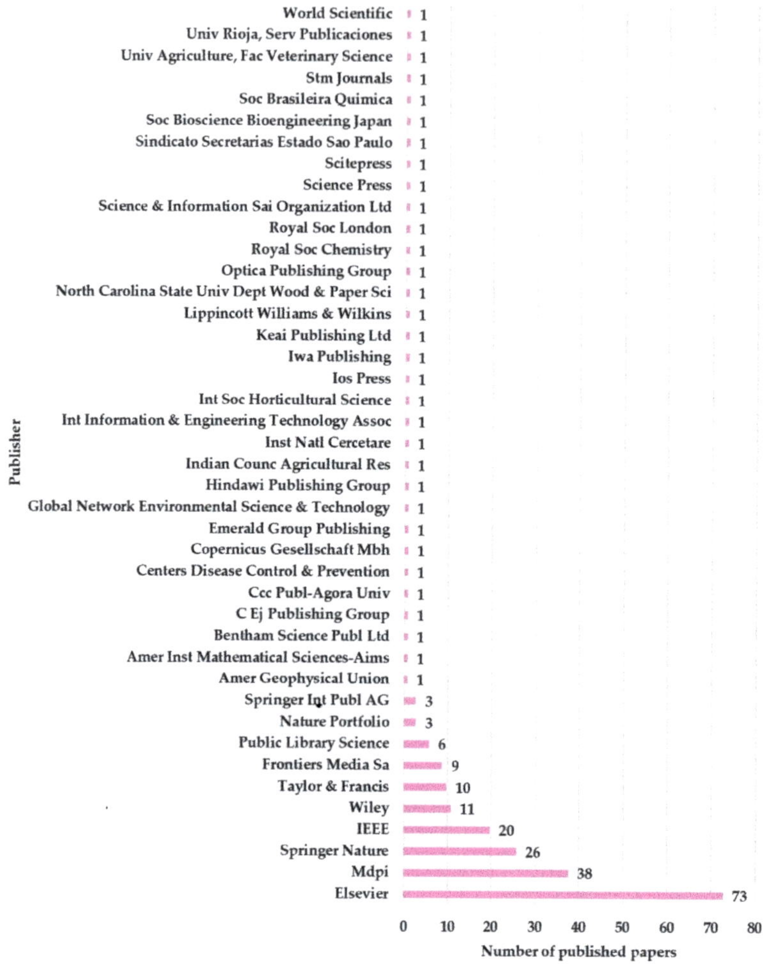

Figure 10. Publishers in AI-integrated soil health research (2020–2025).

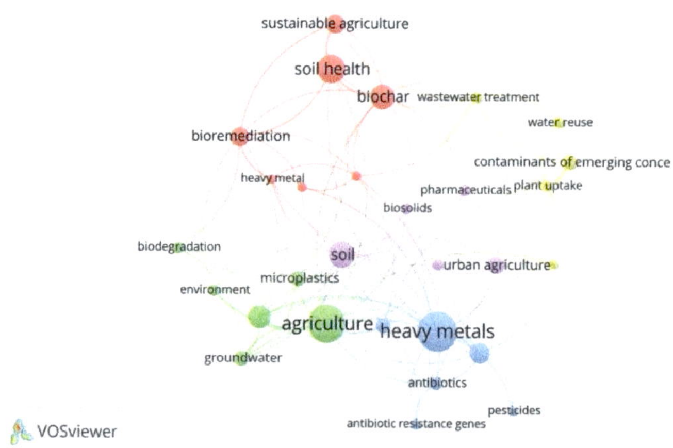

Figure 11. Keyword co-occurrence map in soil health research (2020–2025).

In the second approach, the co-occurrence map of the authors' keywords generated 22 elements grouped into five distinct clusters. These highlight the main directions of research that address soil health through the lens of AI technologies (Figure 12).

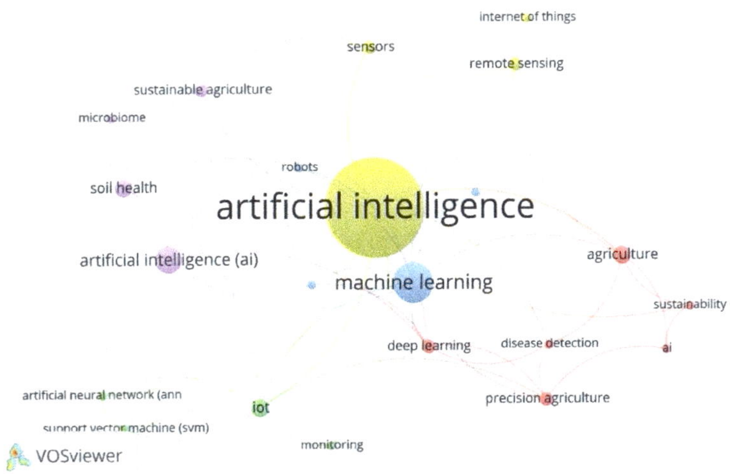

Figure 12. Keyword co-occurrence map for AI-based soil health research (2020–2025).

The yellow cluster focuses on AI, ML, and advanced algorithms. This cluster is associated with remote sensing, sensors, and IoT and demonstrates how digital technologies integrate data from various sources to capture an image of soil health. This approach enables informed decision-making. This cluster demonstrates how AI is a key tool in modern research on soil health management. The red cluster is associated with disease detection from the DL, precision, and sustainability systems perspective. The cluster highlights how AI combined with robotics reduces reliance on manual monitoring methods. The purple cluster corresponds to soil health, microbiome, and bioremediation, with academic relevance highlighting the importance of integrating AI technologies in researching these concepts that directly impact agricultural productivity. The blue cluster focuses on AI algorithms, especially on ML combined with robotics technologies. The green cluster is generated by the convergence of the advanced ML algorithms (Artificial Neural Network—ANN, SVM) with the IoT field, applied in agriculture monitoring applications. The results presented in Figure 12 highlight the trend of research in integrating AI at the level of soil health research. These results reflect the potential of AI transformations in soil health through the collaboration between engineering and ecology.

4.6. AI Branches and Applications in Soil Health

The authors analyzed the results of the WOS search for soil health related to AI and identified seven branches of AI applications. The seven branches are summarized in Table 1 and discussed in the following sections:

1. The first branch focuses on monitoring and predicting soil health. In this category, AI applications use predictive models employing the biological and physical properties of the soil, as well as data from the microbiome, sensors, and satellite images. For example, Kalantzopoulos el al. [95] monitor the soil in real time, allowing for the automatic generation of recommendations regarding decisions aimed at soil actions. Andrade et al. [96] use intelligent indices for soil quality assessment from microbiological data trained with ML models. Papers by Novielli et al. [97,98] employ explainable

AI (XAI) components to predict the sensitivity of microbial respiration to temperature. This is an indicator of soil quality in the context of climate change.

2. The second branch focuses on analyzing the soil microbiome for fertility. Soil microorganisms are responsible for its health and fertility. The analysis of metagenomic data can be performed using AI tools, which build predictive models regarding microbial interactions. AI tools can also analyze the impact of these interactions on nutrient cycles. García el al. [99] study the use of AI to design probiotics and optimize microbial consortia. This research analyzes the impact on the fertility and remediation of degraded soils. Andrade el al. [96] argue for creating a universal soil quality index using the microbiome and ML techniques.
3. Soil organic carbon is also predicted using specific AI techniques. The tools predict soil organic carbon levels with direct implications for resilience. Minasny and McBratney [100] use global-scale soil carbon dynamics forecasting. This forecast optimizes carbon sequestration strategies. The works [97,98,101] identify soils with a higher probability of CO_2 release. The analysis of this probability provides directions for sustainable carbon management.
4. Soil contamination and remediation strategies are associated with the fourth branch of applications. A consistent number of papers in the specialized literature analyze the implications of AI tools in detecting and remedying soil contaminants. Algarni et al. [102] estimate the concentrations of heavy metals such as arsenic, copper, cadmium, and lead using soil reflectance spectra.
5. AI techniques are also used to estimate soil quality from images. These techniques use multispectral images obtained from drones or satellites. The study by Negiş et al. [103] shows how AI tools interpret soil color, vegetation, and other visual indicators to estimate organic matter, moisture, or texture parameters. Lithuania's Soil Data Cube system combines images with AI techniques to create thematic soil maps [104].
6. Estimating nutrient quantities in the soil is also integrated through AI tools in real time. In this way, the growth level of the plants is directly controlled. Hossen el al. [105] propose a system equipped with multispectral sensors that includes an AI component trained with spectroscopy data. This system predicts the total nitrogen content in the soil to optimize the timing and amount of fertilizer application. The study by Kaur and Gupta [106] demonstrates the importance of these applications in increasing food resources.
7. AI and IoT tools are used for soil moisture and water management. Most soil moisture management applications use monitoring and prediction tools. Gaitan et al. [107] integrate both types of tools in the real-time analysis of climatic parameters to generate reports and suggestions for farmers. The works [95,108,109] propose systems that correlate humidity with plant growth for irrigation management and to reduce water stress.

Table 1. Branches of AI applications related to soil health with corresponding references.

Type of AI Application	References
Soil health monitoring and prediction	[95–98,100,104,105,110–112]
Soil microbiome analysis for fertility and health	[96,99,110,113]
Soil organic carbon prediction and carbon management	[97,98,100,101]
Soil contamination and remediation (heavy metals, etc.)	[102,114]
Soil quality estimation from remote sensing (UAV/satellite)	[103–105,115]
Soil nutrient detection and management (N, P, K, etc.)	[105,106,112]
Soil moisture and water management (AI + IoT)	[95,107–109]

Note: UAV—unmanned aerial vehicle; N—nitrogen; P—phosphorus; K—potassium.

This analysis highlights the importance of AI tools in soil health monitoring applications. The results presented in Table 1 highlight the need for further investigations that bring together researchers from modern technology fields, such as AI and IoT, with engineers from the agricultural sector.

4.7. Interdisciplinary Innovation Potential in AI-Soil Studies

The AI technologies identified by the CAI-CIP method indicate their frequency of occurrence in research materials. The results are presented in Table 2.

Table 2. AI Technologies identified in soil health research and their frequency of use.

AI Technology	Frequency of Use
ViT	79
NER	72
OWL	29
PCA	4
Selection	4
SHAP (Explainable AI)	3
ML	2
T5	2
Inception (CNN architecture)	1
NLP	1
BERT (Transformer model)	1
Transformers (general class)	1
Mutation (Genetic Algorithm)	1
RDF	1
Stemming	1
Clustering	1
GloVe (Word Embeddings)	1
GANs	1
LIME (Model interpretability)	1
SfM	1
Topic Modeling	1

Note: ViT—Vision Transformer; NER—Named Entity Recognition; OWL—Web Ontology Language; PCA—Principal Component Analysis; SHAP—SHapley Additive exPlanations; T5—Text-to-Text Transfer Transformer; BERT—Bidirectional encoder representations from transformers; RDF—Resource Description Framework; GloVe—Global Vectors; GAN—Generative Adversarial Network; LIME—Local Interpretable Model-agnostic Explanations; SfM—Structure from Motion.

ViT is a technique for assessing soil quality, having been identified in 79 scientific papers. This method analyzes satellite or drone images to monitor soil conditions visually.

NER and OWL appear as secondary methods, demonstrating the focus on natural language processing and structured knowledge representation. NER is identified 72 times and allows for the automatic extraction of information from textual descriptions. OWL appears 29 times and facilitates the modeling of knowledge and the relationships between concepts such as nutrients, microorganisms, and soil quality.

These results identify 20 AI technologies that are applied in soil quality research. The results demonstrate that the field of image analysis is the preferred one in soil quality analysis. The fact that AI technologies are applied in a fragmented manner, with a high degree of dispersion, but also with varied purposes of monitoring, prediction, remediation, and text analysis, suggests a lack of standardization between the two approaches, AI and soil health. This article highlights this gap through the conceptual map of AI directions in soil health from Table 2. The CAI-CIP results show that the technologies are underutilized. This underutilization does not stem from a lack of potential but from the need for future strategic research directions. The analysis integrated a predefined list of 200 AI technologies, which ensured a level of methodological originality not encountered in other works.

The detailed search for AI technologies yielded 497 papers. This result shows the scientific interest in integrating these modern tools into agriculture. These technologies are used in various applications such as monitoring, prediction, remediation, microbiological analysis, nutrient detection, etc. Although the value of 497 may seem small, the huge number of AI technologies included in the search actually demonstrates the fragmentation of research. Analyzing the value of 497 papers reveals a lack of standardization and fragmentation of approaches, which limits the replicability of the solutions. From these considerations, there is a need for research development that combines AI with holistic assessments of soil health. This result supports the need for future studies that incorporate interdisciplinary collaborations and projects demonstrating the impact of AI technologies in preventing soil contamination or remediating contaminated sites.

The results obtained from the WOS searches on components demonstrate the fragmentation of the research. Most of the research is conducted in relation to ML algorithms, yielding 378 articles. The NLP search yielded three results, while computer vision generated six articles. These modest values once again demonstrate the need to intensify research in the field of AI technologies. Additionally, the robotics sector in relation to AI and soil health or soil contaminants includes only three articles, highlighting the need for future directions in developing automated soil monitoring systems.

Regarding expert systems related to contaminants, only one article was obtained, and it demonstrates a gap that should be addressed by integrating specific knowledge into soil management decisions. The complete absence of results regarding evolutionary computation indicates that computational technologies have not been explored in the context of soil health.

5. Discussion

The results presented in this paper provide an overview of the application of AI-based technologies in determining and predicting soil health. The paper included seven analyses, each analyzing a different perspective.

1. The temporal interest analysis through the distribution of articles over the years shows interest in soil health and contaminants, with the number of articles increasing from 27 in 2020 to 54 in 2024. This increase confirms researchers' interest in studying soil health issues concerning the impact of climate change and intensive agriculture. The analysis dedicated to the AI component also shows an increase from seven articles in 2020 to ninety-two articles in 2024, with thirty articles already published in the first quarter of 2025. The evolution confirms the global trend of integrating modern technologies into agriculture. This paper identified a single article that explicitly combines the themes of soil health and contaminants with AI techniques. This gap outlines a new research direction that the scientific community should be interested in.
2. The productivity of authors and collaboration analysis highlighted two main clusters in soil health research. These clusters show a growing interest, but when the AI component was integrated, the results identified a single cluster with only six primary authors. This identification suggests that the application of AI in soil health requires intense exploration in the coming years. This need should serve as a directive to stimulate interdisciplinary collaborations and attract new researchers to collaborate on soil health studies.
3. The geographical distribution of research confirms that countries with intensive agriculture (China, India, the USA, Pakistan, and Australia) dominate the field. Integrating the AI component in WOS searches regarding soil health has suggested a

strategic opportunity for other countries to develop technological projects concerning agricultural productivity.

4. The publishers' analysis showed differences between traditional research and AI research. Classic publishers, such as Elsevier, MDPI, and Springer Nature, have proven dominant in both cases. Regarding the less popular publishers, it was found that traditional research belongs to technical publishers, while the AI area reflected an openness to multidisciplinary approaches. These aspects are normal, considering the interdisciplinary nature of AI components, and they outline a future development direction, encouraging traditional publications to disseminate works that also contain AI components.
5. The frequency of keywords and thematic directions were used to analyze the co-occurrence of keywords. This analysis identified seven clusters in general research and five clusters in AI research. In traditional research, the central element is represented by classic pollutants (heavy metals and pesticides), sustainable agriculture, water recycling, biosolids, and pharmaceutical pollutants. These themes show the researchers' growing interest in ecological issues. In contrast, AI research has clearly focused on advanced technologies concerning bioremediation. This direction indicates that articles addressing the issue of soil health concerning AI simultaneously aim at the necessity of modernizing agriculture through the integration of modern technologies.
6. The analysis of AI branches and their applications in soil health has led to seven main branches. Thus, branches targeting monitoring, the prediction of microbiological analysis, remediation, and water resource management, as well as other elements reflecting fragmented implementation, were identified. This finding suggests a need for methodological standardization to create specific tools for replicating systems in the agricultural field, such as soil health.
7. The CAI-CIP analysis demonstrated the potential for interdisciplinary innovation. It highlighted the use of ViT as a technology frequently employed in soil health. This stems from the ability of these algorithms to analyze images and process subtle elements identified at the image level. The fact that many of the 200 technologies identified by the authors did not yield results highlights an unexplored area in the specialized literature. This lack of homogenization highlights the increased need for interdisciplinary collaborations in agricultural engineering concerning AI methods.

In this study, 217 original articles published between 1 January 2020 and 30 April 2025 were analyzed. The search conducted in WOS used specific terms related to soil health and contaminants. The second search involved integrating AI into the field of soil health. The result yielded 222 papers. From this collection of results, one article explicitly addresses the relationship between soil contamination, agriculture, and the use of AI technologies.

The results of detailed searches regarding the subfields of AI applied to soil health and contaminants show that ML stands out as the most active branch, with 378 studies published between 2020 and 2025. Although this result suggests that researchers are interested in using ML for prediction, monitoring, remediation, and intervention regarding soil contamination, the number is small compared to the number of existing ML algorithms. Subfields such as NLP, computer vision, robotics, and expert systems have generated lower results, and the interpretation of this low interest is correlated with the need to explore the potential of these technologies in the field. Regarding the evolutionary computation component, it provided zero articles, indicating a gap that could become a research opportunity, considering the potential of genetic algorithms or evolutionary strategies in optimizing soil remediation strategies and reducing associated costs.

This study's main limitation is its exclusive focus on WOS data; however, integrating publications from other alternative databases could have created confusion in identifying future development directives. Additionally, the CAI-CIP analysis may not fully reflect the complexity of the analyzed articles. Another limitation is that AI research is continuously evolving, which means that the analysis window does not reflect the field's maturity for the coming years.

As future development directions, the necessity of expanding international collaborations, diversifying the AI technologies used in soil health studies, standardizing AI analysis procedures in agriculture, and investigating data infrastructure to facilitate the integration of data from satellite sensors or other local sources into a single system, as well as large-scale pilot projects and case studies to demonstrate the impact of AI technologies in soil contamination remediation and prevention, are emerging. This discussion section confirms the progress in soil health and underscores the necessity of further integrating AI technologies to harness this field's potential. The results of this research represent a starting point for a new generation of interdisciplinary studies that will incorporate specific elements of human health protection through soil health into a unique agricultural ecosystem.

6. Conclusions

This work investigated the current state of research dedicated to soil health. This paper emphasizes the integration of AI technologies in the context of soil health. Seven analyses were implemented to measure global scientific interest and identify the academic communities involved in this collaboration to identify the degree of cooperation between the two fields. Additionally, the paper mapped the geographical distribution and analyzed the applied thematic and technological directions.

This study provides an overview of the progress made and identifies the gaps that need to be addressed for a future where soil health is managed using AI techniques. The first conclusion from this research is that interest in soil health has steadily increased over the past five years. This increase reflects the recognition of the importance of protecting soil as a resource for the ecosystem and global food security. The increase from 27 articles in 2020 to 54 in 2024 confirms this trend. Analyses have also shown an increase in the use of AI in the agricultural field. In 2020, there were seven articles, which increased to ninety-two in 2024. This evolution underscores that the academic community perceives modern technologies as viable solutions for contemporary agricultural problems. However, the authors identified only one article that explicitly combines soil health themes with contaminants and AI technologies during the analyzed period. This finding highlights the acute lack of truly interdisciplinary studies.

The analysis of author communities highlighted two main clusters: the first was associated with traditional soil health research, and the second included a smaller circle of six authors who approached the issue of soil health through AI techniques. This finding reveals interest in the field but also highlights the need to strengthen this academic community by attracting a larger number of researchers from complementary domains.

From a geographical perspective, the research has shown that countries with intensive agriculture dominate the research area. This aspect raises a red flag for other countries that have opportunities to improve soil quality and thus intensify agriculture, including the European Union, which has not yet been examined in this field.

The analysis of publishers that have disseminated research in soil health showed that Elsevier, MDPI, and Springer Nature are at the top of researchers' preferences. This research shows a predilection for disseminating works that include AI techniques in technical and multidisciplinary areas, which should encourage other publications to do the same.

The study on keyword frequency showed a thematic diversity regarding traditional research concerning AI techniques. This aspect reflects an awareness of the need for sustainable solutions, and in the area of AI, the confirmation of the potential of these technologies to revolutionize the way soil health can be managed and monitored has been noted.

Another strong point of this research was the analysis of AI branches applied to soil health. Here, seven branches were identified that provide a map of AI applications: monitoring and predicting soil health, microbiome analysis, organic carbon prediction, contaminant detection, soil image analysis, nutrient estimation, and water management. The fact that the applications are fragmented suggests the need to capitalize on these applications. The results of the CAI-CIP analysis demonstrated researchers' interest in technologies such as Vision Transformers; however, this interest highlighted the underutilization of other technologies, indicating a significant opportunity for future research to expand the repertoire of AI technologies used.

Analyzing these results, the future research directions are as follows:

1. Another directive for interdisciplinary collaboration should be to stimulate interdisciplinary collaborations to create partnerships between AI experts and soil science researchers and include agricultural practitioners in the development of integrated solutions applicable on a large scale.
2. The AI technologies used should be diversified so that the entirety of them are inspected in relation to soil health. The authors recommend inspecting ML, neural networks, evolutionary algorithms, generative networks, or fuzzy systems in this category.
3. Methodologies should be standardized regarding the integration of AI in soil monitoring and remediation so that results can be replicated and compared globally. This standardization would allow for a meaningful comparison between research conducted with the same purpose but in different environments.
4. Investments in infrastructure for the development of data platforms allow researchers to study as many AI models as possible using information provided by sensors, satellite images, and local reports so that advancements in the academic environment support farmers.
5. Education should be another directive in which new generations of specialists possess agricultural knowledge corroborated with digital skills, such as those specific to integrating AI tools in agriculture.

In conclusion, this research confirms an increased interest in soil health and highlights the necessity of integrating modern technologies through AI to study soil health. This paper aims to serve as a methodological and thematic benchmark for future research. Also, it seeks to strengthen the relationship between technology and environmental sciences, with the common goal of the benefits generated by advancements in agricultural ecosystems.

Author Contributions: Conceptualization, C.-M.R.; methodology, C.-M.R. and A.S.; software, C.-M.R. and A.S.; validation, C.-M.R. and A.S.; formal analysis, C.-M.R. and A.S.; investigation, C.-M.R. and A.S.; resources, C.-M.R. and A.S.; data curation, C.-M.R. and A.S.; writing—original draft preparation, C.-M.R. and A.S.; writing—review and editing, C.-M.R. and A.S.; visualization, C.-M.R. and A.S.; supervision, C.-M.R. All authors have read and agreed to the published version of the manuscript.

Funding: This research was funded by the Petroleum-Gas University of Ploiesti, Romania.

Institutional Review Board Statement: Not applicable.

Informed Consent Statement: Not applicable.

Data Availability Statement: The raw data supporting the conclusions of this article will be made available by the authors on request.

Conflicts of Interest: The authors declare no conflicts of interest.

Abbreviations

The following abbreviations are used in this manuscript:

AI	Artificial intelligence
ANN	Artificial Neural Network
As	Arsenic
ASGM	Artisanal and small-scale gold mining
ATZ	Atrazine
BERT	Bidirectional encoder representations from transformers
CAI-CIP	Cross-AI Components Innovation Potential
Cd	Cadmium
CNN	Convolutional neural network
Cr	Chromium
Cu	Copper
CV	Computer vision
DEHP	Di(2-ethylhexyl) phthalate
DL	Deep learning
DnBP	Di-n-butyl phthalate
DS	Deinking sludge
DSC	Deinking sludge combined
EC	Evolutionary computation
ES	Expert systems
EU	European Union
GAN	Generative Adversarial Network
GloVe	Global Vectors
HBC	Herbicide
HDPE	High-density polyethylene
Hg	Mercury
IoT	Internet of Things
K	Potassium
LDPE	Low-density polyethylene
LIME	Local Interpretable Model-agnostic Explanations
MC	Microcystins
ML	Machine learning
MP	Microplastic
N	Nitrogen
nB	Nanobiochar
NER	Named Entity Recognition
Ni	Nickel
NLP	Natural Language Processing
nWTR	Nano-water treatment residues
OWL	Web Ontology Language
P	Phosphorus
PAE	Phthalate esters
PAH	Polycyclic aromatic hydrocarbon
Pb	Lead
PCA	Principal Component Analysis
PET	Polyethylene terephthalate
PP	Polypropylene
PTE	potential toxic element
PVC	Polyvinyl chloride
RDF	Resource Description Framework
RF	Random Forest
SfM	Structure from Motion

SHAP	SHapley Additive exPlanations
SVM	Support vector machine
T5	Text-to-Text Transfer Transformer
UAV	Unmanned aerial vehicle
ViT	Vision Transformer
VOC	Volatile organic compound
WOS	Web of Science
XAI	Explainable Artificial Intelligence
Zn	Zinc

References

1. Fernández-González, R.; Puíme-Guillén, F.; Panait, M. A case study of agri-food systems in rural Spain: Impacts, responses and institutional lessons. *Agric. Econ.* **2022**, *68*, 159–170. [CrossRef]
2. Rosca, C.M.; Gortoescu, I.A.; Tanase, M.R. Artificial Intelligence—Powered Video Content Generation Tools. *Rom. J. Pet. Gas Technol.* **2024**, *5*, 131–144. [CrossRef]
3. Popescu, C.; Dissanayake, H.; Mansi, E.; Stancu, A. Eco Breakthroughs: Sustainable Materials Transforming the Future of Our Planet. *Sustainability* **2024**, *16*, 10790. [CrossRef]
4. Griffiths, B.S.; Philippot, L. Insights into the resistance and resilience of the soil microbial community. *FEMS Microbiol. Rev.* **2013**, *37*, 112–129. [CrossRef] [PubMed]
5. Dincă, L.C.; Grenni, P.; Onet, C.; Onet, A. Fertilization and Soil Microbial Community: A Review. *Appl. Sci.* **2022**, *12*, 1198. [CrossRef]
6. Lense, G.H.E.; Servidoni, L.E.; Parreiras, T.C.; Santana, D.B.; Bolleli, T.D.M.; Ayer, J.E.B.; Spalevic, V.; Mincato, R.L. Modeling of soil loss by water erosion in the Tietê River Hydrographic Basin, São Paulo, Brazil. *Semin. Ciências Agrárias* **2022**, *43*, 1403–1422. [CrossRef]
7. Raliya, R. Artificial Intelligence for Precision and Sustainable Agricultural. *ACS Agric. Sci. Technol.* **2024**, *4*, 628–630. [CrossRef]
8. Cappellari, L.D.R.; Santoro, M.V.; Nievas, F.; Giordano, W.; Banchio, E. Increase of secondary metabolite content in marigold by inoculation with plant growth-promoting rhizobacteria. *Appl. Soil Ecol.* **2013**, *70*, 16–22. [CrossRef]
9. Alaba, A.B.; Adefunke, A.T.; Nicholas, O.O. Optimizing Sustainable Agriculture: Harnessing Urine as a Cost-Effective Fertilizer for Enhanced Amaranth Growth and Soil Microbial Health in Tropical Regions. *Dutse J. Pure Appl. Sci.* **2024**, *9*, 249–262. [CrossRef]
10. Imarhiagbe, E.E.; Onwudiwe, C. Pollution indices and microbiological assessment of soil samples from motor parks around New Benin Market, Benin City, Nigeria. *Dutse J. Pure Appl. Sci.* **2022**, *8*, 33–42. [CrossRef]
11. Abu-Qaoud, H.; Al-Fares, H.; Shtaya, M.J.Y.; Shawarb, N. Effect of effective microorganisms on wheat growth under salt stress condition. *Chil. J. Agric. Res.* **2021**, *81*, 351–356. [CrossRef]
12. Hakim, D.L.; Adji, R.; Satwhikawara, R. Analysis of Indigenious Vegetation Diversity in Urban Area of Bekasi Regency. *J. Soc. Econ. Agric.* **2024**, *13*, 32. [CrossRef]
13. Mills, J.G.; Bissett, A.; Gellie, N.J.C.; Lowe, A.J.; Selway, C.A.; Thomas, T.; Weinstein, P.; Weyrich, L.S.; Breed, M.F. Revegetation of urban green space rewilds soil microbiotas with implications for human health and urban design. *Restor. Ecol.* **2020**, *28*, S322–S334. [CrossRef]
14. Rosca, C.-M.; Stancu, A.; Neculaiu, C.-F.; Gortoescu, I.-A. Designing and Implementing a Public Urban Transport Scheduling System Based on Artificial Intelligence for Smart Cities. *Appl. Sci.* **2024**, *14*, 8861. [CrossRef]
15. Xiang, J.; Xu, P.; Chen, W.; Wang, X.; Chen, Z.; Xu, D.; Chen, Y.; Xing, M.; Cheng, P.; Wu, L.; et al. Pollution Characteristics and Health Risk Assessment of Heavy Metals in Agricultural Soils over the Past Five Years in Zhejiang, Southeast China. *Int. J. Environ. Res. Public Health* **2022**, *19*, 14642. [CrossRef]
16. Yang, L.; Wu, P.; Yang, W. Characteristics, Health Risk Assessment, and Transfer Model of Heavy Metals in the Soil—Food Chain in Cultivated Land in Karst. *Foods* **2022**, *11*, 2802. [CrossRef]
17. Wu, D.; Liu, H.; Wu, J.; Gao, X.; Nyasha, N.K.; Cai, G.; Zhang, W. Bi-Directional Pollution Characteristics and Ecological Health Risk Assessment of Heavy Metals in Soil and Crops in Wanjiang Economic Zone, Anhui Province, China. *Int. J. Environ. Res. Public Health* **2022**, *19*, 9669. [CrossRef]
18. Huerta Alata, M.; Alvarez-Risco, A.; Suni Torres, L.; Moran, K.; Pilares, D.; Carling, G.; Paredes, B.; Del-Aguila-Arcentales, S.; Yáñez, J.A. Evaluation of Environmental Contamination by Toxic Elements in Agricultural Soils and Their Health Risks in the City of Arequipa, Peru. *Sustainability* **2023**, *15*, 3829. [CrossRef]
19. Bech, J.; Pradenas, D.; Tume, P.; Cornejo, Ó.; Pedreros, J.; Toledo, S.; Correa, C.; Sepúlveda, B.; Roca, N. Human Health Risk Associated with As, Cu, Pb, and Zn in Soils of the Aconcagua and Casablanca River Basins, Valparaíso Region, Chile. *Appl. Sci.* **2025**, *15*, 2581. [CrossRef]

20. Gao, Z.; Jiang, J.; Sun, G. Evaluation of Heavy Metal Contamination in Black Soil at Sanjiang Plain: From Source Analysis to Health Risk Assessment. *Processes* **2024**, *12*, 2829. [CrossRef]
21. Jiang, Z.; Xiao, X.; Guo, Z.; Zhang, Y.; Huang, X. Impact of Vanadium-Containing Stone Coal Smelting on Trace Metals in an Agricultural Soil–Vegetable System: Accumulation, Transfer, and Health Risks. *Int. J. Environ. Res. Public Health* **2023**, *20*, 2425. [CrossRef] [PubMed]
22. Belanović Simić, S.; Miljković, P.; Baumgertel, A.; Lukić, S.; Ljubičić, J.; Čakmak, D. Environmental and Health Risk Assessment Due to Potentially Toxic Elements in Soil near Former Antimony Mine in Western Serbia. *Land* **2023**, *12*, 421. [CrossRef]
23. Sarim, M.; Jan, T.; Khattak, S.A.; Mihoub, A.; Jamal, A.; Saeed, M.F.; Soltani-Gerdefaramarzi, S.; Tariq, S.R.; Fernández, M.P.; Mancinelli, R.; et al. Assessment of the Ecological and Health Risks of Potentially Toxic Metals in Agricultural Soils from the Drosh-Shishi Valley, Pakistan. *Land* **2022**, *11*, 1663. [CrossRef]
24. Zhang, L.; Yang, Z.; Peng, M.; Cheng, X. Contamination Levels and the Ecological and Human Health Risks of Potentially Toxic Elements (PTEs) in Soil of Baoshan Area, Southwest China. *Appl. Sci.* **2022**, *12*, 1693. [CrossRef]
25. Li, X.; Liu, N.; Meng, W.; He, J.; Wu, P. Accumulation and Health Risk Assessment of Heavy Metal(loid)s in Soil-Crop Systems from Central Guizhou, Southwest China. *Agriculture* **2022**, *12*, 981. [CrossRef]
26. Alsafran, M.; Usman, K.; Al Jabri, H.; Rizwan, M. Ecological and Health Risks Assessment of Potentially Toxic Metals and Metalloids Contaminants: A Case Study of Agricultural Soils in Qatar. *Toxics* **2021**, *9*, 35. [CrossRef]
27. Rosca, C.M.; Popescu, M.; Patrascioiu, C.; Stancu, A. Comparative Analysis of pH Level Between Pasteurized and UTH Milk Using Dedicated Developed Application. *Rev. De Chim.* **2019**, *70*, 3917–3920. [CrossRef]
28. Sharmin, S.; Wang, Q.; Islam, M.R.; Wang, W.; Wang, Y.; Enyoh, C.E.; Rana, M.S. Assessment of Health Risks from Agricultural Soils Contaminated with Polycyclic Aromatic Hydrocarbons (PAHs) Across Different Land-Use Categories of Bangladesh. *Appl. Sci.* **2024**, *15*, 56. [CrossRef]
29. Li, R.; Cheng, M.; Cui, Y.; He, Q.; Guo, X.; Chen, L.; Wang, X. Distribution of the Soil PAHs and Health Risk Influenced by Coal Usage Processes in Taiyuan City, Northern China. *Int. J. Environ. Res. Public Health* **2020**, *17*, 6319. [CrossRef]
30. Li, H.; Liu, H.; Liu, Z.; Su, H.; Simayi, S.; Liu, G. Distribution Features and Health Risk Assessment of Phthalate Pollutants in Facility Soil and Agricultural Products in Xinjiang, China. *Agronomy* **2025**, *15*, 821. [CrossRef]
31. Xing, H.; Yu, X.; Huang, J.; Du, X.; Wang, M.; Sun, J.; Lu, G.; Tao, X. Characteristics and Health Risks of Phthalate Ester Contamination in Soil and Plants in Coastal Areas of South China. *Int. J. Environ. Res. Public Health* **2022**, *19*, 9516. [CrossRef] [PubMed]
32. Rosca, C.-M.; Rădulescu, G.; Stancu, A. Artificial Intelligence of Things Infrastructure for Quality Control in Cast Manufacturing Environments Shedding Light on Industry Changes. *Appl. Sci.* **2025**, *15*, 2068. [CrossRef]
33. Sharmin, S.; Wang, Q.; Islam, M.R.; Wang, W.; Enyoh, C.E. Microplastic Contamination of Non-Mulched Agricultural Soils in Bangladesh: Detection, Characterization, Source Apportionment and Probabilistic Health Risk Assessment. *J. Xenobiotics* **2024**, *14*, 812–826. [CrossRef]
34. Elwaleed, A.; Jeong, H.; Abdelbagi, A.H.; Thi Quynh, N.; Nugraha, W.C.; Agusa, T.; Ishibashi, Y.; Arizono, K. Assessment of Mercury Contamination in Water and Soil from Informal Artisanal Gold Mining: Implications for Environmental and Human Health in Darmali Area, Sudan. *Sustainability* **2024**, *16*, 3931. [CrossRef]
35. Popescu, C.; Gabor, M.R.; Stancu, A. Predictors for Green Energy vs. Fossil Fuels: The Case of Industrial Waste and Biogases in European Union Context. *Agronomy* **2024**, *14*, 1459. [CrossRef]
36. Zoghlami, R.I.; Toukabri, W.; Boudabbous, K.; Hechmi, S.; Barbouchi, M.; Oueriemmi, H.; Moussa, M.; Bahri, H. Assessment of Earthworm Viability and Soil Health after Two Years of Raw and Composted De-Inking Paper Sludge Amendment. *Agriculture* **2023**, *13*, 547. [CrossRef]
37. Moriarity, R.J.; Wilton, M.J.; Tsuji, L.J.S.; Sarkar, A.; Liberda, E.N. Evaluating human health risks from exposure to agricultural soil contaminants using one- and two-dimensional Monte Carlo simulations. *Environ. Res.* **2025**, *265*, 120391. [CrossRef]
38. Ma, J.; Ren, W.; Wang, H.; Song, J.; Jia, J.; Chen, H.; Tan, C.; Teng, Y. Exposure Characteristics and Human Health Risk Assessment of Herbicides in Water in a Typical Region of Northeastern China. *Expo. Health* **2024**, *16*, 1171–1184. [CrossRef]
39. Raimi, L.; Panait, M.; Sule, R. Leveraging Precision Agriculture for Sustainable Food Security in Sub-Saharan Africa: A Theoretical Discourse. In *Shifting Patterns of Agricultural Trade*; Erokhin, V., Tianming, G., Andrei, J.V., Eds.; Springer Nature: Singapore, 2021; pp. 491–509. [CrossRef]
40. Mahmoud, E.; El-Shahawy, A.; Ibrahim, M.; Abd El-Halim, A.E.-H.A.; Abo-Ogiala, A.; Shokr, M.S.; Mohamed, E.S.; Rebouh, N.Y.; Ismail, S.M. Enhancing Maize Yield and Soil Health through the Residual Impact of Nanomaterials in Contaminated Soils to Sustain Food. *Nanomaterials* **2024**, *14*, 369. [CrossRef]
41. Farooqi, Z.U.R.; Ahmad, I.; Abdul Qadir, A.; Murtaza, G.; Rafiq, S.; Jamal, A.; Zeeshan, N.; Murtaza, B.; Javed, W.; Radicetti, E.; et al. Zeolite-Assisted Immobilization and Health Risks of Potentially Toxic Elements in Wastewater-Irrigated Soil under Brinjal (Solanum melongena) Cultivation. *Agronomy* **2022**, *12*, 2433. [CrossRef]

42. Redouane, E.M.; Lahrouni, M.; Martins, J.C.; El Amrani Zerrifi, S.; Benidire, L.; Douma, M.; Aziz, F.; Oufdou, K.; Mandi, L.; Campos, A.; et al. Protective Role of Native Rhizospheric Soil Microbiota Against the Exposure to Microcystins Introduced into Soil-Plant System via Contaminated Irrigation Water and Health Risk Assessment. *Toxins* **2021**, *13*, 118. [CrossRef] [PubMed]
43. Kundu, M.; Krishnan, P.; Prasad, S.; Vashisth, A.; Duhan, S.; Reddy, K.R. Biosensing technology interventions for the detection of nitrate and nitrite contamination in environment and foods. In *Advances in Agronomy*; Sparks, D.L., Ed.; Elsevier: London, UK, 2024; Volume 183, pp. 193–250. [CrossRef]
44. Entwistle, J.; Bramwell, L.; Wragg, J.; Cave, M.; Hamilton, E.; Gardner, A.; Dean, J.R. Investigating the Geochemical Controls on Pb Bioaccessibility in Urban Agricultural Soils to Inform Sustainable Site Management. *Geosciences* **2020**, *10*, 398. [CrossRef]
45. Gao, H.; Wu, M.; Liu, H.; Xu, Y.; Liu, Z. Effect of petroleum hydrocarbon pollution levels on the soil microecosystem and ecological function. *Environ. Pollut.* **2022**, *293*, 118511. [CrossRef] [PubMed]
46. Mahaveerchand, H.; Abdul Salam, A.A. Environmental, industrial, and health benefits of Moringa oleifera. *Phytochem. Rev.* **2024**, *23*, 1497–1556. [CrossRef]
47. Rosca, C.-M.; Stancu, A. Fusing Machine Learning and AI to Create a Framework for Employee Well-Being in the Era of Industry 5.0. *Appl. Sci.* **2024**, *14*, 10835. [CrossRef]
48. Jamieson, L.; Francisco Moreno-García, C.; Elyan, E. A review of deep learning methods for digitisation of complex documents and engineering diagrams. *Artif. Intell. Rev.* **2024**, *57*, 136. [CrossRef]
49. Aykat, Ş.; Senan, S. Using Machine Learning to Detect Different Eye Diseases from OCT Images. *Int. J. Comput. Exp. Sci. Eng.* **2023**, *9*, 62–67. [CrossRef]
50. Rosca, C.M.; Stancu, A.; Ariciu, A.V. Algorithm for child adoption process using artificial intelligence and monitoring system for children. *Internet Things* **2024**, *26*, 101170. [CrossRef]
51. Aldhafeeri, F.M. Perspectives of radiographers on the emergence of artificial intelligence in diagnostic imaging in Saudi Arabia. *Insights Into Imaging* **2022**, *13*, 178. [CrossRef]
52. Chatzimichail, E.; Feltgen, N.; Motta, L.; Empeslidis, T.; Konstas, A.G.; Gatzioufas, Z.; Panos, G.D. Transforming the future of ophthalmology: Artificial intelligence and robotics' breakthrough role in surgical and medical retina advances: A mini review. *Front. Med.* **2024**, *11*, 1434241. [CrossRef]
53. Delanerolle, G.; Bouchareb, Y.; Shetty, S.; Cavalini, H.; Phiri, P. A Pilot Study Using Natural Language Processing to Explore Textual Electronic Mental Healthcare Data. *Informatics* **2025**, *12*, 28. [CrossRef]
54. Dai, W.; Ma, Y.; Fan, Y.; Ma, J. A Multi-Scale Feature Extraction Algorithm for Chinese Herbal Medicine Image Classification. *Appl. Sci.* **2025**, *15*, 4271. [CrossRef]
55. Li, H.; Li, Y.; Xiao, L.; Zhang, Y.; Cao, L.; Wu, D. RLRD-YOLO: An Improved YOLOv8 Algorithm for Small Object Detection from an Unmanned Aerial Vehicle (UAV) Perspective. *Drones* **2025**, *9*, 293. [CrossRef]
56. Rosca, C.M. Comparative Analysis of Object Classification Algorithms: Traditional Image Processing Versus Artificial Intelligence—Based Approach. *Rom. J. Pet. Gas Technol.* **2023**, *4*, 169–180. [CrossRef]
57. Sawicki, P.; Sawicka, H.; Karkula, M.; Zajda, K. Combined Rough Sets and Rule-Based Expert System to Support Environmentally Oriented Sandwich Pallet Loading Problem. *Energies* **2025**, *18*, 268. [CrossRef]
58. Song, H.; Zhao, Y.; Zhang, Y.; Chen, H.; Cui, L. Knowledge Distillation Based Recommendation Systems: A Comprehensive Survey. *Electronics* **2025**, *14*, 1538. [CrossRef]
59. Kingsmore, K.M.; Lipsky, P.E. Recent advances in the use of machine learning and artificial intelligence to improve diagnosis, predict flares, and enrich clinical trials in lupus. *Curr. Opin. Rheumatol.* **2022**, *34*, 374–381. [CrossRef]
60. Kabashkin, I. AI and Evolutionary Computation for Intelligent Aviation Health Monitoring. *Electronics* **2025**, *14*, 1369. [CrossRef]
61. Rosca, C.-M. New Algorithm to Prevent Online Test Fraud Based on Cognitive Services and Input Devices Events. In Lecture Notes in Networks and Systems, Proceedings of the Third Emerging Trends and Technologies on Intelligent Systems, ETTIS 2023, Noida, India, 23–24 February 2023; Noor, A., Saroha, K., Pricop, E., Sen, A., Trivedi, G., Eds.; Springer Nature: Singapore, 2023; Volume 730, pp. 207–219. [CrossRef]
62. Imam, N.; Belda, I.; García-Jiménez, B.; Duehl, A.J.; Doroghazi, J.R.; Almonacid, D.E.; Thomas, V.P.; Acedo, A. Local Network Properties of Soil and Rhizosphere Microbial Communities in Potato Plantations Treated with a Biological Product Are Important Predictors of Crop Yield. *mSphere* **2021**, *6*, 4. [CrossRef]
63. Su, B.; Zhang, H.; Zhang, Y.; Shao, S.; Mouazen, A.M.; Jiao, H.; Yi, S.; Gao, C. Soil C:N:P Stoichiometry Succession and Land Use Effect after Intensive Reclamation: A Case Study on the Yangtze River Floodplain. *Agronomy* **2023**, *13*, 1133. [CrossRef]
64. Shevchenko, V.; Lukashevich, A.; Taniushkina, D.; Bulkin, A.; Grinis, R.; Kovalev, K.; Narozhnaia, V.; Sotiriadi, N.; Krenke, A.; Maximov, Y. Climate Change Impact on Agricultural Land Suitability: An Interpretable Machine Learning-Based Eurasia Case Study. *IEEE Access* **2024**, *12*, 15748–15763. [CrossRef]
65. Chana, A.M.; Batchakui, B.; Nges, B.B. Real-Time Crop Prediction Based on Soil Fertility and Weather Forecast Using IoT and a Machine Learning Algorithm. *Agric. Sci.* **2023**, *14*, 645–664. [CrossRef]

66. Rosca, C.-M.; Carbureanu, M.; Stancu, A. Data-Driven Approaches for Predicting and Forecasting Air Quality in Urban Areas. *Appl. Sci.* **2025**, *15*, 4390. [CrossRef]
67. Condran, S.; Bewong, M.; Islam, M.Z.; Maphosa, L.; Zheng, L. Machine Learning in Precision Agriculture: A Survey on Trends, Applications and Evaluations Over Two Decades. *IEEE Access* **2022**, *10*, 73786–73803. [CrossRef]
68. Hengl, T.; Heuvelink, G.B.M.; Kempen, B.; Leenaars, J.G.B.; Walsh, M.G.; Shepherd, K.D.; Sila, A.; Macmillan, R.A.; Mendes De Jesus, J.; Tamene, L.; et al. Mapping Soil Properties of Africa at 250 m Resolution: Random Forests Significantly Improve Current Predictions. *PLoS ONE* **2015**, *10*, e0125814. [CrossRef]
69. Tripathi, A.; Tiwari, R.K.; Tiwari, S.P. A deep learning multi-layer perceptron and remote sensing approach for soil health based crop yield estimation. *Int. J. Appl. Earth Obs. Geoinf.* **2022**, *113*, 102959. [CrossRef]
70. Yang, X.; Shu, L.; Chen, J.; Ferrag, M.A.; Wu, J.; Nurellari, E.; Huang, K. A Survey on Smart Agriculture: Development Modes, Technologies, and Security and Privacy Challenges. *IEEE/CAA J. Autom. Sin.* **2021**, *8*, 273–302. [CrossRef]
71. Singh, A.K.; Roy, M.L.; Ghorai, A.K. Optimize Fertilizer Use Management through Soil Health Assessment: Saves Money and the Environment. *Int. J. Curr. Microbiol. Appl. Sci.* **2020**, *9*, 322–330. [CrossRef]
72. Rosca, C.-M.; Stancu, A.; Popescu, M. The Impact of Cloud Versus Local Infrastructure on Automatic IoT-Driven Hydroponic Systems. *Appl. Sci.* **2025**, *15*, 4016. [CrossRef]
73. Metwally, M.S.; Shaddad, S.M.; Liu, M.; Yao, R.-J.; Abdo, A.I.; Li, P.; Jiao, J.; Chen, X. Soil Properties Spatial Variability and Delineation of Site-Specific Management Zones Based on Soil Fertility Using Fuzzy Clustering in a Hilly Field in Jianyang, Sichuan, China. *Sustainability* **2019**, *11*, 7084. [CrossRef]
74. Peter-Jerome, H.; Adewopo, J.B.; Kamara, A.Y.; Aliyu, K.T.; Dawaki, M.U. Assessing the Spatial Variability of Soil Properties to Delineate Nutrient Management Zones in Smallholder Maize-Based System of Nigeria. *Appl. Environ. Soil Sci.* **2022**, *2022*, 1–14. [CrossRef]
75. Bhuyan, S.; Patgiri, D.K.; Medhi, S.J.; Chutia, D.; Meena, R.S.; Sandillya, M. Spatial Distribution of Soil Nutrient Status of Biswanath District, Assam, North East India. *Int. J. Plant Soil Sci.* **2023**, *35*, 145–157. [CrossRef]
76. Manimekalai, R.; Vijayashanthi, V.A.; Yogameenakshi, P.; Santhi, P.; Sathish, G. Impact Assessment on Adoption of Soil Health Cards for Fertilizer Management in Tiruvallur District. *Curr. J. Appl. Sci. Technol.* **2021**, *40*, 50–55. [CrossRef]
77. Harris, J.A.; Evans, D.L.; Mooney, S.J. A new theory for soil health. *Eur. J. Soil Sci.* **2022**, *73*, e13292. [CrossRef]
78. Davidovič, D.; Ivajnšič, D. Soil Moisture Index in Pomurje: An Example of Landsat 8 Satellite Data Use. *J. Geogr.* **2020**, *15*, 91–108. [CrossRef]
79. Zhou, M.; Wang, C.; Xie, Z.; Li, Y.; Zhang, X.; Wang, G.; Jin, J.; Ding, G.; Liu, X. Humic substances and distribution in Mollisols affected by six-year organic amendments. *Agron. J.* **2020**, *112*, 4723–4740. [CrossRef]
80. Jordán Vidal, M.M. Criteria for Assessing the Environmental Quality of Soils in a Mediterranean Region for Different Land Use. *Soil Syst.* **2023**, *7*, 75. [CrossRef]
81. Ghazal, S.; Munir, A.; Qureshi, W.S. Computer vision in smart agriculture and precision farming: Techniques and applications. *Artif. Intell. Agric.* **2024**, *13*, 64–83. [CrossRef]
82. Venturini, A.M.; Gontijo, J.B.; Mandro, J.A.; Paula, F.S.; Yoshiura, C.A.; Da França, A.G.; Tsai, S.M. Genome-resolved metagenomics reveals novel archaeal and bacterial genomes from Amazonian forest and pasture soils. *Microb. Genom.* **2022**, *8*, 000853. [CrossRef]
83. Babin, D.; Leoni, C.; Neal, A.L.; Sessitsch, A.; Smalla, K. Editorial to the Thematic Topic "Towards a more sustainable agriculture through managing soil microbiomes". *FEMS Microbiol. Ecol.* **2021**, *97*, fiab094. [CrossRef]
84. Roșca, C.-M.; Cărbureanu, M. A Comparative Analysis of Sorting Algorithms for Large-Scale Data: Performance Metrics and Language Efficiency. In Lecture Notes in Networks and Systems, Proceedings of the Emerging Trends and Technologies on Intelligent Systems, ETTIS 2024, Noida, India, 27–28 March 2024; Springer Nature: Noida, India, 2025; pp. 99–113. [CrossRef]
85. Ballabio, C.; Jones, A.; Panagos, P. Cadmium in topsoils of the European Union—An analysis based on LUCAS topsoil database. *Sci. Total Environ.* **2024**, *912*, 168710. [CrossRef] [PubMed]
86. Fendrich, A.N.; Van Eynde, E.; Stasinopoulos, D.M.; Rigby, R.A.; Mezquita, F.Y.; Panagos, P. Modeling arsenic in European topsoils with a coupled semiparametric (GAMLSS-RF) model for censored data. *Environ. Int.* **2024**, *185*, 108544. [CrossRef]
87. Ballabio, C.; Panagos, P.; Lugato, E.; Huang, J.-H.; Orgiazzi, A.; Jones, A.; Fernández-Ugalde, O.; Borrelli, P.; Montanarella, L. Copper distribution in European topsoils: An assessment based on LUCAS soil survey. *Sci. Total Environ.* **2018**, *636*, 282–298. [CrossRef] [PubMed]
88. Ballabio, C.; Jiskra, M.; Osterwalder, S.; Borrelli, P.; Montanarella, L.; Panagos, P. A spatial assessment of mercury content in the European Union topsoil. *Sci. Total Environ.* **2021**, *769*, 144755. [CrossRef]
89. Wang, J.; Deng, Y.; Huang, Z.; Li, D.A.; Zhang, X. Identification of driving factors for heavy metals and polycyclic aromatic hydrocarbons pollution in agricultural soils using interpretable machine learning. *Sci. Total Environ.* **2025**, *960*, 178384. [CrossRef]
90. Salgado, L.; López-Sánchez, C.A.; Colina, A.; Baragaño, D.; Forján, R.; Gallego, J.R. Hg and As pollution in the soil-plant system evaluated by combining multispectral UAV-RS, geochemical survey and machine learning. *Environ. Pollut.* **2023**, *333*, 122066. [CrossRef]

91. Gantumur, S.; Kharitonova, G.V.; Stepanov, A.S.; Dubrovin, K.N. Assessment of soil contamination using remote sensing data in the Tamsag-Bulag oil field, Mongolia. In Proceedings of the Regions of New Development: The Current State of Natural Complexes and Their Protection, Khabarovsk, Russian, 5–7 October 2021; p. 012013. [CrossRef]
92. Dean, J.R.; Ahmed, S.; Cheung, W.; Salaudeen, I.; Reynolds, M.; Bowerbank, S.L.; Nicholson, C.E.; Perry, J.J. Use of remote sensing to assess vegetative stress as a proxy for soil contamination. *Environ. Sci. Process. Impacts* **2024**, *26*, 161–176. [CrossRef] [PubMed]
93. Wang, Y.; Zou, B.; Zuo, X.; Zou, H.; Zhang, B.; Tian, R.; Feng, H. A remote sensing analysis method for soil heavy metal pollution sources at site scale considering source-sink relationships. *Sci. Total Environ.* **2024**, *946*, 174021. [CrossRef]
94. Anifowose, B.; Anifowose, F. Artificial intelligence and machine learning in environmental impact prediction for soil pollution management—case for EIA process. *Environ. Adv.* **2024**, *17*, 100554. [CrossRef]
95. Kalantzopoulos, G.; Paraskevopoulos, P.; Domalis, G.; Liopa-Tsakalidi, A.; Tsesmelis, D.E.; Barouchas, P.E. The Western Greece Soil Information System (WESIS)—A Soil Health Design Supported by the Internet of Things, Soil Databases, and Artificial Intelligence Technologies in Western Greece. *Sustainability* **2024**, *16*, 3478. [CrossRef]
96. Andrade, V.H.G.Z.D.; Redmile-Gordon, M.; Barbosa, B.H.G.; Andreote, F.D.; Roesch, L.F.W.; Pylro, V.S. Artificially intelligent soil quality and health indices for 'next generation' food production systems. *Trends Food Sci. Technol.* **2021**, *107*, 195–200. [CrossRef]
97. Novielli, P.; Magarelli, M.; Romano, D.; Di Bitonto, P.; Stellacci, A.M.; Monaco, A.; Amoroso, N.; Bellotti, R.; Tangaro, S. Leveraging explainable AI to predict soil respiration sensitivity and its drivers for climate change mitigation. *Sci. Rep.* **2025**, *15*, 12527. [CrossRef]
98. Novielli, P.; Magarelli, M.; Romano, D.; De Trizio, L.; Di Bitonto, P.; Monaco, A.; Amoroso, N.; Stellacci, A.M.; Zoani, C.; Bellotti, R.; et al. Climate Change and Soil Health: Explainable Artificial Intelligence Reveals Microbiome Response to Warming. *Mach. Learn. Knowl. Extr.* **2024**, *6*, 1564–1578. [CrossRef]
99. García, G.; Carlin, M.; Cano, R.D.J. Holobiome Harmony: Linking Environmental Sustainability, Agriculture, and Human Health for a Thriving Planet and One Health. *Microorganisms* **2025**, *13*, 514. [CrossRef] [PubMed]
100. Minasny, B.; McBratney, A.B. Machine Learning and Artificial Intelligence Applications in Soil Science. *Eur. J. Soil Sci.* **2025**, *76*, e70093. [CrossRef]
101. Bhatt, R.; Hossain, A.; Majumder, D.; Chandra, M.S.; Ghimire, R.; Faisal Shahzad, M.; Verma, K.K.; Riar, A.S.; Rajput, V.D.; Oliveira, M.W.; et al. Prospects of artificial intelligence for the sustainability of sugarcane production in the modern era of climate change: An overview of related global findings. *J. Agric. Food Res.* **2024**, *18*, 101519. [CrossRef]
102. Algarni, S.; Tirth, V.; Alqahtani, T.; Kshirsagar, P.R. A Novel Hybrid IOT Based Artificial Intelligence Algorithm for Toxicity Prediction In The Environment And Its Effect On Human Health. *Glob. NEST J.* **2023**, *25*, 12–22. [CrossRef]
103. Negiş, H.; Şeker, C.; Şeker, H.K. Using Artificial Intelligence Algorithms to Analyze Chromatic Attributes for Soil Quality Indicators. *J. Soil Sci. Plant Nutr.* **2025**. [CrossRef]
104. Samarinas, N.; Tsakiridis, N.L.; Kokkas, S.; Kalopesa, E.; Zalidis, G.C. Soil Data Cube and Artificial Intelligence Techniques for Generating National-Scale Topsoil Thematic Maps: A Case Study in Lithuanian Croplands. *Remote Sens.* **2023**, *15*, 5304. [CrossRef]
105. Hossen, M.A.; Diwakar, P.K.; Ragi, S. Total nitrogen estimation in agricultural soils via aerial multispectral imaging and LIBS. *Sci. Rep.* **2021**, *11*, 12693. [CrossRef]
106. Kaur, N.; Gupta, V. Climate Dependent Crop Field Condition Management Through Data Modeling. In *Proceedings of the Third Doctoral Symposium on Computational Intelligence, Lecture Notes in Networks and Systems*; Springer: Singapore, 2023; pp. 651–669. [CrossRef]
107. Gaitan, N.C.; Batinas, B.I.; Ursu, C.; Crainiciuc, F.N. Integrating Artificial Intelligence into an Automated Irrigation System. *Sensors* **2025**, *25*, 1199. [CrossRef] [PubMed]
108. Fuentes-Peñailillo, F.; Gutter, K.; Vega, R.; Silva, G.C. Transformative Technologies in Digital Agriculture: Leveraging Internet of Things, Remote Sensing, and Artificial Intelligence for Smart Crop Management. *J. Sens. Actuator Netw.* **2024**, *13*, 39. [CrossRef]
109. Liang, M.; Mao, K.; Shi, J.; Bateni, S.M.; Meng, F. An AI-Based Nested Large–Small Model for Passive Microwave Soil Moisture and Land Surface Temperature Retrieval Method. *Remote Sens.* **2025**, *17*, 1198. [CrossRef]
110. Pace, R.; Schiano Di Cola, V.; Monti, M.M.; Affinito, A.; Cuomo, S.; Loreto, F.; Ruocco, M. Artificial intelligence in soil microbiome analysis: A potential application in predicting and enhancing soil health—A review. *Discov. Appl. Sci.* **2025**, *7*, 85. [CrossRef]
111. Chinnasamy, P.; Tejaswini, D.; Ayyasamy, R.K.; Dhanasekaran, S.; Kumar, B.S.; Chandran, L. Crop Optimization and Disease Detection using Satellite Imagery; Artificial Intelligence. In Proceedings of the Second International Conference on Intelligent Cyber Physical Systems and Internet of Things, Coimbatore, India, 28–30 August 2024; pp. 1531–1535. [CrossRef]
112. Uddin, M.J.; Sherrell, J.; Emami, A.; Khaleghian, M. Application of Artificial Intelligence and Sensor Fusion for Soil Organic Matter Prediction. *Sensors* **2024**, *24*, 2357. [CrossRef]
113. Garg, S.; Kim, M.; Romero-Suarez, D. Current advancements in fungal engineering technologies for Sustainable Development Goals. *Trends Microbiol.* **2025**, *33*, 285–301. [CrossRef]

114. Haghighizadeh, A.; Rajabi, O.; Nezarat, A.; Hajyani, Z.; Haghmohammadi, M.; Hedayatikhah, S.; Asl, S.D.; Aghababai Beni, A. Comprehensive analysis of heavy metal soil contamination in mining Environments: Impacts, monitoring Techniques, and remediation strategies. *Arab. J. Chem.* **2024**, *17*, 105777. [CrossRef]
115. Kalopesa, E.; Tsakiridis, N.L.; Boletos, G.; Tziolas, N.; Zalidis, G.C. The Greek Soil Data Cube in Support of Generating Soil Related Analysis Ready Data. In Proceedings of the International Geoscience and Remote Sensing Symposium, Pasadena, CA, USA, 16–21 July 2023; pp. 5363–5366. [CrossRef]

agriculture

MDPI

Review

Progress in "Clean Agriculture" for Nitrogen Management to Enhance the Soil Health of Arable Fields and Its Application by Remote Sensing in Hokkaido, Japan

Kiwamu Ishikura [1,*], Nobuhiko Fueki [2] and Katsuhisa Niwa [3]

1 Production Technology Group, Tokachi Agricultural Experiment Station, Hokkaido Research Organization, Memuro 082-0081, Hokkaido, Japan
2 Production Technology Group, Kamikawa Agricultural Experiment Station, Hokkaido Research Organization, Pippu 078-0397, Hokkaido, Japan; fueki-nobuhiko@hro.or.jp
3 Zukosha Co., Ltd., Obihiro 080-0048, Hokkaido, Japan; niwa@zukosha.co.jp
* Correspondence: ishikura-kiwamu@hro.or.jp; Tel.: +81-155-62-9837

Abstract: Soil health has become increasingly important in recent years. The Hokkaido government initiated its original administrative strategy referred to as "Clean Agriculture" in 1991, before the concept of soil health and soil quality evolved in the 1990s. Also, Clean Agriculture has been integrated with remote sensing techniques for spatial application in arable fields. In this review paper, we summarized the scientific progress in Clean Agriculture and the management of soil health using remote sensing. One of the main pillars of Clean Agriculture is the minimal usage of chemical fertilizers and agrochemicals to increase soil fertility through the proper application of organic matter. The other two pillars are the sustainment and enhancement of the natural recycling function in agriculture and the enhancement of a stable production safe and high-quality agricultural products taking into account environmental harmony. These agronomic practices can increase soil fertility, maintain water quality, mitigate climate change, and maintain human health, and are similar to those in North America and the EU. Moreover, soil nitrogen fertility evaluated by autoclaved nitrogen (AC-N) can be estimated in large-scale fields and areas via remote sensing, which can facilitate variable nitrogen fertilization using variable-rate planters or broadcasters. Furthermore, systems comprising the growth sensor and variable-rate broadcaster can determine the additional nitrogen fertilization rates for winter wheat on the fields, which enhances soil health over relatively large areas. Further research is needed to expand the spatial utility of various Clean Agriculture techniques using multiperiod satellite images.

Keywords: autoclaved nitrogen; *Hokkaido Fertilizer Recommendations*; soil diagnosis; variable fertilizer application

1. Introduction

Soil health is an important concept associated with soil conservation, soil function, environmental protection, sustainable agriculture, and human health [1]. The concept of soil health evolved as an offshoot of soil quality in the 1990s [2,3]. Recently, the concept of soil quality has been extended to crop productivity and fertility [4–7], water quality [8–10], climate change and carbon sequestration [11–13], soil biodiversity [14–16], and human health [8,17–19]. Because the functional knowledge and the effective assessment of soil biodiversity are still limited, Lehmann et al. summarized soil health from four viewpoints: sustainable crop production, water quality, climate change, and human health, and they discussed that soil biodiversity is related to all the viewpoints [1]. Nowadays, soil health is a comprehensive concept of ecosystem services related to soils [1,20–22].

Similar to other regions worldwide, there has been considerable research on soil health in Hokkaido, Japan, in terms of sustainable crop production [23] as well as soil and water pollution [24]. Prior to these studies, the Hokkaido government developed "Clean Agriculture" as its original administrative strategy in 1991 [25,26].

Clean Agriculture is defined as "An agriculture that sustains and enhances not only the natural recycling function in agriculture but also a stable production of the feel-relieved safety and high-quality agricultural products taking into account environmental harmony. In this concept, efforts are made to apply soil amendments for example organic application, and minimizes the usage of chemical fertilizers and agrochemicals" [25].

In the 1990s, given the vigorous international agricultural competitions such as the GATT-Uruguay rounds, Clean Agriculture was launched to fortify the international competitiveness and sustainability of Hokkaido agriculture. Its main aims are as follows: (1) the conservation of the clean environment that Hokkaido takes pride in, (2) the promotion of the appeal of Hokkaido agriculture, and (3) the reinforcement of the competitiveness of Hokkaido agriculture [27]. As of 2024, more than 300 relevant techniques have been developed [28], and Clean Agriculture is currently the basis for the agricultural practices relied on to enhance soil health in Hokkaido [26].

In practice, the application of Clean Agriculture techniques depends on spatial variations in the field. Recently, remote sensing [e.g., satellites or unmanned aerial vehicles (UAVs)] has increasingly been used to evaluate spatial variations in soil properties and crop growth. Therefore, the coordinated application of both remote sensing and Clean Agriculture techniques has been adopted, especially for the cultivation of arable crops in Hokkaido. Several studies also pointed out the importance of remote sensing for the assessment of soil health [1,29]. However, studies to manage soil health with remote sensing are still limited.

Recently, many reviews of soil health have been published worldwide [30–32]. Nevertheless, the relationship between Clean Agriculture and soil health, and the practice of soil health using remote sensing have not been summarized yet. The objectives of this paper are (1) to review the scientific progress of Clean Agriculture in Hokkaido from the viewpoints that Lehmann et al. summarized [1] in comparison with other regions, (2) to review the co-application of remote sensing techniques in arable fields in Hokkaido, and (3) to suggest future research directions needed to further enhance soil health mainly for arable fields.

2. Clean Agriculture and *Hokkaido Fertilizer Recommendations*

2.1. History of Clean Agriculture

In 1991, before Doran and Parkin [3] and Doran et al. [2] propounded the concept of soil health, Clean Agriculture was originally advocated by the now deceased Dr. Satoru Souma [27,33]. He established Clean Agriculture mainly on the basis of the concept of "Low Input Sustainable Agriculture" in the United States of America (USA), and emphasized the need for minimizing the usage of chemical fertilizers and agrochemicals by improving soil fertility through the proper application of organic matter. In Clean Agriculture, fertilized soil applied with an adequate amount of organic matter may decrease the need for chemical fertilizers and agrochemicals. In the 1990s, Dr. Souma, the former director of Hokkaido Prefectural Agricultural Experiment Stations, strongly encouraged the scientists he supervised to perform experiments and studies to diagnose soil physicochemical properties, control soil fertility through soil amendments, and minimize the usage of chemical fertilizers and agrochemicals [27,33]. This scientific strategy continues to be applied as of 2025. The research outcomes have been uploaded and updated on the Hokkaido Research Organization website [28], and summarized in *Hokkaido Fertilizer Recommendations* [34], which has been continuously revised almost every five years. These documents have been utilized by farmers, advisors, and professionals in other stakeholders. The Hokkaido government has also promoted Clean Agriculture among farmers and consumers using a cartoon character (Figure 1).

Figure 1. Cartoon character of Dr. Hatakeda (arable field in Japanese) and "clean-dane" (clean seeds in Japanese), which has been used to promote Hokkaido's Clean Agriculture.

Clean Agriculture is "NOT" organic agriculture which excludes any usage of chemical fertilizer and agrochemicals. Clean Agriculture allows their minimal usage [33], aiming for a 30% reduction of chemical fertilizers and agrochemicals used. The political direction of Clean Agriculture resembles the current EU agricultural policy (Farm to Fork Strategy) whose goals are a "20% reduction of chemical fertilizers and 50% reduction of agrochemicals by 2030" [35], which align with the Japanese government's policies of a "30% reduction of chemical fertilizers and 50% reduction of agrochemicals by 2030" [35].

Clean Agriculture is supposed to be a realistic choice when thinking about the failing example of Sri Lanka. The Sri Lankan government imposed organic farming in 2019, with the complete prohibition of the usage of chemical fertilizers and agrochemicals nationwide. However, the policy caused a 30% reduction in crop yield and a serious shortage of food supply [36].

2.2. *Hokkaido Fertilizer Recommendations: Practice of Clean Agriculture Based on Soil Diagnosis*

Hokkaido Fertilizer Recommendations is a guidebook for applying fertilizers on the basis of soil diagnosis results [34]. The first edition was published in 2002, and it has since been revised three times (2010, 2015, and 2020). Before 2002, the guidelines ("Soil Diagnosis-Based Application of Fertilizers in Hokkaido") and the criteria ("Soil and Plant Nutrition") were published separately, but they were combined into one guidebook in 2002.

In *Hokkaido Fertilizer Recommendations*, the criteria for evaluating the soil in arable fields were established (Table 1), such as topsoil thickness, effective soil depth, saturated hydraulic conductivity, bulk density, available moisture, pH, available P_2O_5, and exchangeable K_2O. Similarly, Doran et al. [2] reported the utility of the following soil physical, chemical, and biological factors for assessing soil health: texture, soil depth, infiltration, bulk density, water retention, total organic carbon (C) and nitrogen (N), pH, electrical conductivity (EC), extractable N, phosphorus (P), potassium (K), microbial biomass C and N, potential mineralizable N (anaerobic incubation), and soil respiration. These criteria seem to be the basis for the recent assessment of soil quality and soil health in North America [37] and the EU [38]. Because *Hokkaido Fertilizer Recommendations* referred to the guidelines in the USA, they share similar assessments to those in North America. However, there are some differences. For example, many soil physical properties are assessed in Hokkaido compared with North America and the EU, although these are listed in other soil health assessment guidelines such as USDA-NRCS in the USA [39] and AHDB-BBRO in the UK [40]. In contrast, there are no indices for soil texture, soil organic matter, and cation exchange capacity (CEC) in *Hokkaido Fertilizer Recommendations* because it is difficult for farmers to improve them immediately. Moreover, exchangeable Na_2O is not listed in the *Hokkaido Fertilizer Recommendations* because the annual precipitation is much higher than evapotranspiration in Hokkaido so that the risk of salinity is much less than in North America and the EU. Furthermore, soil biological properties are listed in North America and the EU but are limited in Hokkaido. Instead, autoclaved N (AC-N) content can alternatively be utilized in Hokkaido.

Table 1. Healthy soil criteria for arable fields in Hokkaido [34], North America [37], and the EU [38].

Soil Properties	Hokkaido's Criteria	Notes	North America	EU
Topsoil thickness	20–30 cm			Yes
Effective soil depths	>50 cm			Yes
Penetration resistance	16–20 mm	Yamanaka penetration meter equivalent to 0.94–1.57 MPa		Yes
Solid phase	0.25–0.30 m^3 m^{-3} <0.40 m^3 m^{-3}	Volcanic soils Lowland and upland soils		
Saturated hydraulic conductivity	10^{-3}–10^{-4} cm s^{-1}		Yes	
Bulk density	0.7–0.9 Mg m^{-3} 0.9–1.1 Mg m^{-3}	Volcanic soils Lowland and upland soils	Yes	Yes
Coarse porosity	0.15–0.25 m^3 m^{-3}	Equivalent to a soil water potential of 0–3.2 kPa		
Available moisture	>0.1 m^3 m^{-3}	Equivalent to a soil water potential of 3.2–100 kPa	Yes	
Topsoil tilth/Soil structure	>70%	Weight proportion of soil clods <20 mm		
Groundwater level	>60 cm			Yes
Compaction	<20 mm	Yamanaka penetration meter at the depth of 10 cm below plow horizon		
pH	5.5–6.5	Soil–water = 1:2.5	Yes	Yes
EC	<0.3–0.8 dS m^{-1}	Only for horticultural soils	Yes	
Available P_2O_5	100–300 mg kg^{-1}	Truog method	Yes	Yes
Exchangeable K_2O	150–300 mg kg^{-1}	Ammonium acetate pH 7 extraction	Yes	Yes
Exchangeable MgO	250–450 mg kg^{-1}	Ammonium acetate pH 7 extraction	Yes	Yes
Exchangeable CaO	1700–3500 mg kg^{-1}	Ammonium acetate pH 7 extraction for medium particle size	Yes	
Ratio Ca/Mg	<6	Equivalent charge ratio		
Ratio Mg/K	>2	Equivalent charge ratio		
Reducible Mn	50–500 mg kg^{-1}	Ammonium acetate pH 7 extraction with hydroquinone	Yes	

Table 1. *Cont.*

Soil Properties	Hokkaido's Criteria	Notes	North America	EU
Hot-water extractable B	0.5–1.0 mg kg^{-1}		Yes	
Soluble Zn	2–40 mg kg^{-1}	Hydrochloric acid extraction	Yes	Yes
Soluble Cu	0.3–8.0 mg kg^{-1}	Hydrochloric acid extraction	Yes	Yes
Exchangeable Ni	<5 mg kg^{-1}	Ammonium acetate pH 7 extraction	Yes	
Hg, As, Cd	–	Controlled by Japanese law		Yes
Aggregate stability	–	–	Yes	
Erosion	–	–		Yes
Texture	–	–	Yes	Yes
Organic matter/C	–	–	Yes	Yes
Total N	–	–	Yes	Yes
Mineralizable N	–	Similar to AC-N	Yes	
CEC	–	–	Yes	Yes
Exchangeable Na_2O	–	–	Yes	Yes
Microbial activity or soil respiration	–	Correlated with AC-N	Yes	Yes

AC-N content is a widely used index for available N, with standard values that vary depending on the crop, region, and soil type (as described later). According to the *Hokkaido Fertilizer Recommendations*, the potential mineralizable N (anaerobic incubation) is also applicable for testing the available N in rice paddies. Specifically, a mixture comprising soil–water (10 g:20 mL) is sealed in a container and incubated at 40 °C for 7 days before determining the NH_4-N content. Different diagnosis values for anaerobically incubated N have been set for each region and soil type. It has been reported that AC-N is significantly correlated with microbial biomass C, glucosidase activity, and soil respiration [41,42]. Therefore, AC-N can be utilized as an index of the biological aspect of soil health conditions.

Although there are some differences between *Hokkaido Fertilizer Recommendations* and the soil health assessment in other countries, the former contains relevant information affecting soil health, including soil micronutrients (e.g., Mn, B, Zn, Cu, and Ni), soil nutrient balance indices (Ca/Mg ratio, Mg/K ratio, base saturation, and Ca saturation), and crop nutrition diagnosis values. Thus, the *Hokkaido Fertilizer Recommendations* are a practical guide for maintaining healthy agricultural soil.

2.3. Wide Dissemination of AC-N as the Index for Available N

Clean Agriculture practices necessitate the accurate evaluation of soil N fertility. For arable crops (e.g., wheat, sugar beet, and potato) and horticultural crops, *Hokkaido Fertilizer Recommendations* use AC-N as the index for available N [34]. This index is easily determined by autoclaving a mixture consisting of 8 g of soil and 80 mL of water in a flask at 105 °C for 60 min and then digesting 20 mL of the filtrate according to the Kjeldahl method [43]. Earlier research indicated that AC-N can be easily and rapidly estimated by measuring the absorbance of the filtrate at 280 nm [44] and performing a near-infrared spectroscopic examination [45].

Although AC-N is more likely to overestimate soil N fertility than aerobically incubated N [46], in most cases, the adaptability of AC-N for the N uptake and N fertilization of individual crops was carefully examined before a soil diagnosis technique was disseminated (as described later). Therefore, AC-N is a widely applicable index for analyzing agricultural soil.

3. Sustainable Crop Production

3.1. Sugar Beet

According to *Hokkaido Fertilizer Recommendations*, soil fertility must be enhanced through the proper application of organic matter [34]. This guidebook provides details regarding how to decrease N fertilization rates when organic matter is applied. In particular, the effects of accumulating N due to the consecutive application of manure are thoroughly explained on the basis of long-term experiments for thirty years [47–49]. Moreover, Fueki et al. [50] observed that the amount of N derived from the applied organic matter determined according to *Hokkaido Fertilizer Recommendations* (Table 2) was closely associated with the residual amount of soil mineral N at a soil depth of 0–100 cm. Therefore, Fueki et al. [50] scored the amount of organic matter and field practice in regard to N, called the "N-score". The sum of N-scores showed the highest correlation with the soil mineral N stocks ranging from 44 to 750 kg ha^{-1} in commercial fields ($r = 0.839$, $p < 0.01$) compared with the total soil nitrogen content ($r = 0.388$, $p < 0.05$) or AC-N ($r = 0.397$, $p < 0.01$) [50]. These results suggest that N-score could be representative of soil mineral N stocks in the depth of 0–100 cm (Figure 2). These results correspond to those in the other regions. In the USA, the sugar beet yield was maximized when the amount of N supplied (fertilizer N + spring soil mineral N) was 150–200 kg ha^{-1} [51]. Therefore, Fueki et al. [23] demonstrated that the N-score (+fertilizer N amount) can be used to predict the N uptake by sugar beet (Figure 3). On the other hand, excess organic matter application did not increase N uptake further as shown by × symbols (Figure 3), suggesting that excess organic matter application did not contribute to the further increase in yields and promoted N loss [52]. Thus, the N fertilization technique for sugar beet production was developed and disseminated. Nowadays, farmers can decide N fertilization rates using the following equation:

$$\text{N fertilization rates (kg N ha}^{-1}\text{)} = 210 - \sum \text{N-score} \quad (1)$$

Table 2. Examples of N-scores for the application of organic matter in fields [34].

Organic Matter Application/Field Practice	N-Score
Cattle manure [1], single-year application	1.0 kg N Mg^{-1} manure
Cattle manure [1], more than 10 years consecutive application	3.0 kg N Mg^{-1} manure
Cattle and pig slurry	1.3 kg N Mg^{-1} slurry
Cattle urine	2.5 kg N Mg^{-1} urine
Pig manure	3.7 kg N Mg^{-1} manure
Poultry manure	13 kg N Mg^{-1} manure
N fertilizer amount at sowing green manure	100% of N amount (kg N ha^{-1})
Incorporating sugar beet leaf	40 kg N ha^{-1}
Converting from paddy rice field to arable field within 2 years	10 kg N ha^{-1}

[1] In this study, cattle manure refers to cattle excrement without composting, composted cattle manure (several years), and cattle manure mixed with bark.

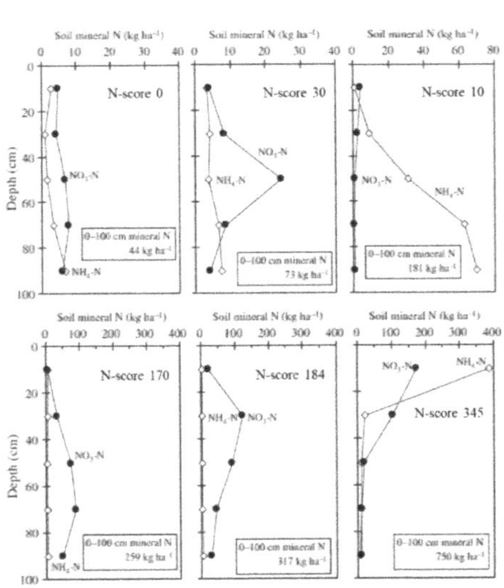

Figure 2. Vertical distribution of soil mineral nitrogen in six typical representative fields modified after [50].

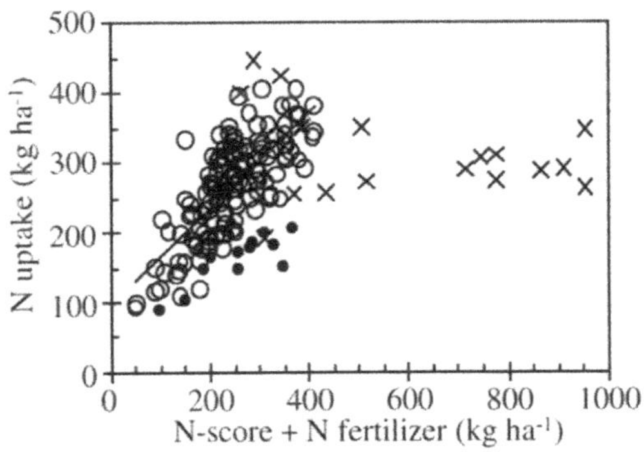

Figure 3. Relationship between N uptake by sugar beet and N-score + N fertilization rate [18]. ×, excess organic matter (N-score exceeds 220 Mg ha^{-1} or consecutive applications more than 30 Mg ha^{-1} yr^{-1}), n = 18; •, confounding problems (diseased sugar beet or the topsoil is too hard), n = 13; ○, the others, n = 140, $y = 0.68x + 100$, $r = 0.75$, $p < 0.001$ [23].

3.2. Potato

In Hokkaido, different potato varieties that are suitable for the production of (1) fresh market potatoes, (2) potatoes that are subsequently processed (e.g., the production of chips, fries, croquettes, and salads), and (3) potatoes as starch resources have been developed. Potatoes that are processed after harvest must have a high starch content, low reducing sugar content, and high dry matter percentage to increase the processing efficiency and production quality at processing plants. Generally, increasing the N uptake by potatoes leads to an increase in tuber yield, but a decrease in starch content [53,54]. Additionally, excessive N uptake adversely affects photosynthetic efficiency and also leads to lodging as well as possibly increased susceptibility to diseases (e.g., late blight and soft rot) [55]. Therefore, optimal N fertilization management practices are needed to improve soil health and increase the starch content.

First, potato plants were cultivated without applying N fertilizers, and soil diagnosis-based techniques were used to measure N uptake. The results indicated that AC-N showed a higher correlation with N uptake ($r = 0.54$, $p < 0.05$) than with the soil mineral N ($r = 0.26$) or incubated mineral N content ($r = 0.47$, $p < 0.05$) [53], reflecting the applicability of AC-N for setting N fertilization rates for potato plants. However, the previous crops in crop rotations are typically sugar beet and pulses, with more N derived from sugar beet residues than from pulses. Accordingly, the previous crops as well as AC-N must be considered when setting N fertilization rates. Another study, which involved fertilizer treatments of potatoes used for processing and various previous crops, revealed that the marketable tuber yields plateaued when the N uptake was 100–110 kg N ha^{-1} regardless of the variety, soil type, AC-N, N fertilization rate, and fertilization method [56]. In addition, the N uptake without N fertilization plateaued when the AC-N was 100 mg kg^{-1} when the previous crops were pulses and green manure. However, the N uptake without N fertilization was much higher when the previous crop was sugar beet rather than pulses or green manure. Accordingly, N fertilization rates for the production of potatoes that will be processed can be determined on the basis of AC-N and previous crops (Table 3) [56].

Table 3. N fertilization rates for potatoes that are processed to produce chips, fries, croquettes, and salads [34].

AC-N	Total N Fertilization Rates (kg N ha^{-1})	
(mg kg^{-1})	Following Sugar Beet	Others
10–20	110	150
30–40	80	120
50–60	40	80
70–80	30	70
90–100	30	50
≥110	30	30

3.3. Winter Wheat

In Hokkaido, winter wheat is mainly cultivated for the production of medium flour (for Japanese "udon" noodles) and hard flour (for bread). Notably, wheat varieties cultivated as a source of medium flour are grown in more than 80% of the total winter wheat cropping area in Hokkaido [57]. Medium flour contains less protein than hard flour. Moreover, the N fertilization rate should be lower for the cultivation of wheat used to produce medium flour. Additionally, excessive N fertilization leads to lodging, which results in a decrease in the harvest efficiency of combine harvesters. Furthermore, because Hokkaido receives snow in winter, soil mineral N is easily lost via infiltration and surface runoff due to snowmelt in spring [24]. Therefore, the split application of N is required for sustainable wheat production while maintaining soil health.

To determine the rates of additional N fertilization between the regrowth stage (after snowmelt) and the panicle formation stage, appropriate soil diagnosis and target depth analyses were conducted. These analyses revealed that N uptake at harvest was related to the abundance of soil nitrate at a depth of 0–60 cm in the regrowth stage [58]. Hence, N fertilization rates can be calculated on the basis of the target yield and the abundance of soil nitrate at a depth of 0–60 cm (Table 4). Similar results were reported in other countries. For example, the optimal N rate was in the range of 71–170 kg N ha^{-1} in the North China Plain [59]. In Denmark, spring N application up to 150 kg N ha^{-1} did not increase the remaining nitrate in the soil at harvest, but N application of more than 150 kg N ha^{-1} retained soil nitrate at harvest [60].

Table 4. Additional N fertilization rates (kg N ha^{-1}) determined according to the target yield of winter wheat and the soil nitrate abundance in a depth of 0–60 cm in the regrowth stage [34].

Target Yield †	Soil Nitrate Abundance in a Depth of 0–60 cm (kg N ha^{-1})								
(Mg ha^{-1})	0	20	40	60	80	100	120	140	160
4.8	80	60	40	20	(20) ‡	(20) ‡	(20) ‡	(20) ‡	(20) ‡
5.4	100	80	60	40	20	(20) ‡	(20) ‡	(20) ‡	(20) ‡
6.0	120	100	80	60	40	20	(20) ‡	(20) ‡	(20) ‡
6.6	(140) ¶	120	100	80	60	40	20	(20) ‡	(20) ‡
7.2	(160) ¶	(140) ¶	120	100	80	60	40	20	(20) ‡
7.8	(180) ¶	(160) ¶	(140) ¶	120	100	80	60	40	20

† Target grain protein content is set at 100 mg g^{-1}. ‡ Minimum additional fertilization rates in the regrowth stage. ¶ Undesirable fertilization rates because they may lead to excessively high protein contents or lodging.

A new high-yielding variety ("Kitahonami") that was subsequently developed responded differently to fertilization compared with previous varieties [61]. More specifically, the additional N fertilization in the regrowth stage increased the yield, but it also resulted in the excessive production of spikes, potentially leading to lodging, especially when the tiller number in the regrowth stage was high. In contrast, the additional N fertilization at the panicle-formation stage resulted in a moderate increase in spike number. Currently, it is recommended that the additional N fertilization rate should be determined according to the tiller number in the regrowth stage (Table 5).

Table 5. Determination of the additional N fertilization rate for winter wheat [34].

Region	Tiller Number at Regrowth Stage	Additional N Fertilization Rates (kg N ha^{-1})	
	(Stem m^{-2})	Regrowth	Panicle Formation
Eastern Hokkaido	>1000	0	Decision from Table 4
	≤1000	20–40 (=A)	(Decision from Table 4) − (A)
Central Hokkaido	≥1300	20	0
	800–1300	60	0 †
	<800	60	40

† 40 kg N ha^{-1} for diluvial upland soils.

Additional N fertilization when the flag leaf appears increases the thousand grain weight [61]. Therefore, a method for setting the additional N fertilization rate in this stage is necessary. Fueki et al. demonstrated a close relationship between N uptake in the flag leaf stage and N uptake at harvest [62]. They also developed a method for estimating the N uptake in the flag leaf stage. Their findings suggested that the additional N fertilization rates at this stage can be calculated using the following equations [34]:

$$\text{N uptake in the flag leaf stage (B, kg N ha}^{-1}) = 0.004 \times \text{superior stem number (stem m}^{-2}) \times \text{leaf color value of the second leaf} - 12 \quad (2)$$

$$\text{Predicted N uptake at harvest without additional N fertilization during the flag leaf stage (C, kg N ha}^{-1}) = 0.58 \times \text{B} + 66 \quad (3)$$

$$\text{Target N uptake at harvest (D, kg N ha}^{-1}) = 0.017 \times \text{target yield (kg ha}^{-1}) + 51 \quad (4)$$

$$\text{N fertilization rates for the flag leaf stage (kg N ha}^{-1}) = (\text{D} - \text{C})/0.7 \quad (5)$$

If farmers did not follow the additional N fertilization rates at the flag leaf stage described above, the N uptake at the mature stage was higher than that when following the recommended fertilization practices, leading to excess protein contents based on Japanese criteria for the medium flour [62]. These wheat grains lead to mineral N remaining in the soil at harvest as well as economic loss to farmers because the grain price decreases.

4. Water Quality

Excess N application results in a high abundance of mineral N remaining in the soil, which lowers crop quality as described in the previous section. Also, the high abundance of mineral N results in N leaching, leading to eutrophication in groundwater, rivers, and oceans [24,63–68]. In Japan, the national upper limit of nitrate concentration in groundwater is 10 mg N L^{-1}. Therefore, it was necessary not to exceed the upper limit of nitrate concentration in agricultural fields.

To maintain water quality, various studies [24,63–68] on the nitrate concentration of leached water in arable fields were conducted. Suzuki and Shiga [24] revealed that the nitrate concentration of leached water did not exceed the upper limit for the N fertilization rates recommended by Clean Agriculture without manure application (Table 6). Also, manure application at 20 Mg ha^{-1} and the reduction of chemical N fertilizer by 20 kg ha^{-1} did not exceed the upper limit on average. However, manure application at 20 Mg ha^{-1} without the reduction of chemical N fertilizer exceeded the upper limit in all years. These results suggest that the nitrate concentration did not exceed the upper limit under the practice of Clean Agriculture. However, the excess N input could exceed the upper limit of nitrate concentration.

Table 6. Nitrate leaching and concentration from arable fields [34], modified to English.

Year	Percolation	No Manure		Manure at 20 Mg ha^{-1}			
		Recommended Chemical Fertilization Rates		No Reduction of Chemical Fertilizer		Reduction by 20 kg N ha^{-1}	
	(mm)	Leaching (kg N ha^{-1})	Conc (mg N L^{-1})	Leaching (kg N ha^{-1})	Conc (mg N L^{-1})	Leaching (kg N ha^{-1})	Conc (mg N L^{-1})
2000	326	25.7	7.9	43.6	13.4	33.3	10.2
2001	391	17.2	4.4	40.6	10.4	32.2	8.2
2002	287	22.6	7.9	34.4	12.0	24.1	8.4
Ave.	335	21.8	6.5	39.6	11.8	29.9	8.9

To expand the previous studies, environmental N capacity and excess N input are defined by the following equations [69]:

$$\text{Environmental N capacity}\left(\text{kg N ha}^{-1}\right) = \text{N output by harvest} + \text{Nitrate-N capacity for groundwater} \quad (6)$$

$$\text{Excess N input} = \text{Total N input} - \text{Environmental N capacity} \quad (7)$$

where the nitrate capacity for groundwater is defined by the Japanese upper limit of nitrate concentration in groundwater (10 mg N L^{-1}) × percolation rate.

Previous studies revealed that the excess N input was significantly correlated with the nitrate concentration in groundwater on the municipal [69] and field [66] scales. Also, these studies suggested that the N leaching was increased when the excess N input exceeded 0 kg N ha^{-1} [66,69], and N fertilization rates equal to the environmental N capacity did not increase the N leaching. Therefore, Clean Agriculture sets the N fertilization rates without surplus N at harvest. A similar concept has been adopted in Poland [70] and the USA [71].

5. Climate Change

Agricultural soils are one of the important sources of greenhouse gas emissions [72]. On the other hand, soils also have the potential to mitigate climate change through appropriate agricultural practices [73,74]. It is expected that the increase in global soil organic matter stocks by 4 per mille can compensate for the global anthropogenic emissions of greenhouse gases [74]. In Hokkaido, the practice of Clean Agriculture did not change soil organic carbon (C) stocks and contents for about 10 years at a depth of 0–0.15 m [75]. This finding suggests that Clean Agriculture can maintain the soil organic C stocks and contents.

Nitrous oxide (N_2O) is another important greenhouse gas, and fertilized soils are an important source of N_2O [72]. Mostly, the N_2O emission is higher with higher surplus N in soils [76–78]. Therefore, it is expected that the same strategy to reduce N leaching can be taken to mitigate climate change.

In arable fields, the annual N_2O emissions of conventional fertilization were 1.85 kg N ha^{-1} yr^{-1} and 1.02 kg N ha^{-1} yr^{-1} for the fields of winter wheat and sugar beet for a three-year average, respectively [52] (Table 7). On the other hand, the annual N_2O emissions of Clean Agriculture by a manure application of 30 Mg ha^{-1} and reduced chemical fertilization of 30 kg N ha^{-1} were significantly decreased [52] (Table 7). In contrast, the excess manure application of 90 Mg ha^{-1} increased N_2O emissions (Table 7). Also, the yields of winter wheat and sugar beet did not change significantly [52].

Table 7. Annual N_2O emissions of conventional chemical fertilization without manure, with Clean Agriculture, and with excess manure application [34], modified to English.

Crop	Treatment	Application Rates		Crop Yield	N_2O Emission
		Chemical N Fertilizer (kg N ha^{-1})	Manure (Mg ha^{-1})	(Mg ha^{-1})	(kg N ha^{-1})
Winter wheat	Conventional	140	0	4.44	1.85
	Clean Agriculture	110	30	6.06	0.22
	Excess manure	110	90	5.45	3.83
Sugar beet	Conventional	210	0	66.43	1.02
	Clean Agriculture	180	30	66.61	0.51
	Excess manure	180	90	76.33	1.21

In grassland, CO_2, CH_4, and N_2O emissions from manure, slurry, and digestive fluid application and reduced chemical fertilizer based on Clean Agriculture were compared [79]. CO_2 (heterotrophic respiration), CH_4, and N_2O emissions did not change among treatments. On the other hand, the net ecosystem C balance (NECB) was significantly lower for manure application (40.5 Mg CO_2-eq ha^{-1} yr^{-1}) than for chemical fertilizer only (64.1 Mg CO_2-eq ha^{-1} yr^{-1}). The greenhouse gas balance (CO_2 equivalent NECB, CH_4, and N_2O) was mitigated 36.2%, 16.5%, and 27.0% for manure, slurry, and digestive fluid, respectively, compared with chemical fertilizer only [79]. Furthermore, grass yield did not change significantly [52,79]. Therefore, Clean Agriculture with manure application and reduced chemical fertilizer can be an important strategy to mitigate climate change without yield loss.

6. Human Health

There are various aspects of the effect of soil health on human health. Lehmann et al. summarized the micronutrients for malnutrition, soil pollutants, and soil host pathogens [1]. In Hokkaido, the main issue related to human health is soil pollutants, especially heavy metals. Although the contents of heavy metals are lower in the parent materials of Andosols than in the other soil groups [80], the control of toxic heavy metals in agricultural soils is important in Hokkaido.

In paddy fields, the cadmium (Cd) contents in rice can be significantly changed by the redox status. A pot experiment revealed that the Cd contents in rice grown (0.01 mg kg^{-1}) with flooding throughout the growing period was significantly lower than that in rice grown (0.24 mg kg^{-1}) without flooding [81]. Moreover, the flooding from transplanting to three weeks before heading can significantly decrease Cd contents in rice from 0.23 mg kg^{-1} to 0.04 mg kg^{-1} in rice grown with the flooding for two weeks after heading [81]. Although all of the results of Cd contents in rice were lower than the Japanese upper limit (0.4 mg kg^{-1}), these results suggest that the Cd contents in rice can be lowered by water management in paddy fields.

Moreover, organic matter application can be a source of heavy metals. Lime-treated sewage sludge compost can be utilized for Clean Agriculture, but it contains zinc (Zn), copper (Cu), arsenic (As), and Cd [82]. Therefore, the application of the compost should be controlled. The application of the compost at 4 Mg ha^{-1} did not change Zn, Cu, and Cd contents in rice compared with the case without compost application for a single year in brown lowland soil and gley lowland soil (Table 8) [82]. Furthermore, the compost application at 2 Mg ha^{-1} for three consecutive years did not change the Zn, Cu, As, and Cd contents in rice compared with the case without compost application in brown lowland soil, gley lowland soil, and peat soil (Table 8). Also, compost application at 2 Mg ha^{-1} did not change acid-soluble Zn (4.9–7.4 mg kg^{-1}), acid-soluble Cu (2.3–5.4 mg kg^{-1}), acid-soluble As (2.0–5.4 mg kg^{-1}), and total Cd (0.19–0.23 mg kg^{-1}) contents in the soil compared with the case without compost application in these three soil types. According to these results, Clean Agriculture allows the application of lime-treated sewage sludge compost with farmers recording the application rates and the actual heavy metal contents.

Table 8. Heavy metal contents in rice grains (mg kg^{-1}) [82], modified to English.

Element	Soil †	Compost (Single Year)				Compost (3 Years)		
		Control	1 Mg	2 Mg	4 Mg	Control	1 Mg	2 Mg
Zn	BL	14.3	12.7	13.2	14.1	10.3	10.7	10.4
	GL	17.9	17.9	17.8	17.7	15.8	15.8	15.4
	PT	17.9	18.2	18.0	–	14.7	14.9	14.0
Cu	BL	0.3	0.2	0.2	0.5	1.2	1.2	0.7
	GL	1.8	1.9	2.4	2.0	3.3	2.9	3.4
	PT	0.7	0.4	0.4	–	1.3	1.1	0.9
As	BL	–	–	–	–	<0.2	<0.2	<0.2
	GL	–	–	–	–	<0.2	<0.2	<0.2
	PT	–	–	–	–	<0.2	<0.2	<0.2
Cd	BL	≤0.02	≤0.02	≤0.02	≤0.02	<0.05	<0.05	<0.05
	GL	≤0.02	≤0.02	≤0.02	≤0.02	<0.05	<0.05	<0.05
	PT	≤0.02	≤0.02	≤0.02	–	<0.05	<0.05	<0.05

† BL, GL, and PT denote brown lowland soil, gley lowland soil, and, peat soil, respectively.

7. Remote Sensing

As mentioned in Section 3, various Clean Agriculture techniques (e.g., N fertilizer management) have been developed in Hokkaido. However, many of these techniques were developed for a single point within the field. To effectively apply Clean Agriculture techniques in the arable fields of Hokkaido where each field exceeds several hectares, it is important to consider the spatial variations in crop growth and soil fertility in many fields. In this section, we introduce some studies that confirmed that remote sensing can facilitate the application of Clean Agriculture techniques in whole fields. Furthermore, we discuss future prospects for remote sensing technology as part of Clean Agriculture initiatives.

7.1. Variable Additional N Application System for Wheat Using Growth Sensors

As shown in Equations (1)–(4), the appropriate additional N application rates for "Kitahonami" wheat during the flag leaf stage can be estimated according to the N uptake at that stage [62]. To apply this method throughout the arable field, the spatial variations in N uptake within the field at a particular stage must be examined. Thus, a "variable additional N application system for wheat using growth sensors" was developed in Hokkaido (Figure 4) [83].

Figure 4. Components of the variable additional N application system for wheat.

This system consists of growth sensors (CropSpec developed by TOPCON Corporation, Tokyo, Japan) mounted on a tractor, a GPS, and a guidance system console. The system is also characterized by sensing crop growth using active optical sensors. Generally, the reflection intensity of crop growth obtained from satellite images is affected by sunlight and aerosols. Therefore, to create an agricultural information map for N uptake using image data, the reflection intensity and the actual measured N uptake for each image must be calibrated. However, the active optical sensors that are used can obtain stable reflection intensities without depending on sunlight. Hence, the reflection intensity can easily be calibrated with the actual crop growth. In other words, once a calibration formula for the reflection intensity and N uptake is created, it can be used to estimate the wheat N uptake in other fields.

The console has a built-in input/output function for sensor values, a calibration formula for sensor values and N uptake, and a formula to convert the N uptake during the flag leaf stage into appropriate additional N application rates. The console performs the following operations in real time: (1) input the sensor values determined by the growth sensors, (2) convert the sensor values into the appropriate additional N application rates, and (3) transfer the N application rates to the console terminal.

Field studies comparing the variable fertilizer (VF) application treatments using this system and the conventional fertilizer (CF) application treatments (uniform fertilization) revealed that the spatial variability in the grain protein contents was lower for the VF treatments than for the CF treatments. Therefore, this system may be useful for restricting the decrease in the wheat quality rank. Furthermore, this system can also be used to calculate variable additional N application rates at the panicle-formation stage of wheat. Specifically, variable application rates are determined on the basis of the difference between the mean N uptake in the field and the N uptake at each location, with data obtained from growth sensors. As shown in Table 9, the studies conducted in six fields indicated that the yield was on average 5% higher for the VF treatments (5.73 t ha^{-1}) than for the CF treatments (5.44 t ha^{-1}). In addition, the coefficient of variance (CV) in the VF treatment was lower than that in the CF treatment for each experiment, suggesting that variable nitrogen application leveled out wheat growth within a field.

Table 9. Wheat yields for the variable and conventional fertilizer application treatments at panicle formation stage [83], edited in English.

Field	Year	Treatment	Yield (t ha^{-1})	CV ¶
A	2003	CF †	6.04	0.214
		VF ‡	6.07	0.204
B	2004	CF †	6.65	0.111
		VF ‡	6.96	0.070
C	2005	CF †	5.38	0.129
		VF ‡	5.91	0.114
D	2010	CF †	4.31	0.085
		VF ‡	4.23	0.078
E	2010	CF †	4.08	0.165
		VF ‡	4.29	0.130
F	2011	CF †	6.17	0.055
		VF ‡	6.89	0.040
Average		CF †	5.44	0.127
		VF ‡	5.73	0.106

† CF: conventional fertilizer application. ‡ VF: variable fertilizer application. ¶ CV: coefficient of variance.

7.2. Variable Basal Nitrogen Application System Using AC-N Maps from Drone Images

As mentioned in Section 2.3, AC-N is used to evaluate soil N fertility in Hokkaido, and methods for estimating the appropriate N application rates according to AC-N have been established for various crops, including sugar beet and potato. In a previous study, AC-N was found to be highly positively correlated with the soil humus content in each field (Figure 5) [84]. In addition, the surface soil humus content in an arable field can be estimated using bare soil images obtained from optical sensors [85]. Thus, a private company in Hokkaido (Zukosha Co., Ltd., Obihiro, Japan) developed a system for determining the variable basal N application rate using the map of AC-N obtained from images captured by the optical camera of a UAV drone [26].

As an example of an AC-N map, Niwa et al. [86] acquired bare soil images of two fields using a visible camera mounted on a UAV drone. These images were separated by color bands of red, green, and blue (RGB) using an image analysis software (PC-MAPPING Version 6.0, Mapcom Inc., Tokyo, Japan). The relationship between RGB and the measured AC-N (Table 10) was investigated with regression analysis for each color band. Then, the color band with the highest R^2 was selected, and the AC-N map was generated by using the images of a single band. The selected color band was not always the same. As shown in Table 11, the blue (R^2 = 0.47) and red (R^2 = 0.79) bands were selected for fields A and B, respectively.

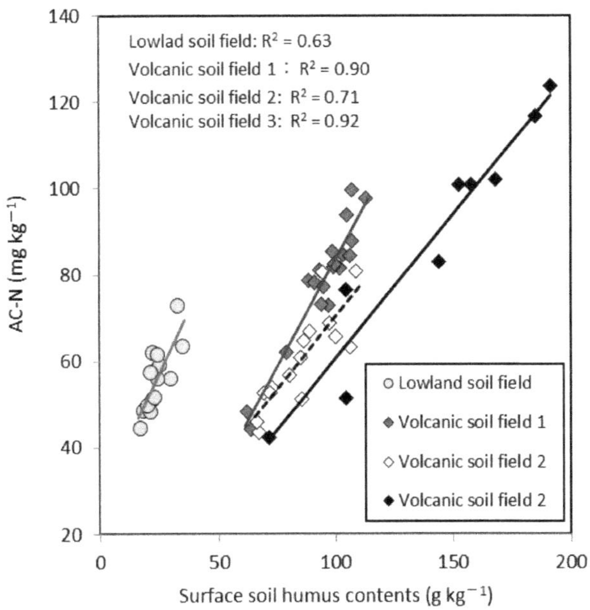

Figure 5. Relationship between AC-N and surface soil humus content under different field conditions [84], edited in English.

Table 10. Actual measured contents of AC-N (mg kg^{-1}) for each field surveyed [86], edited in English.

Field	Maximum	Minimum	Average	CV ¶
A †	110	45	77	0.257
B ‡	145	76	100	0.202

† $n = 16$; ‡ $n = 9$; ¶ CV: coefficient of variance.

Table 11. Results of simple regression analysis of drone-based bare soil image data (red, green, and blue) and AC-N [86], edited in English.

Field	Color	R^2 †	Slope	*p*-Value
A	Red	0.42	−1.08	0.007
	Green	0.44	−0.97	0.005
	Blue	0.47	−1.33	0.003
B	Red	0.79	−0.82	0.001
	Green	0.78	−0.84	0.002
	Blue	0.75	−1.04	0.003

† R^2: coefficient of determination.

The procedure for using this system is used is summarized as follows (Figure 6):

1. An AC-N map of the field is created using bare soil images taken by a UAV-drone and actual AC-N measurements at several points.
2. The map of variable fertilizer application rates is designed on the basis of the AC-N map, the above-mentioned methods for estimating N fertilizer application rates, and the history of organic matter applications and field practices (Table 2). When blended fertilizers are used, variable fertilizer application rates may sometimes be determined after consulting with the user regarding the required phosphate and potassium application amounts.

3. The variable fertilizer application rates are transferred to the application (app) compatible with the Android mobile operating system developed by Zukosha Co., Ltd. (Obihiro, Japan). This app can determine the current location using a GPS and transfer the fertilizer application rates at the current location to the fertilizer applicator terminal via a wireless LAN.

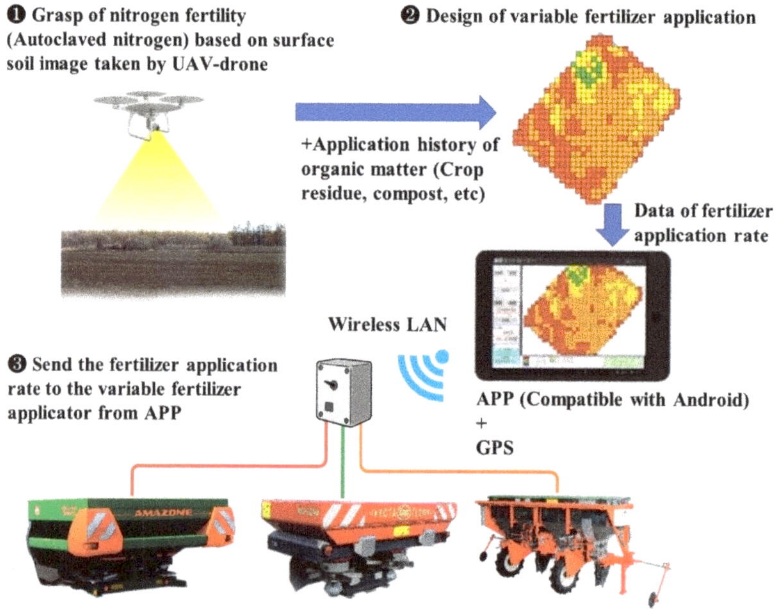

Figure 6. Overview of the variable basal N application system [26].

Recently, ISOBUS fertilizer applicators have been introduced. When using an ISOBUS fertilizer applicator, the app and wireless LAN (Figure 6) are no longer necessary, thereby simplifying the variable fertilizer application.

The utility of this system was investigated in three sugar beet fields comprising volcanic soils in the Tokachi region of Hokkaido in 2015. Briefly, CF and VF treatments were applied to experimental plots established in each field, after which the fertilizer application rates and sugar beet yields for the treatments were compared.

The experimental results for all fields showed that the N fertilizer application rate was on average 25% lower for the VF treatment than for the CF treatment. Moreover, in each field, the sugar beet yield was on average 9% higher for the VF treatment than for the CF treatment (Table 12). Furthermore, in 2017, a similar study was conducted for potatoes in field C using the AC-N field map for variable fertilization in 2015. Compared with the corresponding data for the CF treatment, the N application rate was 18% lower and the marketable potato yield was 15% higher for the VF treatment.

This system has been adapted for various crops, with the AC-N field map applicable for other years. Notably, this system has been introduced in arable fields (more than 1000 ha) in Hokkaido. Furthermore, a local agricultural high school has incorporated this system into their classes.

As mentioned above, in Hokkaido, the technology of variable N application rates has been developed on the basis of crop growth and AC-N so that it is compatible with Clean Agriculture. Similar techniques have been developed on the basis of soil N status [87,88], crop growth detected by sensors [89,90], and their combinations [91,92]. For example, Zhou et al. [87] measured soil total N contents using an on-the-go sensor attached to a tractor. On the basis of the measurements, they generated a map of soil total N, and the N fertilization rates were changed on the basis of the map of soil total N content. The maize grain yield for VF (7275 kg ha^{-1}) was higher than that for CF (6713 kg ha^{-1}).

Table 12. N application rates and sugar beet yields for the variable and conventional fertilizer application treatments [26].

Field	Average N Application Rates [1] (kg N ha^{-1})			Sugar Beet Yields [2] ($Mg\ ha^{-1}$)		
	VF [3]	CF [4]	Ratio (%; VF/CF)	VF [3]	CF [4]	Ratio (%; VF/CF)
A	158	184	14.1	13.3 ± 0.7	12.2 ± 1.1	9.0
B	68	151	55.0	14.6	13.2	10.3
C	108	114	5.3	12.5 ± 0.6	11.7 ± 0.5	6.7
Average	111	150	24.8	13.5	12.4	8.7

[1] N application rates for the VF treatment were 120–200 kg N ha^{-1} for field A, 40–175 kg N ha^{-1} for field B, and 80–140 kg N ha^{-1} for field C. [2] Values represent mean ± standard deviation for fields A and C and one replicate obtained from the harvester for field B. [3] VF: variable fertilizer application. [4] CF: conventional fertilizer application.

There is another approach to delineating the management zone by the apparent electrical conductivity (ECa) [93–95]. ECa is correlated with soil texture, soil moisture, and soil nutrient status. For example, Sanches et al. [95] measured ECa using an electromagnetic induction sensor pulled by a quadricycle. Then, an ECa map was generated by the ordinary kriging interpolation method. The area with higher ECa was attributed to higher base cation and humus contents compared with the area with lower ECa, suggesting that the area with higher ECa was more fertile. Sugarcane yield showed that the area of higher ECa was maximized with lower N fertilization rates compared with the area with lower ECa. Therefore, the potential for the usage of ECa is high for variable N applications.

In contrast, ECa has not been listed in most of the soil health indicators including Clean Agriculture (Table 1). Therefore, the compatibility of variable N application based on ECa would be weak in Hokkaido. For the spatial extension of soil health, it is important to determine N application rates using soil health indicators as well as N fertility. In this context, the variable N application rates based on AC-N are preferable for soil health because AC-N is correlated with soil microbial activity [41,42].

7.3. Determining Suitable Fields for Variable N Application

As shown in Figure 5, the surface humus content is highly correlated with AC-N for each field. In other words, the variability of N fertility within a field can be evaluated on the basis of the surface humus content. Focusing on this point, Niwa et al. [96] investigated the relationship between image data for each wavelength band (RGB and near-infrared) obtained from the satellite image (SPOT) of bare soil in April 2016 and the measured surface humus contents at 64 locations in Hokkaido's arable land. As a result, the highest negative correlation (R^2 = 0.71) was found between the red image data and the measured humus contents, and the surface humus content map (10 m mesh) was generated from the satellite image of the red band. In addition, the normalized difference vegetation index (NDVI) (10 m mesh) of potato and sugar beet was obtained from satellite images (SPOT) taken in early July 2016, which corresponds to the stage of complete foliage cover for sugar beet and the flowering stage for potato. Using the obtained maps, the relationship between the surface humus content and the NDVI was investigated in 123 fields. The year 2016 was characterized by heavy rainfall in June.

As shown in Figure 7, some fields showed a positive correlation, while others showed a negative correlation between surface humus content and NDVI. Approximately 40% of the fields showed positive correlations between these two factors, and variable N application is effective in these fields. However, approximately 20% of the fields showed negative correlations. The soil with high humus content in the study area results from poor drainage during heavy rainfall caused by the excessively high water-holding capacity. For this reason, the drainage in the area with the high surface humus contents should be improved in those fields with negative correlations. The application of a variable N fertilization rate becomes unstable depending on weather conditions.

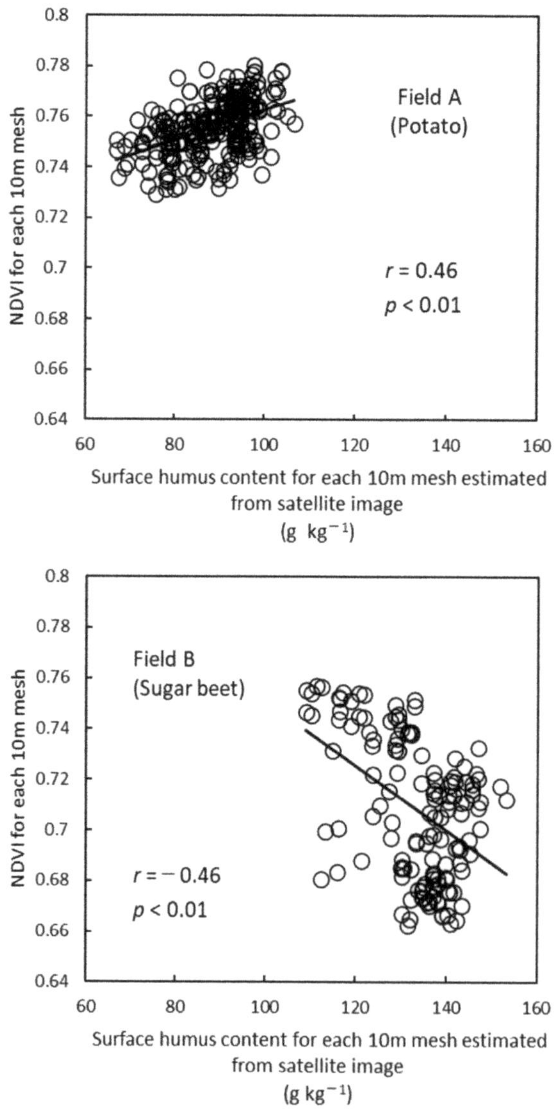

Figure 7. Relationship between surface humus content and crop growth (NDVI) for 10 m mesh estimated from satellite images for each field [96], edited in English.

As described above, variations in crop growth within a field are affected by various factors. Therefore, to conduct variable N application effectively, it is important to understand in advance whether crop growth is affected by N fertility.

7.4. Current Limitations and Future Prospects for Remote Sensing Technology

We have described the technology of variable N application based on remote sensing. In the future, it will be important to estimate various edaphic factors from remote sensing information and to carry out cultivation management and soil improvement based on that information. However, it is difficult to determine various edaphic factors from regression analysis such as those shown in Table 11.

In recent years, more advanced methods such as machine learning have been integrated with remote sensing, and studies on determining various edaphic factors have been conducted [97–100]. For example, Morishita and Ishitsuka [99] took images of bare soils using a UAV drone and attempted to estimate various soil properties by random forest regression. As a result, they showed that particle size, pH (H_2O), EC, and so forth, which were difficult to estimate using conventional regression methods, can be estimated using machine learning.

On the other hand, in recent years, satellite images, including those from Sentinel (European Space Agency) and Landsat (National Aeronautics and Space Administration), have been provided as open data. In terms of Sentinel-2, the 2A and 2B satellites equipped with optical sensors can acquire images with a resolution of 10 m every 5 days. The resolution of Sentinel-2 images is sufficient for revealing the variations in crop growth and soil fertility within arable fields in Hokkaido. In other words, in Hokkaido, agricultural information obtained from multiperiod satellite images can be distributed at a low cost.

Ishikura et al. [101,102] evaluated the variability of NDVI within a field in some wheat fields at the panicle-formation and maturity stages using satellite images. As a result, they showed that areas with relatively low NDVI in both stages showed poor water retention, but areas with low NDVI in the panicle-formation stage and high NDVI in the maturity stage showed poor drainage. Niwa et al. [103] evaluated sugar beet growth within a field in rainy and dry years using NDVI from satellite images. As a result, it was found that some areas of the surveyed fields had low NDVI only in rainy years. In those areas, it was confirmed that the growth of sugar beet was restricted because of waterlogging, which occurred only in a rainy year.

In this way, to understand various edaphic factors from remote sensing, it is important to utilize multiperiod images as well as introduce new analysis methods such as machine learning.

8. Conclusions

In Hokkaido, Clean Agriculture has been applied and refined since the 1990s on the basis of soil diagnosis, appropriate amendment of organic matter, and minimal usage of chemical fertilizers and agrochemicals. Clean Agriculture realizes sustainable crop production, an increase in water quality, the mitigation of climate change, and the maintenance of human health. Although each study has been conducted from the viewpoint of soil quality, implementing Clean Agriculture integrating all of these study outcomes will lead to enhanced soil health as well as soil health indicators in other countries. Moreover, Clean Agriculture has been recommended by the Hokkaido local government, agricultural experiment stations, agricultural extension centers, and other stakeholders. Thus, many farmers have practiced Clean Agriculture in their own arable fields, which contributes to the management of soil health in Hokkaido.

Most farmers request soil diagnosis at only one point in their field. However, soil properties spatially vary within arable fields in Hokkaido. Expanding the use of Clean Agriculture techniques to promote soil health over wide areas will require the incorporation of remote sensing technology into Clean Agriculture practices. Therefore, Clean Agriculture should also be improved to be easily compatible with remote sensing. In the case of N management for winter wheat, the N uptake at the flag leaf stage can be estimated using a handheld NDVI sensor [104], which suggests the possibility of N uptake estimation using satellite images. It is expected that farmers can decide on the N fertilization rate in each field from satellite images in the near future. Although the decision of N fertilization rates by soil diagnosis is accurate, some soil properties have low compatibility with remote sensing. Therefore, further research is required to address the estimation of various soil properties by remote sensing, leading to easy decision-making by farmers. It is expected that these future outcomes of remote sensing and machine learning technologies increasingly enable farmers and other stakeholders to maintain soil health.

Author Contributions: Conceptualization, N.F.; writing—original draft preparation, N.F., K.I. and K.N.; writing—review and editing, K.I.; visualization, K.I.; supervision, N.F. and K.N. All authors have read and agreed to the published version of the manuscript.

Funding: This research received no external funding.

Institutional Review Board Statement: Not applicable.

Data Availability Statement: Not applicable.

Acknowledgments: We appreciate the contributions of the farmers, agricultural extension centers, agricultural cooperatives, and other stakeholders involved in the studies on Clean Agriculture. We also appreciate all of the researchers who helped develop Clean Agriculture techniques.

Conflicts of Interest: Author Katsuhisa Niwa was employed by the company Zukosha Co., Ltd. The remaining authors declare that the research was conducted in the absence of any commercial or financial relationships that could be construed as a potential conflict of interest.

References

1. Lehmann, J.; Bossio, D.A.; Kögel-Knabner, I.; Rillig, M.C. The Concept and Future Prospects of Soil Health. *Nat. Rev. Earth Environ.* **2020**, *1*, 544–553. [CrossRef] [PubMed]
2. Doran, J.W.; Sarrantonio, M.; Liebig, M.A. Soil Health and Sustainability. *Adv. Agron.* **1996**, *56*, 2–55.
3. Doran, J.W.; Parkin, T.B. Defining and Assessing Soil Quality. In *Defining Soil Quality for a Sustainable Environment*; Doran, J.W., Coleman, D.C., Bezdicek, B.F., Stewart, B.A., Eds.; Soil Science Society of America: Madison, WI, USA, 1994; Volume 35, pp. 1–21.
4. Thapa, S.; Bhandari, A.; Ghimire, R.; Xue, Q.; Kidwaro, F.; Ghatrehsamani, S.; Maharjan, B.; Goodwin, M. Managing Micronutrients for Improving Soil Fertility, Health, and Soybean Yield. *Sustainability* **2021**, *13*, 11766. [CrossRef]
5. Silveira, M.L.; Kohmann, M.M. Maintaining Soil Fertility and Health for Sustainable Pastures. In *Management Strategies for Sustainable Cattle Production in Southern Pastures*; Academic Press: New York, NY, USA, 2020; pp. 35–58, ISBN 9780128144749.
6. Chaparro, J.M.; Sheflin, A.M.; Manter, D.K.; Vivanco, J.M. Manipulating the Soil Microbiome to Increase Soil Health and Plant Fertility. *Biol. Fertil. Soils* **2012**, *48*, 489–499. [CrossRef]
7. Karlen, D.L.; Veum, K.S.; Sudduth, K.A.; Obrycki, J.F.; Nunes, M.R. Soil Health Assessment: Past Accomplishments, Current Activities, and Future Opportunities. *Soil. Tillage Res.* **2019**, *195*, 104365. [CrossRef]
8. Lamichhane, S.; Bal Krishna, K.C.; Sarukkalige, R. Polycyclic Aromatic Hydrocarbons (PAHs) Removal by Sorption: A Review. *Chemosphere* **2016**, *148*, 336–353. [CrossRef]
9. Carpenter, S.R.; Caraco, N.F.; Correll, D.L.; Howarth, R.W.; Sharpley, A.N.; Smith, V.H. Nonpoint Pollution of Surface Waters with Phosphorus and Nitrogen. *Ecol. Appl.* **1998**, *8*, 559–568. [CrossRef]
10. Evans, A.E.; Mateo-Sagasta, J.; Qadir, M.; Boelee, E.; Ippolito, A. Agricultural Water Pollution: Key Knowledge Gaps and Research Needs. *Curr. Opin. Environ. Sustain.* **2019**, *36*, 20–27. [CrossRef]
11. Battaglia, M.; Thomason, W.; Fike, J.H.; Evanylo, G.K.; von Cossel, M.; Babur, E.; Iqbal, Y.; Diatta, A.A. The Broad Impacts of Corn Stover and Wheat Straw Removal for Biofuel Production on Crop Productivity, Soil Health and Greenhouse Gas Emissions: A Review. *GCB Bioenergy* **2021**, *13*, 45–57. [CrossRef]

12. Thangarajan, R.; Bolan, N.S.; Tian, G.; Naidu, R.; Kunhikrishnan, A. Role of Organic Amendment Application on Greenhouse Gas Emission from Soil. *Sci. Tot. Environ.* **2013**, *465*, 72–96. [CrossRef]
13. Lal, R. Soil Health and Carbon Management. *Food Energy Secur.* **2016**, *5*, 212–222. [CrossRef]
14. Wall, D.H.; Nielsen, U.N.; Six, J. Soil Biodiversity and Human Health. *Nature* **2015**, *528*, 69–76. [CrossRef]
15. Frąc, M.; Hannula, S.E.; Bełka, M.; Jędryczka, M. Fungal Biodiversity and Their Role in Soil Health. *Front. Microbiol.* **2018**, *9*, 316246. [CrossRef]
16. Wagg, C.; Bender, S.F.; Widmer, F.; Van Der Heijden, M.G.A. Soil Biodiversity and Soil Community Composition Determine Ecosystem Multifunctionality. *Proc. Natl. Acad. Sci. USA* **2014**, *111*, 5266–5270. [CrossRef] [PubMed]
17. Perković, S.; Paul, C.; Vasić, F.; Helming, K. Human Health and Soil Health Risks from Heavy Metals, Micro(Nano)Plastics, and Antibiotic Resistant Bacteria in Agricultural Soils. *Agronomy* **2022**, *12*, 2945. [CrossRef]
18. Morton, C.M.; Pullabhotla, H.; Bevis, L.; Lobell, D.B. Soil Micronutrients Linked to Human Health in India. *Sci. Rep.* **2023**, *13*, 13591. [CrossRef] [PubMed]
19. Barrett, C.B.; Bevis, L.E.M. The Self-Reinforcing Feedback between Low Soil Fertility and Chronic Poverty. *Nat. Geosci.* **2015**, *8*, 907–912. [CrossRef]
20. Kibblewhite, M.G.; Ritz, K.; Swift, M.J. Soil Health in Agricultural Systems. *Philos. T Roy. Soc. B* **2008**, *363*, 685–701. [CrossRef]
21. Adhikari, K.; Hartemink, A.E. Linking Soils to Ecosystem Services—A Global Review. *Geoderma* **2016**, *262*, 101–111. [CrossRef]
22. Bonfante, A.; Basile, A.; Bouma, J. Targeting the Soil Quality and Soil Health Concepts When Aiming for the United Nations Sustainable Development Goals and the EU Green Deal. *Soil* **2020**, *6*, 453–466. [CrossRef]
23. Fueki, N.; Sato, K.; Takeuchi, H.; Sato, H.; Nakatsu, S.; Kato, J. Prediction of Nitrogen Uptake by Sugar Beet (*Beta Vulgaris* L.) by Scoring Organic Matter and Nitrogen Management (N-Score), in Hokkaido, Japan. *Soil. Sci. Plant Nutr.* **2011**, *57*, 411–420. [CrossRef]
24. Suzuki, K.; Shiga, H. The Maximum Permissible Amount of Nitrogen Input into an Andisol Upland Field in Abashiri Area Assessed by Percolate Nitrate Concentrations. *Jpn. J. Soil. Sci. Plant Nutr.* **2004**, *75*, 45–52, (In Japanese with English abstract). [CrossRef]
25. Department of Agriculture Hokkaido Local Government Clean Agriculture of Hokkaido. Available online: https://www.pref.hokkaido.lg.jp/ns/shs/clean/ (accessed on 30 December 2023).
26. Nakatsuji, T.; Shiga, H.; Takeuchi, H.; Tsukamoto, Y.; Goto, E.; Watanabe, Y.; Sakurai, M.; Fueki, N.; Hikasa, Y.; Hayashi, T.; et al. Hokkaido Region. In *The Soils of Japan*; Hatano, R., Shinjo, H., Takata, Y., Eds.; Springer: Singapore, 2021; pp. 135–184.
27. Souma, S. Direction and Technological Agenda of Environmentally Harmonized Agriculture (Clean Agriculture) Aimed in Hokkaido. *J. Agr. Sci.* **1992**, *47*, 193–198. (In Japanese)
28. Department of Agricultural Research Hokkaido Research Organization List of Research Outcomes. Available online: https://www.hro.or.jp/agricultural/center/result/kenkyuseika.html (accessed on 24 December 2023).
29. Tripathi, A.; Tiwari, R.K.; Tiwari, S.P. A Deep Learning Multi-Layer Perceptron and Remote Sensing Approach for Soil Health Based Crop Yield Estimation. *Int. J. Appl. Earth Obs. Geoinf.* **2022**, *113*, 102959. [CrossRef]
30. Hussain, Z.; Deng, L.; Wang, X.; Cui, R.; Liu, G. A Review of Farmland Soil Health Assessment Methods: Current Status and a Novel Approach. *Sustainability* **2022**, *14*, 9300. [CrossRef]
31. Guo, M. Soil Health Assessment and Management: Recent Development in Science and Practices. *Soil. Syst.* **2021**, *5*, 61. [CrossRef]
32. Omer, E.; Szlatenyi, D.; Csenki, S.; Alrwashdeh, J.; Czako, I.; Láng, V. Farming Practice Variability and Its Implications for Soil Health in Agriculture: A Review. *Agriculture* **2024**, *14*, 2114. [CrossRef]
33. Souma, S. Agenda and Perspective of Clean Agriculture. *Hokuno* **1995**, *62*, 314–317. (In Japanese)
34. Department of Agriculture Hokkaido Prefectural Government. *Hokkaido Fertilizer Recommendations 2020*; Department of Agricultural Research Hokkaido Research Organization, Ed.; Department of Agriculture, Hokkaido Prefectural Government: Sapporo, Japan, 2020. (In Japanese)
35. Ministry of Agriculture Forestry and Fisheries MIDORI Strategy for Sustainable Food Systems. Available online: https://www.maff.go.jp/j/kanbo/kankyo/seisaku/midori/ (accessed on 17 March 2025).
36. Hatano, R. Development of International Collaboration and Our Challenges to Regional Issues in Soil Science and Plant Nutrition Research 2. Progress of Internationalization of Soil Science in Japan. *Jpn. J. Soil. Sci. Plant Nutr.* **2022**, *93*, 405–410. (In Japanese) [CrossRef]
37. Norris, C.E.; Mac Bean, G.; Cappellazzi, S.B.; Cope, M.; Greub, K.L.H.; Liptzin, D.; Rieke, E.L.; Tracy, P.W.; Morgan, C.L.S.; Honeycutt, C.W. Introducing the North American Project to Evaluate Soil Health Measurements. *Agron. J.* **2020**, *112*, 3195–3215. [CrossRef]
38. Panagos, P.; Broothaerts, N.; Ballabio, C.; Orgiazzi, A.; De Rosa, D.; Borrelli, P.; Liakos, L.; Vieira, D.; Van Eynde, E.; Arias Navarro, C.; et al. How the EU Soil Observatory Is Providing Solid Science for Healthy Soils. *Eur. J. Soil. Sci.* **2024**, *75*, e13507. [CrossRef]

39. Friedman, D.; Hubbs, M.; Tugel, A.; Seybold, C.; Sucik, M. *Guidelines for Soil Quality Assessment in Conservation Planning*; Joubert, B., Ed.; United States Department of Agriculture, Natural Resources Conservation Service, Soil Quality Institute: Washington, DC, USA, 2001.
40. Agriculture and Horticulture Development Board; British Beet Research Organization. *Soil Health Scorecard Approach Sampling Protocol and Benchmarking Tables England and Wales Version 1.0*; Agriculture and Horticulture Development Board: Coventry, UK, 2022.
41. Okumura, M.; Mino, K.; Miki, N.; Suzuki, K.; Yamagami, M.; Kitagawa, I. Relationship between α-Glucosidase Activity, Microbial Biomass and Physical Properties in the Volcanic Ash Soils Distributed in the Central Part of Hokkaido. *Jpn. J. Soil. Sci. Plant Nutr.* **1998**, *69*, 340–347, (In Japanese with English abstract). [CrossRef]
42. Higashida, S.; Tamura, H.; Yamagami, M. Microbial Activities in Upland Fields and Factors Relating Them. *Bull. Hokkaido Pref. Agr. Exp. Stn.* **1996**, *70*, 17–26, (In Japanese with English abstract).
43. Department of Agriculture Hokkaido Research Organization. *Analytical Methods for Soil and Crop Nutrition Diagnosis*; Department of Agriculture Hokkaido Research Organization: Sapporo, Japan, 2012.
44. Sakaguchi, M.; Sakurai, M.; Nakatsuji, T. Rapid Analysis of Autoclave-Extractable Nitrogen for Assessing Soil Nitrogen Fertility by UV Absorptiometry of L−tryptophan as a Standard Substance. *Jpn. J. Soil. Sci. Plant Nutr.* **2010**, *81*, 130–134. (In Japanese) [CrossRef]
45. Okazaki, T.; Fueki, N.; Koyano, S.; Tanaka, T.; Sato, T.; Ozawa, T.; Ueda, H. Estimation of Soil Autoclaved Nitrogen of Upland and Grassland Soils by Near Infrared Spectroscopy. *Jpn. J. Soil. Sci. Plant Nutr.* **2020**, *91*, 228–231. (In Japanese) [CrossRef]
46. Matsunaga, T.; Moriizumi, M. Methods of Soil Analysis for Available Nitrogen: Past Progress and Future Prospects. *Jpn. J. Soil. Sci. Plant Nutr.* **2012**, *83*, 625–629. (In Japanese) [CrossRef]
47. Nakatsu, S.; Tamura, H. Effects of Thirty Years Continuous Application of Organic Materials (Bark Manure and Crop Residues) on Total Carbon, Total Nitrogen and Physical Characteristics of Upland Field Soil in Light Colored Andosol in Hokkaido. *Jpn. J. Soil. Sci. Plant Nutr.* **2008**, *79*, 139–145, (In Japanese with English abstract). [CrossRef]
48. Nakatsu, S.; Higashida, S.; Yamagami, M. Effects of Continuous Manure Application on Yield and Quality of Upland Crop and Soil in Light Colored Andosol. *Jpn. J. Soil. Sci. Plant Nutr.* **2000**, *71*, 97–100. (In Japanese) [CrossRef]
49. Nakatsu, S.; Tamura, H. Effects of Thirty Years Continuous Application of Organic Matter (Cattle Manure with Bark and Crop Residues) on Yield and Quality of Upland Crops in Light Colored Andosol in Hokkaido. *Bull. Hokkaido Pref. Agr. Exp. Stn.* **2009**, *94*, 81–88, (In Japanese with English abstract).
50. Fueki, N.; Sato, K.; Nakatsu, S. Interpretation of Soil Mineral Nitrogen by Scoring Organic Matter and Nitrogen Management as an "N-Score" in the Fields of Hokkaido before Sugar Beet Planting. *Soil. Sci. Plant Nutr.* **2010**, *56*, 750–759. [CrossRef]
51. Tarkalson, D.D.; Bjorneberg, D.L.; Camp, S.; Dean, G.; Elison, D.; Foote, P. Improving Nitrogen Management in Pacific Northwest Sugarbeet Production. *J. Sugar Beet Res.* **2016**, *53*, 14–38. [CrossRef]
52. Ishikura, K. Reduction of Greenhouse Gas Emission by Clean Agriculture. *Greentechno Inf.* **2023**, *19*, 12–15. (In Japanese)
53. Taniguchi, T. The Effect of Manuring Practice on Processing Quality of Potato. *Jpn. J. Soil. Sci. Plant Nutr.* **1992**, *63*, 723–727. (In Japanese) [CrossRef]
54. Vos, J. Nitrogen Responses and Nitrogen Management in Potato. *Potato Res.* **2009**, *52*, 305–317. [CrossRef]
55. Oka, H. Studies on the High Productivity of Potato in the Tokachi District of Hokkaido Part II Relation between the Growth of Potatoes and the Use of Applied Nitrogen under Heavy Application of Phosphoric Acid. *Res. Bull. Hokkaido Natl. Agr. Exp. Stn.* **1969**, *95*, 46–52, (In Japanese with English abstract).
56. Fueki, N.; Otsuka, S.; Tamura, H.; Nakamoto, H.; Watanabe, Y. Nitrogen Fertilization Method for Potato Processing for Chips, Fries, Croquettes and Salads Based on Relationships between Vendible Tuber Yield, Nitrogen Uptake by Potatoes, and Soil Autoclaved Nitrogen. *Jpn. J. Soil. Sci. Plant Nutr.* **2020**, *91*, 341–350, (In Japanese with English abstract). [CrossRef]
57. Ministry of Agriculture Forestry and Fisheries Trends of Production and Distribution of Domestic Wheat. Available online: https://www.maff.go.jp/j/seisan/boueki/mugi_zyukyuu/attach/pdf/index-10.pdf (accessed on 3 December 2023).
58. Sato, K.; Nakatsu, S.; Miki, N.; Nakamura, R.; Fueki, N.; Shiga, H. Recommendation of Nitrogen Fertilizer Application for Winter Wheat Based on Using Diagnosis of Soil Nitrate-N in Early Spring in Hokkaido. *Jpn. J. Soil. Sci. Plant Nutr.* **2008**, *79*, 45–51, (In Japanese with English abstract). [CrossRef]
59. Cui, Z.; Zhang, F.; Chen, X.; Dou, Z.; Li, J. In-Season Nitrogen Management Strategy for Winter Wheat: Maximizing Yields, Minimizing Environmental Impact in an over-Fertilization Context. *Field Crops Res.* **2010**, *116*, 140–146. [CrossRef]
60. Rasmussen, I.S.; Dresbøll, D.B.; Thorup-Kristensen, K. Winter Wheat Cultivars and Nitrogen (N) Fertilization-Effects on Root Growth, N Uptake Efficiency and N Use Efficiency. *Eur. J. Agron.* **2015**, *68*, 38–49. [CrossRef]
61. Fueki, N.; Nakamura, R.; Sawaguchi, A.; Watanobe, K.; Suzuki, T.; Uchida, T.; Onodera, M. Effect of Timing of Additional N Fertilization on Spike Number, Grain Yield, Grain Protein and N Use Efficiency of Winter Wheat Cultivar "Kitahonami. " *Bull. Hokkaido Res. Organization Agr. Exp. Stn.* **2015**, *99*, 61–72.

62. Fueki, N.; Nakamura, R.; Sawaguchi, A.; Watanobe, K.; Suzuki, T.; Uchida, T.; Onodera, M. Prediction of Nitrogen Uptake by Winter Wheat (*Triticum Aestivum* L.) by Measurement of Superior Stem Number and Leaf Color Value, for Decision-Making Regarding Additional Nitrogen Fertilization. *Soil. Sci. Plant Nutr.* **2015**, *61*, 769–774. [CrossRef]
63. Hayashi, Y.; Hatano, R. Annual Nitrogen Leaching to Subsurface Drainage Water from a Clayey Aquic Soil Cultivated with Onions in Hokkaido, Japan. *Soil. Sci. Plant Nutr.* **1999**, *45*, 451–459. [CrossRef]
64. Hatano, R.; Nagumo, T.; Hata, H.; Kuramochi, K. Impact of Nitrogen Cycling on Stream Water Quality in a Basin Associated with Forest, Grassland, and Animal Husbandry, Hokkaido, Japan. *Ecol. Eng.* **2005**, *24*, 509–515. [CrossRef]
65. Yang, X.; Lu, Y.; Ding, Y.; Yin, X.; Raza, S.; Tong, Y. Optimising Nitrogen Fertilisation: A Key to Improving Nitrogen-Use Efficiency and Minimising Nitrate Leaching Losses in an Intensive Wheat/Maize Rotation (2008–2014). *Field Crops Res.* **2017**, *206*, 1–10. [CrossRef]
66. Nakatsuji, T.; Fueki, N.; Nakatsu, S.; Suzuki, K.; Shiga, H. Validity of a Concept of Permissible Nitrogen Input as an Index for Risk Assessment of Nitrate Pollution in Groundwater. *Jpn. J. Soil. Sci. Plant Nutr.* **2016**, *87*, 360–364. (In Japanese) [CrossRef]
67. Liang, L.; Nagumo, T.; Hatano, R. Nitrogen Flow in the Rural Ecosystem of Mikasa City in Hokkaido, Japan. *Pedosphere* **2006**, *16*, 264–272. [CrossRef]
68. Nagumo, T.; Hatano, R. Regional Characteristics of Stream Water Quality during the Snow-Melting Season in Hokkaido. *Jpn. J. Soil. Sci. Plant Nutr.* **2001**, *72*, 41–48, (In Japanese with English abstract). [CrossRef]
69. Matsumoto, T.; Tou, S. Risk Assessment of Nitrate Pollution in Groundwater Based on the Environmental Nitrogen-Assimilation Capacity of Agricultural Lands in Hokkaido. *Jpn. J. Soil. Sci. Plant Nutr.* **2006**, *77*, 17–24, (In Japanese with English abstract). [CrossRef]
70. Dybowski, D.; Dzierzbicka-Glowacka, L.A.; Pietrzak, S.; Juszkowska, D.; Puszkarczuk, T. Estimation of Nitrogen Leaching Load from Agricultural Fields in the Puck Commune with an Interactive Calculator. *PeerJ* **2020**, *8*, e8899. [CrossRef]
71. Meisinger, J.J.; Delgado, J.A. Principles for Managing Nitrogen Leaching. *J. Soil. Water Conserv.* **2002**, *57*, 485–498. [CrossRef]
72. Calvin, K.; Dasgupta, D.; Krinner, G.; Mukherji, A.; Thorne, P.W.; Trisos, C.; Romero, J.; Aldunce, P.; Barrett, K.; Blanco, G.; et al. *Climate Change 2023: Synthesis Report. Contribution of Working Groups I, II and III to the Sixth Assessment Report of the Intergovernmental Panel on Climate Change*; Core Writing Team, Lee, H., Romero, J., Eds.; IPCC: Geneva, Switzerland, 2023.
73. Lal, R. Soil Carbon Sequestration Impacts on Global Climate Change and Food Security. *Science* **2004**, *304*, 1623–1627. [CrossRef] [PubMed]
74. Minasny, B.; Malone, B.P.; McBratney, A.B.; Angers, D.A.; Arrouays, D.; Chambers, A.; Chaplot, V.; Chen, Z.S.; Cheng, K.; Das, B.S.; et al. Soil Carbon 4 per Mille. *Geoderma* **2017**, *292*, 59–86. [CrossRef]
75. Iwasaki, S.; Endo, Y.; Hatano, R. The Effect of Organic Matter Application on Carbon Sequestration and Soil Fertility in Upland Fields of Different Types of Andosols. *Soil. Sci. Plant Nutr.* **2017**, *63*, 200–220. [CrossRef]
76. Van Groenigen, J.W.; Velthof, G.L.; Oenema, O.; Van Groenigen, K.J.; Van Kessel, C. Towards an Agronomic Assessment of N_2O Emissions: A Case Study for Arable Crops. *Eur. J. Soil. Sci.* **2010**, *61*, 903–913. [CrossRef]
77. Shcherbak, I.; Millar, N.; Robertson, G.P. Global Metaanalysis of the Nonlinear Response of Soil Nitrous Oxide (N_2O) Emissions to Fertilizer Nitrogen. *Proc. Natl. Acad. Sci. USA* **2014**, *111*, 9199–9204. [CrossRef]
78. Nagatake, A.; Mukumbuta, I.; Yasuda, K.; Shimizu, M.; Kawai, M.; Hatano, R. Temporal Dynamics of Nitrous Oxide Emission and Nitrate Leaching in Renovated Grassland with Repeated Application of Manure and/or Chemical Fertilizer. *Atmosphere* **2018**, *9*, 485. [CrossRef]
79. Kitamura, R.; Sugiyama, C.; Yasuda, K.; Nagatake, A.; Yuan, Y.; Du, J.; Yamaki, N.; Taira, K.; Kawai, M.; Hatano, R. Effects of Three Types of Organic Fertilizers on Greenhouse Gas Emissions in a Grassland on Andosol in Southern Hokkaido, Japan. *Front. Sustain. Food Syst.* **2021**, *5*, 649613. [CrossRef]
80. Mizuno, N.; Amano, Y.; Mizuno, T.; Nanzyo, M. Changes in the Heavy Mineral Content and Element Concentration of Tarumae-a Tephra with Distance from the Source Volcano. *Soil. Sci. Plant Nutr.* **2008**, *54*, 839–845. [CrossRef]
81. Nakatsu, S.; Nakamoto, H.; Matsumoto, T.; Igarashi, T.; Sugawara, A. Factors Affecting Cadmium Concentration in Lowland Rice in Hokkaido and Countermeasures for the Control of Cadmium Uptake. *Jpn. J. Soil. Sci. Plant Nutr.* **2010**, *81*, 514–517. (In Japanese)
82. Sugikawa, Y.; Goto, E.; Asaka, D. Application Standard of Lime-Treated Sewage Sludge Compost for Paddy Rice in Hokkaido. *Jpn. J. Soil. Sci. Plant Nutr.* **2009**, *80*, 530–533. (In Japanese) [CrossRef]
83. Hara, K.; Suda, T.; Watanobe, K. Evaluation of Sensor-Based Variable Rate Nitrogen Fertilization for Winter Wheat Production. *J. Jpn. Soc. Agr. Mach. Food Eng.* **2015**, *77*, 485–493, (In Japanese with English abstract). [CrossRef]
84. Niwa, K.; Yokobori, J.; Yoneyama, A.; Shinagawa, H. The Monitoring of Soil Information within an Agricultural Field Using Low Altitude Remote Sensing. *Water Land. Environ. Eng.* **2016**, *84*, 752–794. (In Japanese) [CrossRef]
85. Niwa, K.; Seino, N.; Akashi, N.; Kikuchi, K. Drawing of Large-Scaled Soil Map on Volcanic Acid Soil Area in Tokachi District of Hokkaido. *Jpn. J. Soil. Sci. Plant Nutr.* **2004**, *75*, 69–78, (In Japanese with English abstract). [CrossRef]

86. Niwa, K.; Yokobori, J.; Hara, K.; Fueki, N.; Wakabayashi, M. Use of Remote-Sensing Data on Soil Nitrogen Availability and Wheat Growth to Estimate Factors That Affect within-Field Variation in Wheat Growth. *Jpn. J. Soil. Sci. Plant Nutr.* **2018**, *89*, 544–551. (In Japanese with English abstract) [CrossRef]
87. Zhou, P.; Ou, Y.; Yang, W.; Gu, Y.; Kong, Y.; Zhu, Y.; Jin, C.; Hao, S. Variable-Rate Fertilization for Summer Maize Using Combined Proximal Sensing Technology and the Nitrogen Balance Principle. *Agriculture* **2024**, *14*, 1180. [CrossRef]
88. Shahandeh, H.; Wright, A.L.; Hons, F.M. Use of Soil Nitrogen Parameters and Texture for Spatially-Variable Nitrogen Fertilization. *Precis. Agric.* **2011**, *12*, 146–163. [CrossRef]
89. Mirzakhaninafchi, H.; Singh, M.; Bector, V.; Gupta, O.P.; Singh, R. Design and Development of a Variable Rate Applicator for Real-Time Application of Fertilizer. *Sustainability* **2021**, *13*, 8694. [CrossRef]
90. Basso, B.; Fiorentino, C.; Cammarano, D.; Schulthess, U. Variable Rate Nitrogen Fertilizer Response in Wheat Using Remote Sensing. *Precis. Agric.* **2016**, *17*, 168–182. [CrossRef]
91. Guerrero, A.; De Neve, S.; Mouazen, A.M. Data Fusion Approach for Map-Based Variable-Rate Nitrogen Fertilization in Barley and Wheat. *Soil. Tillage Res.* **2021**, *205*, 104789. [CrossRef]
92. Guerrero, A.; De Neve, S.; Mouazen, A.M. Current Sensor Technologies for in Situ and On-Line Measurement of Soil Nitrogen for Variable Rate Fertilization: A Review. *Adv. Agron.* **2021**, *168*, 1–38. [CrossRef]
93. Heiß, A.; Paraforos, D.S.; Sharipov, G.M.; Griepentrog, H.W. Real-Time Control for Multi-Parametric Data Fusion and Dynamic Offset Optimization in Sensor-Based Variable Rate Nitrogen Application. *Comput. Electron. Agric.* **2022**, *196*, 106893. [CrossRef]
94. Kazlauskas, M.; Šarauskis, E.; Lekavičienė, K.; Naujokienė, V.; Romaneckas, K.; Bručienė, I.; Buragienė, S.; Steponavičius, D. The Comparison Analysis of Uniform-and Variable-Rate Fertilizations on Winter Wheat Yield Parameters Using Site-Specific Seeding. *Processes* **2022**, *10*, 2717. [CrossRef]
95. Sanches, G.M.; Faria, H.M.; Otto, R.; Neto, A.S.; Corá, J.E. Using Soil Apparent Electrical Conductivity (ECa) to Assess Responsiveness of Nitrogen Rates and Yield in Brazilian Sugarcane Fields. *Agronomy* **2025**, *15*, 606. [CrossRef]
96. Niwa, K.; Yokobori, J.; Ishikura, K.; Hara, K.; Fueki, N.; Imada, S. The Possibility of Introducing Variable-Rate Nitrogen Application in Uplands of Andosol Areas as Evaluated by Surface Soil and Crop Growth Attributes Estimated from Satellite Images. *Jpn. J. Soil. Sci. Plant Nutr.* **2021**, *92*, 249–254, (In Japanese with English abstract). [CrossRef]
97. Takata, Y.; Yamada, H.; Kanuma, N.; Ise, Y.; Kanda, T. Digital Soil Mapping Using Drone Images and Machine Learning at the Sloping Vegetable Fields in Cool Highland in the Northern Kanto Region, Japan. *Soil. Sci. Plant Nutr.* **2023**, *69*, 221–230. [CrossRef]
98. Morishita, M.; Ishitsuka, N. Mapping of Soil Management Zones by Unsupervised Classification of Drone Aerial Images. *Jpn. J. Soil. Sci. Plant Nutr.* **2023**, *94*, 254–262, (In Japanese with English abstract). [CrossRef]
99. Morishita, M.; Ishitsuka, N. Estimation of Soil Properties Distribution Using UAV Observation and Machine Learning-Application of Data Augmentation to Soil Physicochemical Properties. *J. Jpn. Agr. Syst. Soc.* **2021**, *37*, 21–28. (In Japanese with English abstract) [CrossRef]
100. Morishita, M.; Ishitsuka, N. Estimation of Soil Moisture Distribution in Soybean Field Using UAV-Application of Machine Learning by Data Augmentation of Ground Truth. *J. Jpn. Agr. Syst. Soc.* **2020**, *36*, 55–61. (In Japanese with English abstract) [CrossRef]
101. Ishikura, K.; Fueki, N.; Hara, K.; Niwa, K.; Seshimo, T. Evaluation of Soil Physical Properties in Agricultural Fields Using Satellite Images and Topographic Information I. Poor Water Retention. *Jpn. J. Soil. Sci. Plant Nutr.* **2024**, *95*, 11–20. (In Japanese with English abstract) [CrossRef]
102. Ishikura, K.; Fueki, N.; Hara, K.; Niwa, K.; Seshimo, T. Evaluation of Soil Physical Properties in Agricultural Fields Using Satellite Images and Topographic Information II. Poor Drainage. *Jpn. J. Soil. Sci. Plant Nutr.* **2024**, *95*, 21–29. [CrossRef]
103. Niwa, K.; Yokobori, J.; Imada, S. Estimation of Factors Affecting Sugar Beet Growth in Lowland Soils Based on Satellite Imagery Data from Two Different Years. *Jpn. J. Soil. Sci. Plant Nutr.* **2025**, *in press* (In Japanese with English abstract).
104. Ishikura, K.; Fueki, N.; Suda, T.; Sugikawa, Y.; Tou, S. Estimation of Nitrogen Uptake and Tiller Number of Winter Wheat Using a Handheld Optical Sensor in Hokkaido, Japan. *Soil. Sci. Plant Nutr.* **2020**, *66*, 828–836. [CrossRef]

agriculture

MDPI

Review

Impacts of Biochar Application on Inorganic Phosphorus Fractions in Agricultural Soils

Liwen Lin [1], Yutao Peng [1,2], Lin Zhou [2], Baige Zhang [3], Qing Chen [2] and Hao Chen [1,*]

[1] School of Agriculture and Biotechnology, Shenzhen Campus of Sun Yat-sen University, Shenzhen 518107, China; linlw5@mail2.sysu.edu.cn (L.L.); pengyt39@mail.sysu.edu.cn (Y.P.)

[2] Beijing Key Laboratory of Farmyard Soil Pollution Prevention-Control and Remediation, College of Resources and Environmental Sciences, China Agricultural University, Beijing 100193, China; s20223030430@cau.edu.cn (L.Z.); qchen@cau.edu.cn (Q.C.)

[3] Vegetable Research Institute, Guangdong Academy of Agricultural Sciences, Guangzhou 510640, China; plantgroup@126.com

* Correspondence: chenh626@mail.sysu.edu.cn

Abstract: Inorganic phosphorus (P) is a key component of soil P pools, influencing their availability and mobility. Although studies on biochar's effect on inorganic P fractions in various soils are growing, a critical review of these findings is lacking. Herein, we conducted a quantitative meta-analysis of 74 peer-reviewed datasets, drawing general conclusions and confirming the absence of publication bias through funnel plot statistics. The results showed that biochars can influence soil inorganic P fractions, with their effects depending on biochar (i.e., feedstock, pyrolysis temperature and time, C:N ratio, pH, ash and P content) and soil-related properties (i.e., pH, texture, P content). Specifically, the addition of biochar significantly enhanced the diverse soil inorganic P fractions and P availability (as indicated by Olsen-P). Only biochars produced from wood residues and having high C/N ratios (>200) did not significantly increase the labile P fractions (water extracted soil phosphorus (H_2O-P), Olsen-P, and soil calcium compounds bound phosphorus (Ca_2-P)). The application of biochars derived from crop residues significantly increased the soil P associated with iron and aluminum oxides, while there was no significant effect on manure- and wood residue-derived biochars. In addition, applications of low temperature biochars and manure residue-derived biochars could increase the proportions of soil highly stable P. We identified knowledge gaps in biochar production and its potential for soil phosphorus regulation. Due to the complex processes by which biochar affects soils, more systematic evaluations and predictive methods (e.g., modeling, machine learning) are needed to support sustainable agriculture and environmental practices.

Keywords: soil phosphorus form; biochar properties; quantative analysis; soil texture; phosphorus availability

1. Introduction

Phosphorus (P) plays a role in the cell development of all organisms as an indispensable element [1]. P supply in soil is essential for natural as well as managed ecosystems because P often limits primary productivity [2,3]. This is due to the fact that only a small fraction (<1%, predominantly as orthophosphate anion form) of the total P in the soil solution is readily available for plant assimilation [3–5]. To meet the plant's P requirements, a variety of P-containing materials have been used as soil amendments to improve soil P availability, among which the use of P fertilizer derived from non-renewable phosphate rock has increased rapidly and has become dominant since the 1950s [6–8]. While around

15 million tons of P fertilizer is globally applied to agricultural soils each year [9], less than 30% of the P in fertilizers can be taken up by crops in the year after application [10]. Thus, more efficient P management in agroecosystems is required because P deposits are finite and losses of excess P to natural environments can cause adverse effects such as eutrophication [11] and biodiversity variation [12].

Soil P can be supplied not only by mineral fertilizers but also by organic materials from sources such as agroforestry, the breeding industry, and human habitats. Recycling these P resources has great potential to supplement or replace traditional fertilizers, promoting more sustainable agriculture [9,13]. Among these potential new strategies, biochar has gained significant interest in recent years because of its improvements in P management and many other merits for soil quality enhancement [14–16]. Biochar is a carbon-rich material produced through the process of pyrolysis, which involves heating organic biomass—such as crop residues, wood, manure, or other plant materials—in a low-oxygen environment. It is also worth mentioning that as carbon-neutral and environmental-friendly technologies, biochar production and application also meet the trend of using wastes following the concept of "waste to wealth" and are closely related to the themes of the sustainable development goals (SDGs) [17,18].

Numerous studies have demonstrated that biochar has the potential to increase P availability in soils [19–23]. Biochar itself is a relatively P-rich product because of the recovery and enrichment of P during thermochemical treatment and thus shows a similar fertilization effect as slow-release P fertilizer [24]. It was reported that the total P in a biochar produced from animal bones could be 152.0 g kg^{-1} and that the water-extractable P was 6.6 g kg^{-1} [25]. On the other hand, soil P availability is controlled by dynamics of various transformation processes (e.g., sorption/desorption, precipitation/dissolution, immobilization/mineralization) among different P forms that biochar application can affect via direct and indirect ways [26–30]. Biochar may release P from Fe and Al oxides, hydroxides, calcium carbonate ($CaCO_3$), and Ca, Fe, and Al phosphates [31]. Phosphates that are bound to free cations like Ca^{2+}, Mg^{2+}, Fe^{3+}, and Al^{3+} also have great potentials to feed the available P pool via dissolution processes, as stimulated by biochar application [9]. Although the enhancement of the soil available P pool by biochar has been well documented and reviewed [7,9,23,32,33], a better understanding of P dynamics governing P availability under biochar application requires separate investigations into the different forms of P in soils.

Soil P occurs in inorganic and organic forms, both of which are associated with the above-mentioned elements or compounds [28]. Compared to natural soils such as grassland and forest, inorganic P pools and their transformation processes are dominant for P supply in agricultural soils [4]. Thus, the investigation of diverse inorganic P fractions in agricultural soils under biochar applications has received more attention. For example, previous studies have indicated that straw-derived biochar enhances the proportion of labile P in lateritic soils [34,35], while straw biochar promotes the transformation of labile P into moderate soil P in calcareous soil [36]. A similar phenomenon was also found in manure biochar application cases. Troy et al. [37] stated that 10 g kg^{-1} swine manure biochar can increase the soil Morgan's P content and total P leaching, but Laird et al. [38] demonstrated that 20 g kg^{-1} swine manure biochar treatments enhanced moderately stable P proportion and reduced total dissolved P leaching by 69%. The above cases implied that the effect of biochar application on soil inorganic P may vary depending on biochar and soil properties. It is noteworthy that biochars are fabricated from a variety of feedstocks using diverse thermochemical conditions (e.g., duration time, pyrolysis temperature, atmosphere) and have diverse properties (e.g., pH, hydrophilicity, aromaticity, and nutrient content) [39].

Also, soils greatly differ in basic properties (i.e., pH, texture) and native P status, which can substantially mediate the biochar effect on inorganic P fractions.

To date, two meta-analysis articles have revealed biochars' effects on soil P availability [7,23], while results regarding biochar's effect on diverse inorganic P fractions in different soils have not been critically reviewed yet. Herein, we propose the scientific question, which is how do biochar properties and feedstocks affect soil fractions and ultimately influence P availability? Therefore, a quantitative review is necessary to fill this knowledge gap. Meta-analysis is an effective tool to achieve general conclusions as it is a statistical method used to systematically integrate the published results. By using this method, with comprehensive data analyses derived from 74 collected articles (incl. 673 independent observations) published from 1980 to 2022, we aim to (i) identify the impacts of biochar on the concentration of P in different P fractions; (ii) explain the relationships among P availability, fraction content, and soil properties; and (iii) reveal the potential mechanisms of soil P fraction transformation processes as affected by biochar application. This research is important for optimizing biochar's role in improving soil P availability, enhancing nutrient efficiency, and promoting sustainable agricultural practices while reducing environmental impact.

2. Materials and Methods

2.1. Database and Data Collection Criteria

The target publications were collected using the Web of Science (WOS), Google Scholar (GS), and the Chinese National Knowledge Infrastructure (CNKI). Papers published from 1980 to 2022 were included in these databases. The search keywords included soil AND phosph* AND avail* AND *char. The collected publications were further refined by adopting the following categories: (i) articles which set un-amended soil as a control; (ii) articles where all treatments were without P fertilizer addition; (iii) articles which featured agricultural soils only; and (iv) articles where each treatment had at least three replications and the data had standard errors. As a result, in total, 74 articles were collected for further analysis. The software of GetData (version 2.24) was adopted for extracting the data shown in the figure columns. Data on the soil variables measured in these studies (Olsen-P, H_2O-P, Al-P, Fe-P, Ca_2-P, Ca_8-P, Ca_{10}-P, Residual P (R-P)) were obtained from publications and included mean and standard errors of all treatments. "Olsen-P" refers to $NaHCO_3$ extractable P. Olsen-P primarily represents available P, which refers to P in the soil that is loosely bound to soil particles and readily available for plant uptake. The Olsen-P test extracts calcium-bound P (Ca-P), aluminum-bound P (Al-P), and some iron-bound P (Fe-P), which are active forms of inorganic P that can be easily taken up by plants [36]. For labile P, "H_2O-P" refers to water-soluble P; "Ca_2-P" refers to P bound to calcium in a low-oxidation state. For moderate labile P, "Al-P" refers to P bound to aluminum; "Fe-P" refers to P bound to iron; and "Ca_8-P" refers to P bound to calcium in a medium-oxidation state. For highly stable P, "Ca_{10}-P" refers to P bound to calcium in a high-oxidation state, and "Residual P (R-P)" refers to P that is not easily extracted by the methods listed above [40]. Specifically, a total of 6 P fractions representing low stable soil P (H_2O-P, Ca_2-P), moderately stable soil P (Al-P, Fe-P, Ca_8-P), and highly stable soil P (Ca_{10}-P, R-P) were categorized in this meta-analysis. Besides the soil P availability and diverse P fraction contents, the soil properties (i.e., soil texture, pH, Olsen-P), experimental duration, and biochar properties (i.e., pyrolysis duration, pyrolysis temperature, feedstock type, C/N ratio, pH, ash content, biochar application rate and duration) were carefully recorded in the meta-analysis database. While general biochar properties, such as feedstock type and pyrolysis conditions, were reported across the studies, specific chemical composition data were not available for biochar types. Notably, the P quantities in this meta-analysis refer to the P contents in the soils amended with(out) biochar. P changes reflect both the original soil content and the P added through biochar, capturing the overall impact of biochar on P levels. Its effect on P dynamics is assessed by considering both soil and biochar-derived P.

Data were normalized to the same units for comparison. By using the bulk density of the soil and the depth of the soil where biochar was applied, the data of biochar application rate were converted into mass percentage content (%). The subgroups were grouped as follows. The classification criteria used were primarily based on previous meta-analyses (e.g., Glaser et al. [7]; Yuan et al. [39]; Jeffery et al. [41]; Gao et al. [42]) and adapted to align with the dataset obtained in this study. The biochar raw materials were (1) crop residue; (2) manure residue; and (3) wood residue. The pyrolysis temperatures were (1) <400 °C; (2) 400–500 °C; and (3) >500 °C. The biochar C:N was (1) <50; (2) 50–200; and (3) >200. The soil pH was (1) <5; (2) 5–7; and (3) >7. The application rate was (1) <1%; (2) 1–2.5%; and (3) >2.5%. The experiment's duration was (1) <3 month; (2) 3 months to 2 years; and (3) >2 years. The pyrolysis time was (1) <1 h; (2) 1 h–2 h; and (3) >2 h. The biochar pH was (1) <6; (2) 6–8; and (3) >8. The biochar ash content was (1) <15%; (2) 5–30%; and (3) >30%. The soil Olsen-P was (1) <9 ppm; (2) 9–40 ppm; and (3) >40 ppm. The soil texture was (1) coarse (sandy loam, sandy clay loam, loam); (2) medium (clay loam, loam, silty clay loam, silty or silty loam); and (3) fine (clay, silty clay, sandy clay).

2.2. Meta-Analysis

The response ratio (*RR*) was employed to appraise the impacts of biochar amendment on P availability and diverse soil P fractions' content, which is the mean of the biochar-treated soil divided by the mean of the control group without biochar. In addition, to avoid the poor statistical properties of ratios, transformation of *RR* into the natural log of the response ratio ($ln(RR)$) was performed as

$$lnRR = ln\left(\frac{X_E}{X_C}\right) \quad (1)$$

where X_E and X_C represent the means of the treatment (with biochar application) and control (without biochar application) groups, respectively.

The effect size was converted to %*Change* by using the following equation [39]:

$$\%Change = \left(e^{lnRR} - 1\right) \times 100\% \quad (2)$$

The effect sizes and 95% confidence intervals (CIs) were calculated using the Metawin 2.1 software of all the categorical groups with the random effect model. In total, 9999 iterations were performed in Resampling tests [41]. We assessed the potential publication bias and the stability of our meta-analysis results using the Egger test and fail-safe N test, as detailed in Table S1 [39,43]. The importance and interactions of the variables explaining the changes in each soil P fraction (shown in Tables S3–S10) were calculated using the boosted regression tree (BRT) model, implemented with the 'gbm' package in R Studio version 1.2.5042 [44].

3. Results and Discussion

3.1. General Trend

Overall, biochar application remarkably enhanced the content of diverse soil inorganic P fractions, except R-P, Ca_8-P, and Ca_{10}-P (Figure 1). As R-P, Ca_8-P, and Ca_{10}-P are relatively stable forms of P in soil, as they usually combine with calcium to form insoluble minerals, they are less susceptible to biochar addition in the short term. Biochar may affect more water-soluble P and metal oxides loosely bound P [30,31]. To garner deeper insights into critical factors influencing soil availability and fractions, BRT model results were used to processes the In RR data (Table 1). The results implied that biochar P content and soil pH were important factors influencing the soil Olsen-P and H_2O-P, respectively. Further, in terms of moderate labile soil P, Al-P was mainly influenced by application duration, while Fe-P and Ca_2-P were greatly influenced by pyrolysis time and biochar pH, respectively. Lastly, pyrolysis temperature, biochar pH, and soil pH were the most important factors inducing changes in stable P (Ca_8-P, Ca10-P, and R-P, respectively).

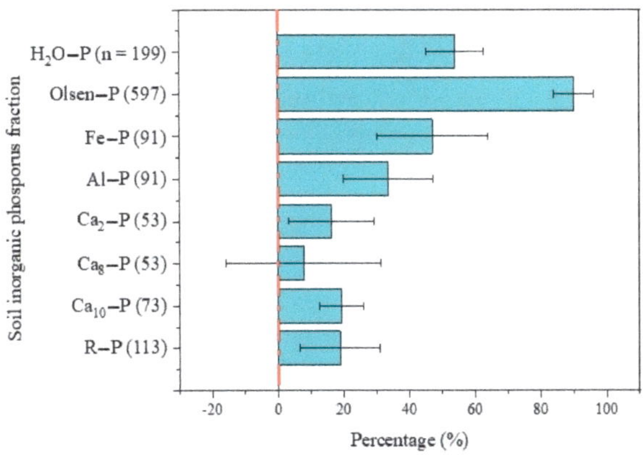

Figure 1. Grand mean of all cases for diverse inorganic P fractions when biochar was applied regardless of experimental conditions.

Table 1. Contribution rate (%) of explanatory variables in the boosted regression tree (BRT) model for explaining the variation in soil P fractions.

Explanatory Variable	Olsen-P	H_2O-P	Al-P	Fe-P	Ca_2-P	Ca_8-P	Ca_{10}-P	R-P
Feedstock	0.48	0.00	0.04	0.00	0.09	0.43	2.06	0.00
Pyrolysis temperature	10.3	3.94	0.66	18.9	0.03	21.8	1.54	0.00
Pyrolysis time	6.00	0.69	0.66	17.5	1.34	6.38	2.38	0.00
Biochar C:N ratio	4.29	6.81	5.74	9.86	5.57	6.31	5.07	0.18
Biochar pH	7.59	7.68	7.11	1.69	30.4	9.59	18.1	0.07
Biochar ash	2.64	0.00	0.00	12.2	0.00	3.09	0.04	0.00
Application rate	8.39	8.42	7.46	23.9	0.08	10.8	15.2	0.25
Application duration	6.78	17.6	18.0	0.78	0.12	7.45	8.89	0.14
Biochar total P	19.8	13.6	10.9	0.69	0.04	0.00	10.2	0.13
Biochar Olsen-P	0.00	1.95	2.39	2.10	4.12	0.00	17.8	0.02
Soil pH	12.7	22.7	26.8	4.62	58.3	14.7	10.2	67.7
Soil Olsen-P	18.5	7.64	6.87	2.10	0.00	16.6	3.29	0.08
Soil texture	2.42	8.98	9.97	0.00	0.00	2.71	5.17	31.3

In addition, the results of funnel plot statistics showed that there was generally no publication bias, suggesting that the results were not distorted by selective reporting. Additionally, the fail-safe numbers greater than 5N + 10 for all subgroups further verified the robustness and reliability of the meta-analysis (Table S1) [43].

Specifically, the percentage (%) of H_2O-P was significantly improved (Figure 1), as biochar has certain alkalinity and can introduce anions such as hydroxide and chloride into soil to compete with P on soil P adsorption sites [35]. On the other hand, the activity of P-soluble microorganisms (e.g., *Lysinobacteria*) enhanced by biochar application could promote the conversion of organic P to water-soluble phosphate [6,35]. As shown in Figure 1, the percentage change in moderately stable soil P (Al-P, Fe-P, Ca_8-P) and highly stable soil P (Ca_{10}-P, R-P) was smaller than that of more labile fractions like H_2O-P and Olsen-P, with the change generally decreasing as the stability of the P fraction increased. Olsen-P is the most common indicator of soil P availability. Our results show that biochar significantly enhances soil P availability, consistent with previous meta-analyses [23,34].

Although some studies have focused on biochar amendments used to improve plant available P content [36,45], current research has emphasized the potential transformations among inorganic P fractions. Biochar affects soil P both directly and indirectly. Directly, it influences P availability through its nutrient content, which may result from mineral changes during production (e.g., the formation of white lockite) and be gradually released into the soil. Indirectly, biochar can alter soil properties like pH, cation exchange capacity (CEC), and metal concentrations, which in turn affect P availability. However, these impact processes could be implicated depending on experimental conditions including biochar and soil properties and experimental durations (Figure 2, Table S2), which will be further discussed in the following sections.

3.2. *Biochar Properties*

3.2.1. Feedstock

Our results showed that wood residue-derived biochar, compared to crop and manure residue-derived biochars, produced no significant improvement in H_2O-P and Olsen-P (Figure 2). This may be attributed to the lower ash content and higher specific surface area of wood residue-derived biochar, meaning that it can exhibit more adsorption and fixation capacities of water-soluble phosphate [33]. Among these three types of biochar feedstocks, manure residue-derived biochar enhanced H_2O-P and Olsen-P to the greatest extent. This may be due to the high P content of manure, which can be retained in biochar during the pyrolysis process [46]. Other studies have also reported that, compared to cotton straw- and corncob-derived biochars, chicken manure biochar has a significantly greater effect on improving soil H_2O-P and Olsen-P content [47,48]. Wang et al. [49] pointed out that manure-derived biochar itself contains a large amount of soluble P, and the application of 1% pig manure biochar can increase soil Olsen-P by 1.89 times. Furthermore, they found that the content of Olsen-P gradually increases during the 30-day cultivation period, which is related to the release of P from biochar.

Crop residue-derived biochar significantly improved Al-P and Fe-P, likely due to its higher content of inorganic elements like Al, Fe, and Mg, which promote the formation of these compounds in soil [50–52]. It was also found that all types of biochar can significantly improve residue P (R-P) content, with the most obvious improvement effect of the manure residue-derived biochar (Figure 2). This could be attributed to the relatively high P concentration and more sites for P sorption within the manure residue-derived biochar. Further, the application of manure residue-derived biochars may stimulate microbial P fixations in soil, as they could provide favorable conditions of pH and nutrients, as well as more habitat positions for microorganisms [53,54].

The above results indicate that the feedstock indeed has great impacts on soil P fractions. This would happen even within a certain feedstock type. For instance, Han et al. [55] showed that biochar produced from soybean pods and straw, but not corncob biochar, could significantly improve soil H_2O-P content, because corncob biochar has a larger specific surface area and stronger anion exchange performance. Given the diversity of organic waste and biochar's impact on soil P, future biochar production for agriculture should balance both economic and ecological benefits.

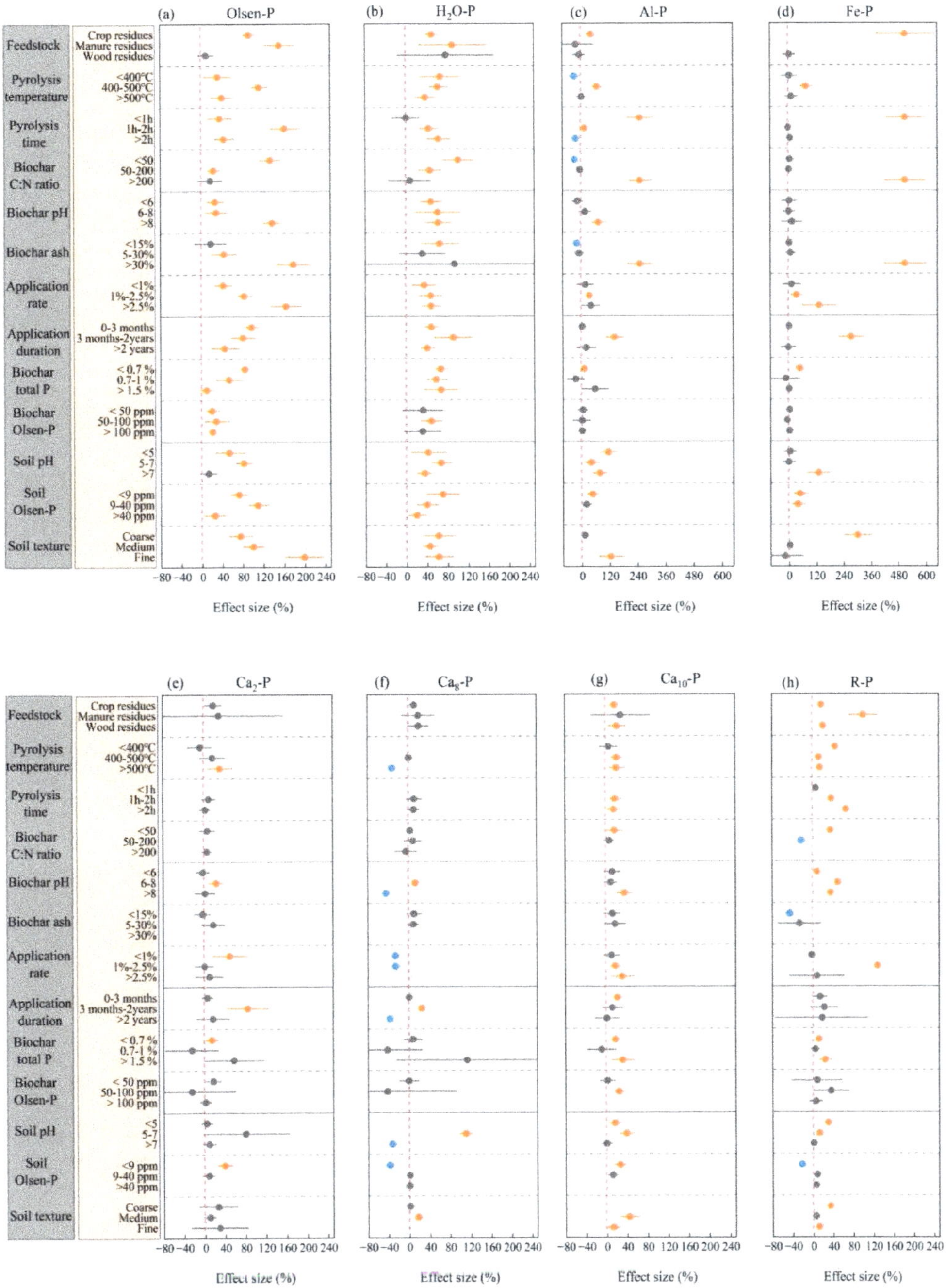

Figure 2. Effect of explanatory variables on content of Olsen-P (**a**), H_2O-P (**b**), Al-P (**c**), Fe-P (**d**), Ca_2-P (**e**), Ca_8-P (**f**), Ca_{10}-P (**g**) and R-P (**h**) in soil. Symbols indicate the mean % change in effect size with 95% confidence interval. The number after the name of group indicates the amount of pairwise comparison. The red dotted line indicates the zero line. Orange, blue, and grey dots represent the positive, negative, and no significant effects.

3.2.2. Pyrolysis Temperature and Duration

Biochar pyrolyzed at relatively low and medium temperatures (<500 °C) was more conducive to improving soil H_2O-P, and medium temperature (400–500 °C) biochar had the best improvement effect on Olsen-P (Figure 2). Since P volatilizes around 700 °C and biochar is typically produced at lower temperatures, the P contained in biochar remains similar to its original feedstock [56]. It is noteworthy that these P transformation processes under different pyrolysis temperatures are also time-dependent (Figure 2). Our results showed that as pyrolysis time increased, the percentage of soil Al-bound P decreased, while soil H_2O-P and R-P increased. This suggests that the transformation processes of P contained in biochar may occur between different pools, and these processes are dependent on pyrolysis temperature and time. Our results revealed that biochar prepared at a relatively low temperature can enhance soil available P more effectively, which may be because relatively stable P compounds were formed from other 'labile' P fractions at higher pyrolysis temperatures. In addition, biochar fabricated at high pyrolysis temperatures may exhibit high ion-binding strength through chemical adsorption, which may immobilize nutrients into unusable forms [57]. Biochars produced at low to medium temperatures have more oxygen-containing functional groups (e.g., hydroxyl, carboxyl, carbonyl) and higher cation exchange capacity (CEC), making it easier for them to release P into the soil's available P pool [55,58]. Our study emphasizes a significant role of the pyrolysis conditions of biochar in mediating P availability in soil since they greatly affect biochar properties and in turn affect soil P pools and dynamics after application via direct and indirect pathways.

3.2.3. C/N Ratio

Biochar with a lower C/N ratio showed a more profound improvement effect on soil H_2O-P and Olsen-P, which may be related to the unstable C and relatively sufficient N supply within biochar (Figure 2). Biochars with more available C and N compounds are beneficial to soil microbe growth. Many studies have pointed out that the soil P solubilizing bacteria (e.g., *Pseudomonas* and *Bacillus* species) can be improved after biochar addition [35,59]. Biochars with a higher C/N ratio are usually produced at a higher temperature and have higher contents of mineral compounds, which would strengthen P immobilization and therefore have greater potentials for decreasing soil P availability [55,59]. In contrast, biochars with lower C/N ratios, often derived from non-woody materials (e.g., manure, straw), contain relatively higher soluble N and P [60].

3.2.4. pH

Biochar with alkaline pH (>8) had a greater effect in improving H_2O-P and Olsen-P than biochar with acidic (<6) and neutral pHs (6–8). Most biochars are alkaline, as the surface acidic groups (e.g., carboxyl, hydroxyl, and phenolic groups) will decrease and the surface basic groups (e.g., lactones) will increase with increasing pyrolysis temperature [61]. In addition, mineral elements such as Na, K, Mg, and Ca remain as oxides or carbonates during biochar production [62]. Zhou et al. [63] produced sawdust biochar at 300 °C and 600 °C, resulting in a pH of 4.05 and 7.96, respectively. They applied the same dosage of biochar to soil (pH 4.34) and found similar increases in available P at 7 days. However, after 80 days, the available P in the alkaline biochar treatment was significantly lower than in the acid biochar treatment. This may be due to the alkaline biochar promoting P precipitation by minerals and/or P fixation by microorganisms in the acidic soil.

3.2.5. Ash Content

The high-ash-content biochar showed significantly higher improvement effects on Al- and Fe-bound P, as well as Olsen-P (Figure 2). In addition to phosphate, the ash in biochar mainly consists of minerals such as sulfate, silicate, and chloride [64]. Thus, the ash content in biochar may serve as a direct contributor to the availability of P; otherwise, the remaining P would have high potential to be bound to Al- and Fe-containing minerals, as indicated by the results of the present meta-analysis. Further, wood residue-derived biochar usually has a low ash content. It was reported that the ash content of corn straw and rice husk biochars is about 40%, which is much higher than the ash content of pine biochar (11.6%) [65]. Thus, high-ash-content biochar has a similar effect on inorganic P fractions as biochars pyrolyzed from cellulose-rich materials, rather than lignin.

3.2.6. Biochar Application Rate and Duration

Biochar application rate can generally magnify the effects of other biochar properties mentioned above on soil inorganic P fractions. For example, Ippolito et al. [66] conducted a 6-month incubation experiment by applying switchgrass biochar (pH 5.8) to a soil with a pH of 7.6, and the soil pH decreased by 0.2–0.4 units and the soil available P, as well as the soil microbial growth, increased with increasing biochar application rate. The biochar application improved the availability of Olsen-P, which decreased over time. In contrast, stable P fractions (e.g., Al-P, Fe-P, Ca_2-P, Ca_8-P) increased from short-term (<3 months) to medium-term (3 months–2 years), suggesting that precipitation and sorption processes may occur over time. It was evidenced in another study that biochar application increased the contents of Ca_2-P and Al-P in soil after one rice season [67]. A longer application duration may strengthen the biochar effect, which was supported by findings that 5-year biochar application increased the contents of Ca_2-P, Ca_8-P, Al-P, Fe-P, and Ca_{10}-P by 1.92 times, 2.61 times,1.87 times, and 0.43 times, respectively [68]. In addition to the time effect, it is noteworthy that agronomy activities such as planting, irrigation, fertilization, and tillage may also substantially affect the soil P dynamics and fractions.

3.3. *Soil Properties*

Soil properties can substantially mediate the biochar effect on P fractions. It was found that soils with neutral pH showed higher soil water-soluble P content after biochar application than alkaline and acid soils. This is probably due to the fact that high alkalinity stimulates P fixation by metal oxides [42,69]. In addition, the precipitation of Ca-P in alkaline soils could form series of products that reduce P solubility, resulting in low P availability. Under relatively acidic conditions, the P newly introduced by biochar application may be adsorbed on soil minerals or precipitate with aluminum and iron oxides [34]. There is a significant difference in soil P retention in acidic and alkaline soils amended with biochar as a P carrier [8,70]. Also, biochar can mediate other physicochemical and biological processes that affect P availability in soils with different pHs. For instance, the competition between dissolved organic matter from biochar and P in soil solution at soil P adsorption sites will vary with soil pH. On the one hand, biochar can induce changes in soil enzyme activities and/or microbial population dynamics that are sensitive to soil pH change [71]. These together suggest that biochar should be used cautiously to retain P in soils with different acidity. It is also noteworthy that there are exceptions to this situation. For example, in fertile soils, when the soil is already saturated for P adsorption sites, biochar application can cause the release of P from the fertile soil, thereby rapidly increasing soil P availability [35]. While studying the pH effects of biochar on soil P, it is necessary to consider this in conjunction with other relevant factors (e.g., soil P adsorption saturation degree, soil and biochar characteristics, and the duration of biochar in soil).

The application of biochar in fine-textured soils can significantly increase Olsen-P content by promoting aggregate formation, enhancing P adsorption and fixation, and reducing the risk of P loss [72]. The introduction of biochar can improve the stability of soil aggregates and increase the content of large aggregates in soil [73].

3.4. Soil P Dynamics Influenced by Biochar

Previous studies have proved that most P is present as orthophosphate, pyrophosphate, and hydroxyapatite (Ca_{10}-P) [74]. Studies have noted that biochar application could commonly induce a rapid increase in the soil labile P and Ca_{10}-P, which is in line with our results (Figure 1). However, the less significant increase in moderate labile P (Fe/Al/Ca_8-P) is probably induced by indirect methods after biochar application. Manure-derived biochars are particularly effective at enhancing soil available P fractions such as H_2O-P and Olsen-P, owing to their high intrinsic P content [20,37]. This biochar type also fosters microbial activity that can facilitate P mineralization and solubilization [46]. In contrast, wood-derived biochars, characterized by their low ash content, large surface area, and higher porosity, exhibit superior P adsorption capacity [36]. Crop residue-derived biochars contribute to improvements in aluminum-bound (Al-P) and iron-bound (Fe-P) P fractions, due to their relatively high inorganic element content, which interacts with soil minerals to stabilize P [35].

Pyrolysis conditions further modulate biochar's effects. Low to medium pyrolysis temperatures (<500 °C) produce biochars rich in oxygen-containing functional groups, higher cation exchange capacity (CEC), and soluble P, which enhance soil P availability [33,37]. Conversely, biochars generated at higher temperatures (>500 °C) result in the formation of more stable and less soluble P forms, favoring long-term P sequestration but potentially reducing short-term availability [35]. Biochars with lower C/N ratios release more soluble nutrients, promoting microbial growth and enzymatic activity, which facilitate P solubilization [27]. The biochar pH also plays a critical role. Alkaline biochars (>pH 8) can increase H_2O-P and facilitate P release in neutral to acidic soils but may cause P precipitation or immobilization over time in highly acidic soils [51].

High-ash-content biochars contribute to the retention of P in mineral-bound forms, such as calcium phosphates, particularly in alkaline soil environments [35,36]. Application rates and durations also influence these effects. In the short term, biochar addition tends to increase soil available P by enhancing desorption and solubilization processes. Over time, however, the dynamics shift toward the stabilization of P in forms that are less prone to leaching, improving P retention [52]. The interplay between biochar and soil properties, such as soil pH, texture, and mineral composition, further dictates the outcomes. For example, in sandy soils with low P buffering capacity, biochar can significantly improve P retention, while in clay-rich soils, its effects may be moderated by the native P dynamics and existing mineral interactions [30,31].

4. Future Research Recommendations

The following aspects warrant attention in future research and field applications: (1) the endogenous P contents of biochar produced from manure and sludge residues were higher than that of other biochars, and attention should be given to the potential risks associated with the soil P in manure or sludge biochar-amended agricultural soils. However, studies evaluating P fractions and mobility in soils amended with these biochars are not sufficient yet. (2) The duration of biochar application in soil could significantly affect soil P forms, availability, and mobility through indirect ways. In addition, some reports have claimed that applying biochar to soil is risky because it contains some harmful substances such as heavy metals, organophosphorus, and inorganic pollutants, although the concentrations of these toxins are usually below the permitted levels included in the regulations of various organizations. Therefore, the effect of biochar duration/aging time on soil P fractions and mobility after application into different soils requires more systematic investigations. (3) The physicochemical properties of biochar are highly different depending on feedstock and pyrolysis conditions, and these series of properties play joint roles in biochar's effects on soil P fractions and dynamics. More comprehensive methods such as model simulations and machine learning, together with verification of field experiments, should be used for better understanding and practice.

5. Conclusions

From the current quantitative meta-analysis results, it can be seen that biochar properties, such as feedstock type, pyrolysis temperature, C:N ratio, pH, and ash content, influence soil P fractions by interacting with soil minerals and organic matter. Biochars from different feedstocks (e.g., wood, crop residues, manure) can affect the availability of labile P (like H_2O-P, Olsen-P) and more stable forms (such as P bound to iron, aluminum, or calcium (hydr)oxides) in varying ways. High C:N ratio biochars, for instance, may have limited impacts on labile P fractions, while biochars from crop residues can enhance P associated with iron and aluminum oxides. In summary, the application of biochar has shown an overall increase in the proportion of various P fractions, with the most significant enhancement observed in the active P fraction. However, only slight improvements in the moderately stable and stable P fractions were found. The effects of biochar on the enhancement and transformation of P fractions are driven by a combination of biological and abiotic processes. Further systematic and in-depth research is needed to better understand these mechanisms.

Supplementary Materials: The following supporting information can be downloaded at: https://www.mdpi.com/article/10.3390/agriculture15010103/s1. Table S1: The results of publication bias used in the present study; Table S2. Summary of the averaged relative change (%) in diverse soil fractions in response to biochar addition. Significance of Wilcoxon signed rank tests: * $p < 0.05$, no symbol following the number indicates not statistically significant. N/A indicates data not available. The numbers in brackets represent the 95% confidence intervals (CI); Table S3: Pairwise interactions of Olsen-P modeled using the boosted regression tree (BRT) approach; Table S4: Pairwise interactions of H_2O-P modeled using the boosted regression tree (BRT) approach; Table S5 Pairwise interactions of Fe-P modeled using the boosted regression tree (BRT) approach; Table S6: Pairwise interactions of Al-P modeled using the boosted regression tree (BRT) approach; Table S7: Pairwise interactions of Ca_2-P modeled using the boosted regression tree (BRT) approach; Table S8: Pairwise interactions of Ca_8-P modeled using the boosted regression tree (BRT) approach; Table S9: Pairwise interactions of Ca_{10}-P modeled using the boosted regression tree (BRT) approach; Table S10: Pairwise interactions of R-P modeled using the boosted regression tree (BRT) approach.

Author Contributions: Conceptualization, H.C. and L.L.; methodology, L.L., Y.P., L.Z., B.Z. and Q.C.; validation, L.L.; investigation, L.L. and Y.P.; resources, Y.P.; data curation, Y.P.; writing—original draft preparation, L.L., Y.P. and H.C.; writing—review and editing, L.L., L.Z., B.Z., Q.C. and H.C.; supervision, H.C.; project administration, L.L.; funding acquisition, Y.P. and H.C. All authors have read and agreed to the published version of the manuscript.

Funding: This study was supported by the National Natural Science Foundation of China (No. 42007047, 42207015), post-doctorate research funding to Shenzhen (szbo202207, szbo202323), the modern agricultural innovation center, and the Henan Institute of Sun Yat-sen University (N2021-003, N2021-002).

Data Availability Statement: All data generated or analyzed during this study are included in this published article [and its supplementary information files].

Conflicts of Interest: The authors declare no conflicts of interest.

References

1. Wrage, N.; Chapuis-Lardy, L.; Isselstein, J. Phosphorus, Plant Biodiversity and Climate Change. In *Sociology, Organic Farming, Climate Change and Soil Science*; Lichtfouse, E., Ed.; Springer: Dordrecht, The Netherlands, 2010; Volume 3.
2. Westheimer, F.H. Why nature chose phosphates. *Science* **1987**, *235*, 1173–1178. [CrossRef] [PubMed]
3. Peñuelas, J.; Poulter, B.; Sardans, J.; Ciais, P.; van der Velde, M.; Bopp, L.; Boucher, O.; Godderis, Y.; Hinsinger, P.; Llusia, J.; et al. Human-induced nitrogen-phosphorus imbalances alter natural and managed ecosystems across the globe. *Nat. Commun.* **2013**, *4*, 2934. [CrossRef] [PubMed]
4. Bünemann, E.K. Assessment of gross and net mineralization rates of soil organic phosphorus-A review. *Soil Biol. Biochem.* **2015**, *89*, 82–98. [CrossRef]
5. Zhu, J.; Li, M.; Whelan, M. Phosphorus activators contribute to legacy phosphorus availability in agricultural soils: A review. *Sci. Total Environ.* **2018**, *612*, 522–537. [CrossRef]
6. Chen, S.; Zhang, S.; Yan, Z.; Peng, Y.; Chen, Q. Differences in main processes to transform phosphorus influenced by ammonium nitrogen in flooded intensive agricultural and steppe soils. *Chemosphere* **2019**, *226*, 192–200. [CrossRef]
7. Glaser, B.; Lehr, V.-I. Biochar effects on phosphorus availability in agricultural soils: A meta-analysis. *Sci. Rep.* **2019**, *9*, 9338. [CrossRef]
8. Li, H.; Cui, S.; Tan, Y.; Peng, Y.; Gao, X.; Yang, X.; Ma, Y.; He, X.; Fan, B.; Yang, S. Synergistic effects of ball-milled biochar-supported exfoliated LDHs on phosphate adsorption: Insights into role of fine biochar support. *Environ. Pollut.* **2021**, *294*, 118592. [CrossRef] [PubMed]
9. Zhang, H.; Chen, C.; Gray, E.M.; Boyd, S.E.; Yang, H.; Zhang, D. Roles of biochar in improving phosphorus availability in soils: A phosphate adsorbent and a source of available phosphorus. *Geoderma* **2016**, *276*, 1–6. [CrossRef]
10. Price, G. *Australian Soil Fertility Manual*, 3rd ed.; CSIRO Pub: Collingwood, Australia, 2006.
11. Carpenter, S.R. Eutrophication of aquatic ecosystems: Bistability and soil phosphorus. *Proc. Natl. Acad. Sci. USA* **2005**, *102*, 10002–10005. [CrossRef]
12. Ceulemans, T.; Merckx, R.; Hens, M.; Honnay, O. A trait-based analysis of the role of phosphorus vs. nitrogen enrichment in plant species loss across Northwest European grasslands. *J. Appl. Ecol.* **2011**, *48*, 1155–1163. [CrossRef]
13. Martiny, T.R.; Avila, L.B.; Rodrigues, T.L.; Tholozan, L.V.; Meili, L.; de Almeida, A.R.F.; da Rosa, G.S. From waste to wealth: Exploring biochar's role in environmental remediation and resource optimization. *J. Clean. Prod.* **2024**, *453*, 142237. [CrossRef]
14. Purakayastha, T.J.; Bera, T.; Bhaduri, D.; Sarkar, B.; Mandal, S.; Wade, P.; Kumari, S.; Biswas, S.; Menon, M.; Pathak, H.; et al. A review on biochar modulated soil condition improvements and nutrient dynamics concerning crop yield: Pathways to climate change mitigation and global food security. *Chemosphere* **2019**, *227*, 345–365. [CrossRef] [PubMed]
15. Hossain, M.Z.; Bahar, M.M.; Sarkar, B.; Donne, S.W.; Ok, Y.S.; Palansooriya, K.N.; Kirkham, M.B.; Chowdhry, S.; Bolan, N. Biochar and its importance on nutrient dynamics in soil and plant. *Biochar* **2020**, *2*, 379–420. [CrossRef]
16. Qian, S.X.; Zhou, X.R.; Fu, Y.K.; Song, B.; Yan, H.C.; Chen, Z.X.; Sun, Q.; Ye, H.Y.; Qin, L.; Lai, C. Biochar-compost as a new option for soil improvement: Application in various problem soils. *Sci. Total Environ.* **2023**, *870*, 162024. [CrossRef]
17. Ghorbani, M.; Amirahmadi, E.; Cornelis, W.; Zoroufchi, B.K. Understanding the physicochemical structure of biochar affected by feedstock, pyrolysis conditions, and post-pyrolysis modification methods–A meta-analysis. *J. Environ. Chem. Eng.* **2024**, *12*, 114885. [CrossRef]
18. Xu, Q.; Zhang, T.; Niu, Y.Q.; Mukherjee, S.; Abou-Elwafa, S.F.; Nguyen, N.S.H.; Aboud, N.M.A.; Wang, Y.K.; Pu, M.J.; Zhang, Y.R.; et al. A comprehensive review on agricultural waste utilization through sustainable conversion techniques, with a focus on the additives effect on the fate of phosphorus and toxic elements during composting. *Sci. Total Environ.* **2024**, *942*, 173567. [CrossRef] [PubMed]
19. Wang, T.; Camps-Arbestain, M.; Hedley, M.; Bishop, P. Predicting phosphorus bioavailability from high-ash biochars. *Plant Soil* **2012**, *357*, 173–187. [CrossRef]
20. Liang, Y.; Cao, X.; Zhao, L.; Xu, X.; Harri, W. Phosphorus release from dairy manure, the manure-derived biochar, and their amended soil: Effects of phosphorus nature and soil property. *J. Environ. Qual.* **2014**, *43*, 1504. [CrossRef]
21. Uchimiya, M.; Hiradate, S.; Antal, M.J. Dissolved phosphorus speciation of flash carbonization, slow pyrolysis, and fast pyrolysis biochars. *ACS Sustain. Chem. Eng.* **2015**, *3*, 1642–1649. [CrossRef]
22. Zhao, L.; Cao, X.D.; Zheng, W.; Scott, J.W.; Sharma, B.K.; Chen, X. Copyrolysis of biomass with phosphate fertilizers to improve biochar carbon retention, slow nutrient release, and stabilize heavy metals in soil. *ACS Sustain. Chem. Eng.* **2016**, *4*, 1630–1636. [CrossRef]
23. Gao, S.; DeLuca, T.H.; Cleveland, C.C. Biochar additions alter phosphorus and nitrogen availability in agricultural ecosystems: A meta-analysis. *Sci. Total Environ.* **2019**, *654*, 463–472. [CrossRef]
24. Wang, Y.; Lin, Y.; Chiu, P.C.; Imhoff, P.T.; Guo, M. Phosphorus release behaviors of poultry litter biochar as a soil amendment. *Sci. Total Environ.* **2015**, *512*, 454–463. [CrossRef] [PubMed]

25. Siebers, N.; Leinweber, P. Bone char: A clean and renewable phosphorus fertilizer with cadmium immobilization capability. *J. Environ. Qual.* **2013**, *42*, 405–411. [CrossRef]
26. Frossard, E.; Condron, L.M.; Oberson, A.; Sinaj, S.; Fardeau, J.C. Processes governing phosphorus availability in temperate soils. *J. Environ. Qual.* **2000**, *29*, 15–23. [CrossRef]
27. Turner, B.L. Storage-induced changes in phosphorus solubility of air-dried soils. *Soil Sci. Soc. Am. J.* **2005**, *29*, 630–633. [CrossRef]
28. Tiessen, H.; Moir, O. Characterization of available P by sequential extraction. In *Soil Sampling and Methods of Analysis*, 2nd ed.; Carter, M.R., Gregorich, E.G., Eds.; CRC Press: Boca Raton, FL, USA, 1993; pp. 293–306.
29. Vitousek, P.M.; Porder, S.; Houlton, B.Z.; Chadwick, O.A. Terrestrial phosphorus limitation: Mechanisms, implications, and nitrogen-phosphorus interactions. *Ecol. Appl.* **2010**, *20*, 5–15. [CrossRef] [PubMed]
30. Barrow, N. A mechanistic model for describing the sorption and desorption of phosphate by soil. *Eur. J. Soil Sci.* **2015**, *66*, 9–18. [CrossRef]
31. Mehlich, A. Amodification of Mehlich 2 extractant. *Commun. Soil Sci. Plant Anal.* **1984**, *15*, 1409–1416. [CrossRef]
32. Gul, S.; Whalen, J.K.; Thomas, B.W.; Sachdeva, V.; Deng, H. Physico-chemical properties and microbial responses in biochar-amended soils: Mechanisms and future directions. *Agric. Ecosyst. Environ.* **2015**, *206*, 46–59. [CrossRef]
33. Gao, S.; DeLuca, T.H. Wood biochar impacts soil phosphorus dynamics and microbial communities in organically-managed croplands. *Soil Biol. Biochem.* **2018**, *126*, 144–150. [CrossRef]
34. Zhang, B.; Yin, R.; Wei, Q.; Qin, S.; Peng, Y.; Zhang, B. Effects of Combined Applications of Biogas Slurry and Biochar on Phosphorus Leaching and Fractionations in Lateritic Soil. *Sustainability* **2022**, *14*, 7924. [CrossRef]
35. Peng, Y.; Zhang, B.; Guan, C.-Y.; Jiang, X.; Tan, J.; Li, X. Identifying biotic and abiotic processes of reversing biochar-induced soil phosphorus leaching through biochar modification with MgAl layered (hydr)oxides. *Sci. Total Environ.* **2022**, *843*, 157037. [CrossRef] [PubMed]
36. Peng, Y.; Sun, Y.; Fan, B.; Zhang, S.; Bolan, N.S.; Chen, Q.; Tsang, D.C. Fe/Al (hydr) oxides engineered biochar for reducing phosphorus leaching from a fertile calcareous soil. *J. Clean. Prod.* **2021**, *279*, 123877. [CrossRef]
37. Troy, S.M.; Lawlor, P.G.; O'Flynn, C.J.; Healy, M.G. The impact of biochar addition on nutrient leaching and soil properties from tillage soil amended with pig manure. *Water Air Soil Poll.* **2014**, *225*, 1900. [CrossRef]
38. Laird, D.; Fleming, P.; Wang, B.; Horton, R.; Karlen, D. Biochar impact on nutrient leaching from a Midwestern agricultural soil. *Geoderma* **2010**, *158*, 436–442. [CrossRef]
39. Yuan, C.; Gao, B.; Peng, Y.; Gao, X.; Fan, B.; Chen, Q. A meta-analysis of heavy metal bioavailability response to biochar aging: Importance of soil and biochar properties. *Sci. Total Environ.* **2021**, *756*, 144058. [CrossRef]
40. Zhang, S.; Wei, L.; Trakal, L.; Wang, S.; Shaheen, S.M.; Rinklebe, J.; Chen, Q. Pyrolytic and hydrothermal carbonization affect the transformation of phosphorus fractions in the biochar and hydrochar derived from organic materials: A meta-analysis study. *Sci. Total Environ.* **2024**, *906*, 167418. [CrossRef]
41. Jeffery, S.; Verheijen, F.G.A.; Kammann, C.; Abalos, D. Biochar effects on methane emissions from soils: A meta-analysis. *Soil Biol. Biochem.* **2016**, *101*, 251–258. [CrossRef]
42. Gao, X.; Peng, Y.; Guo, L.; Wang, Q.; Guan, C.-Y.; Yang, F.; Chen, Q. Arsenic adsorption on layered double hydroxides biochars and their amended red and calcareous soils. *J. Environ. Manag.* **2020**, *271*, 111045. [CrossRef]
43. Rosenthal, R.; Rosnow, R.L. *Essentials of Behavioral Research: Methods and Data Analysis*; McGraw-Hill: New York, NY, USA, 2008.
44. Elith, J.; Leathwick, J.R.; Hastie, T. A working guide to boosted regression trees. *J. Anim. Ecol.* **2008**, *77*, 802–813. [CrossRef] [PubMed]
45. Parvage, M.M.; Ulén, B.; Eriksson, J.; Strock, J.; Kirchmann, H. Phosphorus availability in soils amended with wheat residue char. *Biol. Fert. Soils* **2013**, *49*, 245–250. [CrossRef]
46. Jin, Y.; Liang, X.; He, M.; Liu, Y.; Tian, G.; Shi, J. Manure biochar influence upon soil properties, phosphorus distribution and phosphatase activities: A microcosm incubation study. *Chemosphere* **2016**, *142*, 128–135. [CrossRef]
47. Bohara, H.; Dodla, S.; Wang, J.J.; Darapuneni, M.; Kongchum, M.; Fromme, D.D.; Harrell, D. Impacts of N-stabilizers and biochar on nitrogen losses, nitrogen phytoavailability, and cotton yield in poultry litter-fertilized soils. *Agron. J.* **2018**, *110*, 2016–2024. [CrossRef]
48. Ahmad, S.; Ghaffar, A.; Rahman, M.H.U.; Hussain, I.; Iqbal, R.; Haider, G.; Khan, M.A.; Ikram, R.M.; Hussnain, H.; Bashir, M.S. Effect of application of biochar, poultry and farmyard manures in combination with synthetic fertilizers on soil fertility and cotton productivity under arid environment. *Commun. Soil Sci. Plant Anal.* **2021**, *52*, 2018–2031. [CrossRef]
49. Wang, Z.; Chen, L.; Liu, C.; Jin, Y.; Li, F.; Khan, S.; Liang, X. Reduced colloidal phosphorus loss potential and enhanced phosphorus availability by manure-derived biochar addition to paddy soils. *Geoderma* **2021**, *402*, 115348. [CrossRef]
50. Liu, L.; Tan, Z.; Gong, H.; Huang, Q. Migration and transformation mechanisms of nutrient elements (N, P, K) within biochar in straw–biochar–soil–plant systems: A review. *ACS Sustain. Chem. Eng.* **2018**, *7*, 22–32. [CrossRef]
51. Jing, Y.; Zhang, Y.; Han, I.; Wang, P.; Mei, Q.; Huang, Y. Effects of different straw biochars on soil organic carbon, nitrogen, available phosphorus, and enzyme activity in paddy soil. *Sci. Rep.* **2020**, *10*, 8837. [CrossRef]

52. Yang, C.; Lu, S. The dynamic changes of phosphorus availability in straw/biochar-amended soils during the rice growth revealed by a combination of chemical extraction and DGT technique. *J. Soil. Sediments* **2022**, *22*, 957–967. [CrossRef]
53. Wu, H.; Zhang, J.; Li, C.; Fan, J.; Zou, Y. Mass balance study on phosphorus removal in constructed wetland microcosms treating polluted river water. *CLEAN–Soil Air Water* **2013**, *41*, 844–850. [CrossRef]
54. Akmal, M.; Maqbool, Z.; Khan, K.S.; Hussain, Q.; Ijaz, S.S.; Iqbal, M.; Aziz, I.; Hussain, A.; Abbas, M.S.; Rafa, H.U. Integrated use of biochar and compost to improve soil microbial activity, nutrient availability, and plant growth in arid soil. *Arab. J. Geosci.* **2019**, *12*, 232. [CrossRef]
55. Han, Y.; Chen, X.; Wang, E.; Xia, X. Optimum biochar preparations enhance phosphorus availability in amended Mollisols of Northeast China. *Chil. J. Agr. Res.* **2019**, *79*, 153–164. [CrossRef]
56. Li, F.; Liang, X.; Niyungeko, C.; Sun, T.; Liu, F.; Arai, Y. Effects of biochar amendments on soil phosphorus transformation in agricultural soils. *Adv. Agron.* **2019**, *158*, 131–172.
57. Peng, Y.; Sun, Y.; Hanif, A.; Shang, J.; Shen, Z.; Hou, D.; Zhou, Y.; Chen, Q.; Ok, Y.S.; Tsang, D.C.W. Design and fabrication of exfoliated Mg/Al layered double hydroxides on biochar support. *J. Clean. Prod.* **2021**, *289*, 125142. [CrossRef]
58. Yu, L.U.O.; Jiao, Y.-j.; Zhao, X.-r.; Li, G.-t.; Zhao, L.-x.; Meng, H.-b. Improvement to maize growth caused by biochars derived from six feedstocks prepared at three different temperatures. *J. Integr. Agr.* **2014**, *13*, 533–540.
59. Azeem, M.; Hassan, T.U.; Tahir, M.I.; Ali, A.; Jeyasundar, P.G.S.A.; Hussain, Q.; Bashir, S.; Mehmood, S.; Zhang, Z. Tea leaves biochar as a carrier of Bacillus cereus improves the soil function and crop productivity. *Appl. Soil Ecol.* **2021**, *157*, 103732. [CrossRef]
60. Li, Y.; Zhao, Y.; Cheng, K.; Yang, F. Effects of biochar on transport and retention of phosphorus in porous media: Laboratory test and modeling. *Environ. Pollut.* **2022**, *297*, 118788. [CrossRef]
61. Silber, A.; Levkovitch, I.; Graber, E.R. PH-dependent mineral release and surface properties of cornstraw biochar: Agronomic implications. *Environ. Sci. Technol.* **2010**, *44*, 9318–9323. [CrossRef] [PubMed]
62. Jia, M.; Wang, F.; Bian, Y.; Jin, X.; Song, Y.; Kengara, F.O.; Xu, R.; Jiang, X. Effects of pH and metal ions on oxytetracycline sorption to maize-straw-derived biochar. *Bioresour. Technol.* **2013**, *136*, 87–93. [CrossRef] [PubMed]
63. Zhou, C.; Heal, K.; Tigabu, M.; Xia, L.; Hu, H.; Yin, D.; Ma, X. Biochar addition to forest plantation soil enhances phosphorus availability and soil bacterial community diversity. *Forest Ecol. Manag.* **2020**, *455*, 117635. [CrossRef]
64. Xu, G.; Sun, J.; Shao, H.; Chang, S.X. Biochar had effects on phosphorus sorption and desorption in three soils with differing acidity. *Ecol. Eng.* **2014**, *62*, 54–60. [CrossRef]
65. Lv, Y.; Zhao, X.; Shu, Y.; Chang, H.; Zhao, S.; Liu, S. Effect of biochar on the migration and leaching of phosphorus in black soil. *Paddy Water Environ* **2021**, *19*, 1–9. [CrossRef]
66. Ippolito, J.A.; Ducey, T.F.; Cantrell, K.B.; Novak, J.M.; Lentz, R.D. Designer, acidic biochar influences calcareous soil characteristics. *Chemosphere* **2016**, *142*, 184–191. [CrossRef]
67. Wei, W.; Zhang, S.; Wu, L.; Cui, D.; Ding, X. Biochar and phosphorus fertilization improved soil quality and inorganic phosphorus fractions in saline-alkaline soils. *Arch. Agron. Soil Sci.* **2021**, *67*, 1177–1190. [CrossRef]
68. Gao, T.; Gao, M.; Peng, J.; Li, N. Effects of different amount of biochar on nitrogen, phosphorus and potassium nutrients in soil. *IOP Conf. Ser. Mater. Sci. Eng.* **2018**, *394*, 022043. [CrossRef]
69. Gao, X.; Peng, Y.; Zhou, Y.; Adeel, M.; Chen, Q. Effects of magnesium ferrite biochar on the cadmium passivation in acidic soil and bioavailability for packoi (*Brassica chinensis* L.). *J. Environ. Manag.* **2019**, *251*, 109610. [CrossRef]
70. Peng, Y.; Zhang, T.; Tang, B.; Li, X.; Cui, S.; Guan, C.-Y.; Zhang, B.; Chen, Q. Interception of fertile soil phosphorus leaching with immobilization materials: Recent progresses, opportunities and challenges. *Chemosphere* **2022**, *308*, 136337. [CrossRef] [PubMed]
71. Palansooriya, K.N.; Wong, J.T.F.; Hashimoto, Y.; Huang, L.; Rinklebe, J.; Chang, S.X.; Bolan, N.; Wang, H.; Ok, Y.S.J.B. Response of microbial communities to biochar-amended soils: A critical review. *Biochar* **2019**, *1*, 3–22. [CrossRef]
72. Richards, B.K.; Steenhuis, T.S.; Peverly, J.H.; McBride, M.B. Effect of sludge-processing mode, soil texture and soil pH on metal mobility in undisturbed soil columns under accelerated loading. *Environ. Pollut.* **2000**, *109*, 327–346. [CrossRef]
73. Burrell, L.D.; Zehetner, F.; Rampazzo, N.; Wimmer, B.; Soja, G. Long-term effects of biochar on soil physical properties. *Geoderma* **2016**, *282*, 96–102. [CrossRef]
74. Adhikari, S.; Gascó, G.; Méndez, A.; Surapaneni, A.; Jegatheesan, V.; Shah, K.; Paz-Ferreiro, J. Influence of pyrolysis parameters on phosphorus fractions of biosolids derived biochar. *Sci. Total Environ.* **2019**, *695*, 133846. [CrossRef]

agriculture

MDPI

Review

Fire Impacts on Soil Properties and Implications for Sustainability in Rotational Shifting Cultivation: A Review

Noppol Arunrat [1,*], Praeploy Kongsurakan [2], Lemlem Wondwossen Solomon [3] and Sukanya Sereenonchai [1]

1 Faculty of Environment and Resource Studies, Mahidol University, Nakhon Pathom 73170, Thailand; sukanya.ser@mahidol.ac.th
2 Graduate School of Fisheries and Environmental Sciences, Nagasaki University, Nagasaki 852-8102, Japan; praeploy.kong@hotmail.com
3 College of Natural Resources and Environmental Science, Oda Bultum University, Chiro 226, Ethiopia; wondowsenlemlem@gmail.com
* Correspondence: noppol.aru@mahidol.ac.th

Abstract: Fire, a prevalent land management tool in rotational shifting cultivation (RSC), has long been debated for its immediate disruption of surface soil, vegetation, and microbial communities. While low-intensity and short-duration slash-and-burn techniques are considered beneficial for overall soil function, the dual nature of fire's impact warrants a comprehensive exploration. This review examines both the beneficial and detrimental effects of fire on soil properties within the context of RSC. We highlight that research on soil microbial composition, carbon, and nitrogen dynamics following fire events in RSC is gaining momentum. After fires, soil typically shows decreases in porosity, clay content, aggregation, and cation exchange capacity, while sand content, pH, available phosphorus, and organic nitrogen tend to increase. There remains ongoing debate regarding the effects on bulk density, silt content, electrical conductivity, organic carbon, total nitrogen, and exchangeable ions (K^+, Ca^{2+}, Mg^{2+}). Certain bacterial diversity often increases, while fungal communities tend to decline during post-fire recovery, influenced by the soil chemical properties. Soil erosion is a major concern because fire-altered soil structures heighten erosion risks, underscoring the need for sustainable post-fire soil management strategies. Future research directions are proposed, including the use of advanced technologies like remote sensing, UAVs, and soil sensors to monitor fire impacts, as well as socio-economic studies to balance traditional practices with modern sustainability goals. This review aims to inform sustainable land management practices that balance agricultural productivity with ecological health in RSC systems.

Keywords: fire; slash-and-burn; prescribed burning; rotational shifting cultivation; soil property; soil microbial

1. Introduction

Fire is a significant ecological disturbance that can profoundly affect terrestrial ecosystems. It plays a crucial role in shaping vegetation patterns [1,2], influencing nutrient cycling [3–5], and altering soil physicochemical properties [6–8]. In many regions, fire is also an integral component of traditional agricultural practices, such as rotational shifting cultivation (RSC) [9]. Fire plays a multifaceted role beyond nutrient release; it aids in controlling pests and diseases and shapes the landscape by influencing plant succession and diversity [10]. Over time, repeated applications of ash contribute to preserving or enhancing soil nutrient levels and organic matter content in agricultural systems that employ crop rotation [11,12]. When these substances are combined with topsoil layers through the process of rainfall, they become easily accessible for uptake by crops [13]. Additionally, burning makes the process of clearing land faster and easier for farmers compared to using their own hands. After numerous cycles of cultivation and burning, the soil's fertility declines, at which time farmers abandon that plot and shift to a new place, resuming the cycle.

RSC also referred to as slash-and-burn or swidden agriculture, is a traditional farming technique that has been practiced by indigenous and rural communities for millennia [14–17]. Historically, shifting agriculture has been prevalent in various tropical regions of South and Southeast Asia [18–20], as well as parts of Africa and South America [16,21–23]. This traditional practice is crucial for many communities, as it ensures food security, preserves cultural heritage, and supports biodiversity. However, population growth in many regions has altered fallow periods and led to the overexploitation of agricultural resources [24,25]. While shifting cultivation offers significant agricultural benefits, it also sparks debate over its ecological impact, particularly concerning soil sustainability and long-term environmental health [26–30]. RSC involves cyclically clearing land—primarily using fire—cultivating crops, and then allowing the land to lie fallow to recover. While fire can enhance soil fertility in the short term by releasing nutrients from burned vegetation, its repeated use can lead to long-term soil degradation, including the loss of organic matter, deterioration of soil structure, and disruption of essential microbial communities.

In the context of RSC, the use of fire has evolved significantly due to technological advancements and policy changes. Over time, technological progress has introduced new methods that reduce dependence on fire, such as the application of fertilizers, mulching, and incorporation [31,32], minimizing the need for extensive burning. Furthermore, policy changes and population pressures have increasingly influenced fire usage in RSC. Many countries have implemented regulations to control or limit slash-and-burn as well as prescribed burning, motivated by concerns over deforestation, soil degradation, and environmental pollution. However, fire still serves as a crucial tool for clearing vegetation and preparing the land for planting, a traditional agricultural method employed by some indigenous and rural communities [33–35]. In Thailand, for example, the government prohibited slash-and-burn in the 1960s [36], but enforcement of the ban was sporadic until the 1980s. Currently, open burning is banned for short periods as part of hazardous air pollution mitigation measures [37,38], particularly in Northern Thailand. However, burning in RSC fields still occurs in some areas [39–43]. These fields are managed in fallow cycles, where one plot is temporarily cultivated and then abandoned, while the cultivator moves on to another plot. The village remains a permanent settlement without relocating to new sites.

To achieve this, the impact of fire on soil physicochemical properties and soil microorganisms was discussed. For this review, existing peer-reviewed articles were searched using electronic databases such as ISI Web of Science, Scopus, and Google Scholar. Articles published up to May 2024 were collected, focusing on the impact of fire on soil properties and soil microorganisms. Keywords used for the search included 'fire', 'burning', 'swidden cultivation', 'slash-and-burn', 'shifting cultivation', 'rotational shifting cultivation', 'prescribed fire', 'soil properties', 'microorganisms', and 'post-fire management'. We selected only articles written in English and included those reporting both field and laboratory studies. This review delineates the role of fire in RSC, its effects on soil properties, soil erosion, and soil microorganisms, as well as soil recovery post-fire and mitigation and management strategies. The aim of this review is to explore and discuss the dual nature of fire's impact on soil within the context of RSC. By examining both the beneficial and detrimental effects, the review seeks to provide a comprehensive understanding of how fire influences soil properties and to highlight sustainable practices that can mitigate negative outcomes while leveraging positive effects.

Recent international research on the impact of fire on soil, particularly in the context of RSC system, distribution, and ecological effects, has significantly increased over the past two decades, as shown in Figure 1. The overall growth trend in publications reflects the evolving patterns in research findings over time. A total of 75 publications were analyzed, and Figure 1 illustrates the dynamics of publication numbers over the past 20 years.

Between 1998 and 2005, the annual number of published papers remained below 10 papers. However, there was a notable increase from 2006 to 2010, with publications rising from 5 to 13 papers. From 2010 to 2024, the growth curve shows an exponential trend, with the number of publications increasing from 13 to 37 papers. This time series analysis provides insights into research trends and highlights specific phases of research, particularly in areas such as soil properties, microbial composition, soil carbon, and soil nitrogen. Notably, from 2006 to 2020, the majority of publications focused on soil properties. Since 2016, there has been a significant rise in research topics related to microbial composition, soil carbon, and soil nitrogen, indicating an emerging trend in these areas (Figure 2).

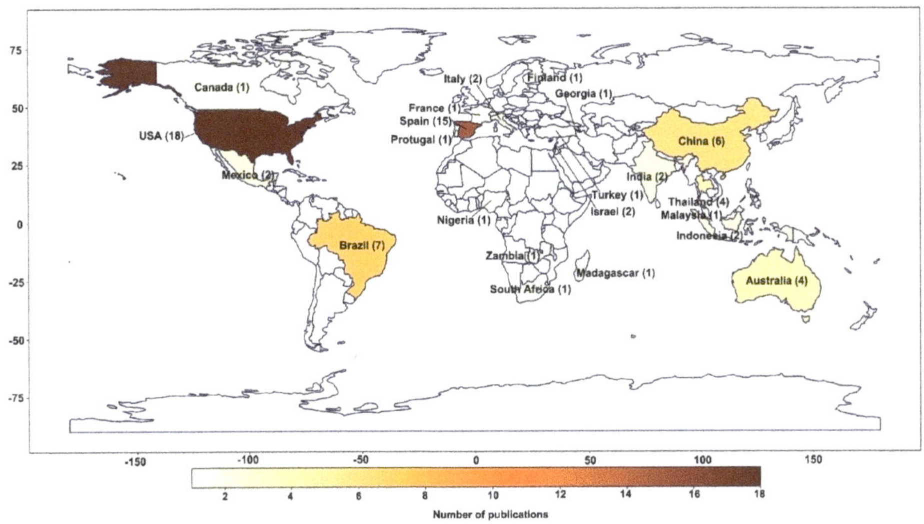

Figure 1. Global distribution of publications on the impact of fire in RSC across major countries (n = 75). The numbers in bracket represents the number of papers from each country.

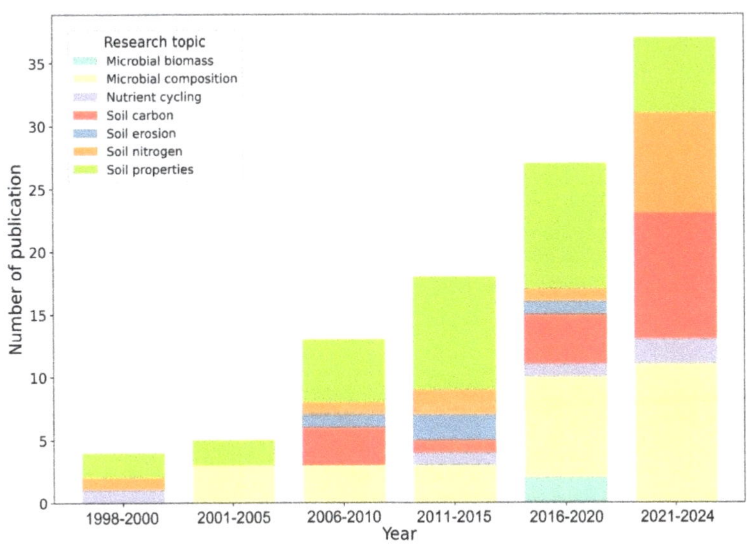

Figure 2. Number of publications on the impact of fire in RSC over 5-year intervals, color-coded by research topic (n = 75).

In our review, we identified a total of 422 keywords in the literature on this research field from 1998 to 2024. Figure 3 visualizes the 30 keywords with the highest co-occurrence and frequency. The most frequent keywords were 'microbial composition' (11 occurrences), 'soil nutrient' (10), 'wildfire' (9), 'carbon' (9), 'prescribed burning' (9), 'fire ecology' (8), 'soil properties' (7), 'nitrogen' (6), 'soil chemical' (5), and 'soil physical' (5). This analysis highlights that current research priorities and emerging topics are centered around soil microorganisms, feedback processes post-fire, and soil properties within the context of agricultural management ecosystems.

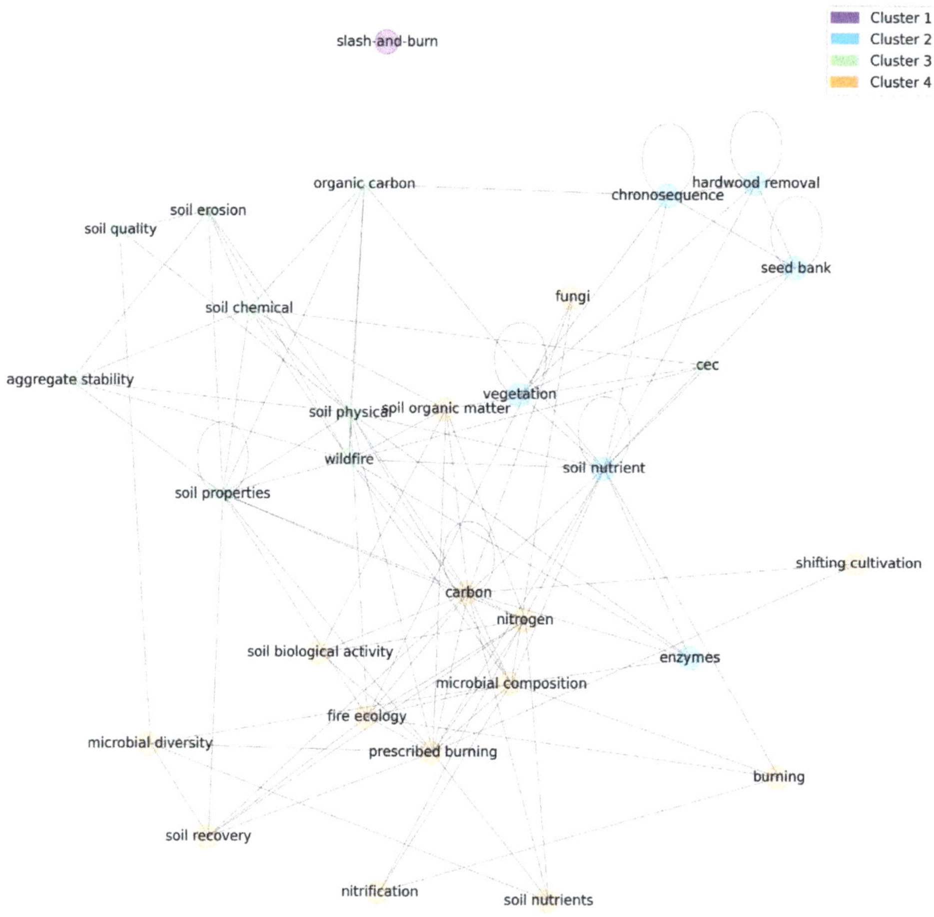

Figure 3. Network of keywords based on the co-occurrence method for the study of the impact of fire on soil in RSC (1998–2024).

The cycle of RSC begins with the selection of a suitable site in a previously fallow area. Initially, vegetation cover is cleared using tools such as axes, machetes, and chainsaws (Figure 4a). After clearing, the residues are left to dry for approximately one week to two months, followed by burning to eliminate the vegetation residues, weeds, and plant pathogens (Figure 4b). This burning process aims to release soil nutrients in the form of ash (Figure 4c,d), temporarily enhancing soil fertility and supporting crop growth. Subsequently, crops such as upland rice, tubers, vegetables, and flowers are planted and tended until harvest time. After the harvest, the land is left fallow to regenerate its fertility, with this period varying from a few months to several years. The cycle then repeats with the selection of a new site for cultivation.

Figure 4. Land clearing of RSC in Northern Thailand. (**a**) cutting, (**b**) burning, (**c**) after burning, and (**d**) remaining ash and charcoal. Photos were taken by Noppol Arunrat.

2. Effects of Fire on Soil Properties

The impact of fire on soil texture, structure, and porosity can vary depending on the severity of the fire, resulting in both positive and negative effects [44]. Fire can profoundly alter soil structure by combusting organic matter [45] and directly affecting its physical, chemical, and biological characteristics through heating and combustion [5,46–48]. Indirect effects include modifications to biological, pedological, and hydrological processes [5,49,50].

In a fire, organic materials such as wood, soil, and leaves rapidly oxidize, releasing various gasses into the atmosphere, including carbon dioxide and water vapor. This process accelerates the conversion of carbon contained in organic matter into atmospheric carbon dioxide, surpassing the rate at which natural decomposition processes [51]. Here, we illustrate the physical changes in topsoil (0–30 cm) over various periods, from before burning to after burning, in an RSC field where ash covers the surface soil. The residual ash, visible as a black substrate, remained on the surface soil 6 months and 12 months post-burning (Figure 5). Intense fires can completely destroy the organic layer on the surface, while low-intensity fires may promote decomposition by releasing nutrients. However, large and intense fires frequently have the opposite effect [52]. Although fires can enhance nutrient availability for decomposition by releasing them from organic matter, uncontrolled and repeated burning can lead to nutrient loss through volatilization or leaching [53].

Nutrient-poor conditions following a fire can reduce microbial activity and nutrient cycling within the ecosystem, ultimately impacting soil health and productivity. These dynamics highlight the complex interplay between fire intensity, nutrient cycling, and microbial activity in post-fire soil recovery.

Figure 5. Soil profile (0–30 cm) at different periods of RSC in Northern Thailand. (**a**) before burning, (**b**) 5 min after burning, (**c**) 6 months after burning, and (**d**) 12 months after burning. Photos were taken by Noppol Arunrat.

Alcañiz et al. [54] reviewed the effects of prescribed fires on soil properties, highlighting that these fires have a less severe impact compared to wildfires due to lower soil heating and fire intensity. Intense heat from fires can also lead to the breakup of soil aggregates, which are groupings of soil particles contributing to its crumbly texture [49]. The combustion process leads to the collapse of organic-mineral aggregates and the destruction of organic matter, reducing both soil bulk density and porosity [55–59]. Despite this, bulk density may also increase due to soil compaction and the loss of organic matter, while porosity typically decreases as soil pores collapse [60]. Fire also affects soil structural stability, particularly altering the distribution and stability of soil aggregates $\geq$ 4 mm, described by Thomaz [61] as the formation of fire-hardened aggregates through rapid physical-chemical processes like mineral fusion and recrystallization. Moreover, prescribed burning conducted on moist soil significantly reduces heat penetration compared to dry soil conditions [62,63]. However, intermittent low-intensity fires can promote soil aggregation under specific conditions. The residual ash generated from combustion may contain oxides, such as aluminum and silica, which act as binding agents among mineral particles. This process facilitates the formation of soil aggregates, enhancing the cohesion and clumping of soil particles over time [64].

Post-fire, organic matter, and volatile substances burning alter soil composition, often increasing sand or silt content [65]. In general, fire tends to reduce the organic content of the soil and disturb its structure, temporarily giving the soil a sandier and less aggregated appearance. However, in specific ecosystems, long-term aggregation can be promoted by periodic low-intensity fires. The texture of the soil can be influenced by factors such as the frequency of fires, their intensity, and the characteristics of the surrounding environment [52]. Studies on slash-and-burn practices reveal varying impacts on soil texture: an increase in sand content due to aggregate breakdown and particle loss during combustion [33,55,56,66], fluctuating silt content influenced by fire severity and erosion processes [33,52,57,67], and a decrease in clay content attributable to physical and chemical transformations induced by fire [45,68].

Both RSC and prescribed burning significantly alter soil chemical properties, notably nutrient availability, pH levels, electric conductivity (EC), and cation exchange capacity (CEC). Fire can significantly affect soil EC due to the release of soluble ions from ash and burned organic matter. The combustion of organic materials releases nutrients such as nitrogen, phosphorus, potassium, calcium, and magnesium into the soil, temporarily increasing their availability [42,45,68–70]. While low-intensity and short-duration fires may not significantly alter some soil properties, they do result in notable increases in pH and EC [41]. The ash produced from burning is rich in alkaline elements [71], which raises soil pH and subsequently increases CEC [72]. The rise in pH is one of the most beneficial effects of prescribed burning, as it enhances the availability of essential nutrients, particularly in acidic soils [69]. The degree of pH alteration is contingent upon the severity of the fire, as more intense burns result in higher amounts of ash, hence leading to more substantial pH rises. The most significant elevations generally manifest in the upper layers of soil, where there is a buildup of ash [45]. The EC often rises in the medium to long term as soluble salts from ash accumulate, although there may be a short-term decrease as these salts leach away [73]. The formation of biochar is also observed in cases of extremely high temperatures, leading to the development of a more enduring alkaline nature compared to ash alone [44]. Moreover, the high-porosity physical structure of biochar can increase the soil's water content [74].

Long-term impacts arise from the interactions between ash and carbonates, which facilitate the retention of pH and induce changes in the soil microbial community responsible for organic matter breakdown [75–77]. This enhanced cation exchange capacity improves the soil's ability to retain essential nutrients, making them more accessible to plants.

Fires can have dual effects on soil nutrients: initially causing losses through volatilization and combustion of organic matter or releasing nutrients like organic nitrogen from burned plant material, which becomes available to soil microorganisms and plants. The breakdown of organic matter post-fire can release organic nitrogen [58] and increase the availability of nitrogen, phosphorus, and potassium [12,33,54,55,58,67,78]. Generally, fire temperatures vary from moderate to very high at the soil surface but have a short residence time and rapidly decrease with soil depth [79]. Despite high temperatures, there is no depletion of carbon content in the topsoil [58]. While combustion may lead to the loss of some organic carbon, the charred residues left in the soil can contribute to long-term soil carbon pools [8,80,81]. Organic and total carbon levels initially increase with the addition of charred residues [41,55,80]. However, the volatilization of nutrients, which varies across regions, can result in the depletion of labile soil carbon, nitrogen, and potassium pools over time due to weathering processes [4,57,82]. Nitrogen dynamics are also significantly affected by fire. Organic nitrogen often experiences a temporary increase following events like fires [55,70,80], but total nitrogen tends to decline over time, largely due to processes such as volatilization and soil erosion. While available nitrogen may show short-term increases as fire releases nitrogen from organic matter, it typically declines in the long term [55,66,82–84]. Phosphorus and potassium levels usually increase after a fire as these nutrients are released from organic matter and minerals [12,33,56,67,78,84], although their availability can fluctuate over time [55,85]. This loss of key nutrients in specific areas can lead to decreased nutrient turnover in the long term, as observed in studies such as that by Wang et al. [35]. Table 1 provides an overview of the effects of fire on changes in soil physicochemical properties, as observed in studies of both slash-and-burn and prescribed burning systems.

Table 1. Effect of fire on soil physicochemical properties.

Soil Properties	Changes	Post-Fire Period	References
Bulk density	increase decrease	5–15 years 5–15 years	[56,60] [59,66]
Porosity	decrease	0–7 years	[55,59]
%Sand	increase	0–2 years	[33,66]
%Silt	increase decrease	10–12 years 5–15 years	[57] [66]
%Clay	decrease	0–15 years	[33,66]
Aggregation	decrease	0–1 year	[67,86]
pH	increase	0–15 years	[12,33,55,56,60,66,67,69,70,75,76,80,82,84–87]
EC	increase decrease	5–15 years 0–3 years	[33,55,56,66,67,69,80,84] [86]
CEC	decrease	12 h after fire	[67,88]
Organic carbon	increase decrease	0–1 year 0–15 years	[47,55,88] [33,56,89]
Total C	increase decrease	12 h after fire 0–15 years	[80,87] [12,66,69,75,82,86,90,91]
Organic nitrogen	increase	0–13 years	[83]
Total N	increase decrease	0–1 year 1–7 years	[55,70,80] [55,66,75,82–84,86]
Available N	increase decrease	0–7 years 0–15 years	[55,80] [33,56,57,82,89]
Available P	increase	0–15 years	[12,33,56,67,76,78,84,85,88,90]
Available K	increase decrease	0–20 years 0–7 years	[56] [55,85,92]
Exchangeable ion (K^+, Ca^{2+}, Mg^{2+})	increase decrease	12 h after fire 0–3 years	[66,67,69,80] [70,85]

Fire significantly influences microbial activities (Table 2) that play crucial roles in soil processes, including the decomposition of organic matter, nutrient mineralization, and enzyme production. The impacts of fire on microbial activity are complex and varied, with extreme temperatures and changes in nutrient availability substantially affecting microbial dynamics [93]. High temperatures from fire can directly damage microbial cells, leading to a significant decrease in overall microbial biomass immediately following the fire [94]. However, the extent of the impact varies depending on fire severity and subsequent environmental conditions [95,96]. Low-to-moderate severity fires often stimulate diverse microbial activities involved in decomposition, enzyme production, and nutrient mineralization. Small, low-intensity fires can stimulate microbial activity and facilitate nutrient mineralization by releasing organic nutrients such as nitrogen, phosphorus, and sulfur into mineralized forms that are more available for microbial and plant consumption [97,98]. However, in surface soils, microorganisms are often killed and organic matter is consumed by combustion during intense fire events. Following such events, nutrient mineralization rates slow down, and microbial activity is reduced [99]. Low-level fires may improve microbial enzyme production by making enzymes bound to soil particles easier to absorb. The heat from low-intensity fires causes structural alterations and increased mobility of enzymes such as cellulases, proteases, and chitinases, allowing them to interact with substrates and catalyze reactions vital for nutrient cycling [100]. Nevertheless, these benefits are limited as the intensity of the fire increases. Temperatures over 200 °C can eradicate enzyme activity within surface soils. Conversely, high-severity fires can temporarily inhibit several microbial processes until recovery [97]. Extreme burns not only eliminate the existing enzymes but also negatively impact the quantity, variety, and activity of soil microbial communities [101].

Fire also affects soil fungi and nitrogen-fixing bacteria. The intensity and duration of flames, along with the fire regime—frequency and return interval—determine the extent of this impact [102]. Arunrat et al. [103] found that bacteria exhibited greater sensitivity to fire compared to fungi. Bacterial richness and diversity increased significantly and recovered more rapidly than fungi one month after burning in RSC fields in Northern Thailand. Some fungi, such as pioneer fungi (*pyrophilous* fungi), are fire-adapted and can survive post-fire by colonizing and breaking down charred organic matter [104]. Mycorrhizal fungi, which form symbiotic relationships with plant roots, are also affected by fire, altering their diversity and abundance. Actinobacteria and Proteobacteria often become more widespread after a fire, while other bacterial species may decline due to competition or sensitivity to post-fire conditions [105]. High temperatures from fire can damage or reduce the population of nitrogen-fixing bacteria, which are vital for nitrogen cycling within ecosystems [106]. However, certain nitrogen-fixing bacteria adapted to fire conditions, such as those associated with fire-adapted plants or residing in fire-resistant structures, can persist or even increase in number following a fire event, including species like *Clostridium* and *Paenibacillus* [107]. These changes influence plant establishment and growth during post-fire recovery, ultimately altering the composition and dynamics of soil microbial communities [104].

Table 2. Effect of fire on soil bacteria and fungi.

Microbial Parameter	Post-Fire Recovery	Relate Factors	References
Bacterial diversity and richness	increase	higher C source	[88]
Actinobacteria	increase	higher N source	[108]
Acidobacteria	increase	higher soil pH	[75,77]
Proteobacteria	increase	higher P source	[75–77]
Firmicutes	increase	higher soil pH	[76]

Table 2. *Cont.*

Microbial Parameter	Post-Fire Recovery	Relate Factors	References
Fungal community composition	decrease	lower C source	[109]
Arbuscular Mycorrhizal Fungi (AMF)	decrease	Lower MBC	[110,111]
Ectomycorrhizal Fungi	decrease	lower C and N source	[91,109]
Cellulolytic Fungi	decrease	lower C source	[112]
Enzyme activities			
Urease	decrease	denatured/lower N source	[55,86,113]
Phosphatase	decrease	lower P source	[55,89,113,114]
β-glucosidase	decrease	denatured, lower MBC	[86,89,90,113,114]
Microbial C utilization	decrease	Lower labile C	[86,110,115]
Microbial Biomass Carbon (MBC)	increase	higher DOC	[47,89,110]
	decrease	denatured/lower DOC	[75,86,87]

DOC refers to soil dissolved organic carbon

3. Impacts of Fire on Soil Erosion

Fire disrupts soil structure by breaking down soil aggregates, resulting in a looser and more granular soil texture prone to erosion. This is compounded by the formation of water-repellent soil layers [7,116], which further exacerbates erosion by reducing water infiltration and increasing surface runoff [117]. Are et al. [118] documented significant reductions in structural stability, saturated hydraulic conductivity, sorptivity, and infiltration rate following slash-and-burn practices. In prescribed burning, fires influence soil natural density, bulk density, porosity, water repellency, and permeability, predominantly in the topsoil within a 5 cm depth [119]. Immediately following a fire, the soil's susceptibility to erosion increases dramatically due to the loss of vegetation and changes in soil properties [6]. Heating from fires can induce significant changes in water repellency and structural stability, influenced by fire intensity and initial soil characteristics [46,52,67,120,121]. Moreover, fire can increase the soil's susceptibility to wind erosion, which is associated with soil hydrophobicity. This is due to water-repellent compounds released by burning vegetation [122,123]. Over the long term, fire-induced changes in soil properties can have lasting effects on soil stability and landscape morphology [54,69,124]. The degree of this effect is influenced by various factors such as soil texture, slope grade, and rainfall intensity, with particles like clays and ashes being the most susceptible to loss. The absence of proper gaps for regeneration between repeated burns can accelerate erosion by inhibiting the complete restoration of root systems and ground cover, which are essential for absorbing and dispersing runoff energy [125]. Repeated fires can lead to persistent changes in soil structure and composition, making the soil more prone to erosion even years after the initial fire event. The loss of topsoil and nutrients can hinder vegetation regrowth, further perpetuating the cycle of erosion [116,126].

In agricultural areas with sloping topography, the absence of proper management practices can lead to uncontrollable runoff, causing erosion and the transport of both particulate and dissolved nutrients downslope. Arunrat et al. [33] reported that average soil surface loss ranged from 1.6 to 3.1 cm, with the highest loss observed during the rainy season on the upper slope in RSC in Northern Thailand. The implementation of RSC practices reduces fallow periods, limiting the time available for vegetation regeneration and the restoration of soil organic carbon levels before subsequent rotations. Consequently, erosion gradually reduces soil fertility by causing oxidation and the subsequent loss of topsoil over time within these dynamic smallholder systems [9].

4. Post-Fire Recovery, Successional Changes after Fire

Post-fire recovery and successional changes following agricultural burning are critical for understanding the resilience and long-term health of affected ecosystems. The soil recovery process after a fire is complex and varies based on several factors, including soil characteristics, fire intensity, microbial communities, and prevailing environmental conditions post-fire [105,127]. Creech et al. [128] found that after a fire with RSC, soil properties may take at least 6 years to return to pre-burn conditions, with changes in nutrient levels and soil pH persisting post-burn. Aboim et al. [129] observed that maintaining fallow plots for periods longer than 5 years can conserve soil quality in RSC in the Atlantic forest region of Rio de Janeiro. In highly vulnerable ecosystems, such as pine stands, the most significant changes in species composition and the lowest rates of post-fire plant recovery were observed [130,131]. However, in revegetated woodland communities in southeastern Australia, post-fire recovery has shown promising results. Pickup et al. [132] found high survival rates of revegetation plantings and substantial recovery of soil function to pre-fire levels within 5 years. Kutiel and Shaviv [133] observed that both bulk density and aggregate stability experienced long-lasting impairments exceeding 15 years due to the slow replenishment of organic matter. Similarly, Murdiyarso et al. [134] documented persistent issues with bulk density and compaction in Indonesian soils over a duration of 10–15 years, attributing these problems to the extended loss of organic matter caused by repeated fires in RSC areas. Piché and Kelting [135] observed that surface soils recover physical properties such as lower bulk density and higher macroporosity within 5–10 years. However, subsoils displayed a legacy effect of agricultural compaction even 55–60 years later. Arunrat et al. [43] reported that the total nitrogen stocks in soil under RSC in Northern Thailand significantly decreased after burning and had not returned to pre-burning levels even after 2 years.

Prescribed burning has minimal short-term effects on soil microbial community composition, likely due to limited soil heating and rapid post-fire vegetative recovery. Post-fire impacts on soil properties can induce short-term microbial responses and shift soil nutrient limitations [136–139]. Soil microbial activity plays a crucial role in nutrient recovery, highlighting its importance in post-fire ecosystem restoration. Studies have revealed diverse temporal recovery patterns in soil microbial communities following burns associated with RSC.

Leal et al. [51] observed that microbial biomass and activity in Amazon soils decreased for 10–15 years post-fire. Similarly, Kutiel and Shaviv [133] found microbial recovery in Israeli shifting plots extended beyond 15 years. In Thailand, Arunrat et al. [41] reported significant changes in certain bacterial phyla, specifically Proteobacteria and Acidobacteria, with soil bacterial communities beginning to recover during the rainy season despite declining nutrient availability. Alpha diversity decreased immediately after the fire but increased from the early rainy season until summer, with bacterial richness and community diversity returning to pre-fire levels within a year.

Some species, such as mycorrhizal fungi, exhibit resistance to surface fires and contribute to recovery [140]. Though severe burns can have long-lasting impacts, persisting for 10 years or more, Zhu et al. [141] found that burning led to relatively high bacterial diversity but low fungal diversity, while mowing increased the abundance of *Nitrospirae* bacteria. These findings underscore the complex and varied responses of soil microbial communities to fire, emphasizing the need for long-term monitoring and tailored management practices in different ecosystems.

Vegetation and plant communities play a crucial role in the soil recovery process after fire, significantly influencing soil structure, nutrient cycling, and microbial activity. Post-fire vegetation regrowth helps stabilize the soil, reducing erosion and promoting the retention of nutrients [142]. For instance, Qiu et al. [143] found that vegetation restoration, including the planting of trees and grasses, improved soil hydraulic properties and increased soil infiltration capacity on the Loess Plateau in China. Additionally, the presence of plant roots can enhance soil microbial activity by providing organic matter and root exudates, which serve as energy sources for soil microbes. Wang et al. [144] observed that coarse root biomass and soil organic matter were strong predictors of soil infiltration capacity, showing a significant positive correlation with infiltration rates after a severe forest fire in Daxing'anling, Northeast China. Different plant species contribute varying amounts and types of organic matter to the soil, influencing microbial community structure and function. Randriamalala et al. [145] studied the slow recovery of endangered xerophytic thicket vegetation after fire in Madagascar, highlighting the significant impact on soil as a driver of plant biodiversity and the key role of shrub species growth in influencing diversity and floristic composition along secondary succession stages. In RSC, the vegetation cover is primarily designed by the farmer. Therefore, the successional process of biodiversity in these areas is influenced by cropping practices and land management, as exemplified in Northern Thailand (Figures 6 and 7). Overall, the complex and prolonged nature of soil biological recovery after fire highlights the need for adequate fallow periods and targeted management strategies to facilitate the complete restoration of soil biota and maintain ecosystem functionality.

Figure 6. Post-fire rotational plots of RSC in Northern Thailand. (**a**) 3 months after burning, (**b**) 6 months after burning, (**c**) 9 months after burning, and (**d**) 12 months after burning. Photos were taken by Noppol Arunrat.

Figure 7. Examples of vegetables and flowers in RSC in Northern Thailand. Photos were taken by Noppol Arunrat.

5. Implications for Sustainability: Mitigation and Management Strategies

Effective mitigation and management strategies are crucial to minimizing the negative impacts on soil health, biodiversity, and carbon emissions while promoting sustainable agricultural practices. The practice of burning in RSC presents significant challenges and opportunities for sustainability. This method involves clearing land by systematically cutting down vegetation, followed by the controlled use of fire. Implementing careful burn management strategies, such as regulated application and maintenance of buffer zones, can help mitigate the risks of erosion caused by fires. The loss of vegetation that protects soils from fire creates at least a short-term window of increased erosion risk, which can alter long-term soil quality and productivity until cover is restored [146]. Arunrat et al. [43] recommended three approaches for post-fire management in RSC: (1) leaving weeds and grasses on the soil surface during vegetation cutting, (2) conducting burns in late winter or early summer to reduce the complete combustion, and (3) constructing contour-felled log erosion barriers using the trunks left after the fire to trap sediment and slow surface runoff.

Controlled burns in RSC fields are intentional and planned fires set by farmers as part of their agricultural practices. Land zoning and establishing protected areas are crucial for conserving valuable ecosystems and mitigating fire risks. Allocating distinct zones for agriculture, forestry, and conservation helps prevent the expansion of RSC into ecologically vulnerable areas. Protected areas act as buffer zones, mitigating fire spread and safeguarding biodiversity. This methodology necessitates a thorough understanding of the appropriate location, limited area, and vegetation cover to prepare the land for cultivation. Additionally, conducting controlled burns in a sustainable manner requires taking into account weather conditions and implementing safety procedures [147,148]. While burning practices in RSC pose significant environmental risks, they also offer potential benefits if managed with sustainable strategies [149,150]. Key measures include careful planning of burn locations, maintaining buffer zones, monitoring weather conditions, and following safety protocols.

Rapid recovery, genetic adaptation, nutrient cycling, and organic matter decomposition are essential for restoring and functioning fire-affected ecosystems [151]. Some soil microorganisms can quickly recolonize burned areas from less impacted or unburned nearby regions [103,152]. Dispersal mechanisms like wind, water, or animal vectors aid in moving microbes to fire-affected areas. This rapid rehabilitation preserves vital ecosystem functions, facilitating plant regeneration. In post-fire pasture, soil organic matter is primarily contributed by *Brachiaria*. At depths of 40–50 cm, alkyl and hydroaromatic compounds accumulate in the pasture post-fire, while unspecific aromatic compounds (UACs) accumulate in the pasture after burning. UACs and polycyclic aromatic hydrocarbons (PAHs) are abundant in RSC practices, likely air-transported from the burn sites [48]. Over time, soil microorganisms can genetically adapt to fire disturbances. Through natural selection, some microorganisms develop traits that confer resistance or tolerance to fire, such as heat resistance or the ability to metabolize fire-altered compounds. These genetic adaptations enhance their survival and recovery in fire-affected soils, contributing to ecosystem resilience [153].

Improperly managed RSC areas can lead to significant deforestation and biodiversity loss. However, when executed with sustainable principles, these practices can enhance soil fertility and biodiversity conservation. Conservation agriculture techniques, such as no-tillage, cover cropping and mulching, and soil erosion protection, can reduce the need for chemical substances. No-tillage reduces soil disturbance and improves moisture retention, while cover cropping enhances soil fertility, decreases weed spread, and increases organic matter content, promoting sustainable land management [154]. Extending the fallow periods between agricultural cycles allows for the restoration and recovery of soil. Longer fallow periods promote vegetation regeneration, replenishing organic matter and nutrient levels, and enhancing microbial activity [150,155–157].

Alternative land management strategies can effectively minimize the adverse effects of fire in these systems. A pivotal strategy involves transitioning from traditional practices to agroforestry systems, integrating tree maintenance with crop production. Agroforestry, integrating trees with crops, not only provides shade and mitigates soil erosion but also contributes to biodiversity conservation and carbon sequestration, aligning with climate action goals. Chowdhury et al. [158] demonstrated that agroforestry holds greater potential for soil restoration after RSC compared to reforestation, showing significantly higher concentrations of soil organic matter, available phosphorus, and exchangeable potassium in agroforestry plots. Moreover, agroforestry extends the rotation period, thereby reducing the frequency of land clearing and burning, and enhances carbon sequestration [149,159].

In conjunction with agroforestry, intercropping and crop rotation are effective agricultural practices that contribute to soil conservation, nutrient cycling, and overall soil health. These techniques have been extensively studied, revealing numerous ecological benefits. Intercropping, the practice of growing multiple crop species in the same field, offers significant advantages. It enhances soil carbon and nitrogen content, leading to increased soil organic carbon and nitrogen levels compared to sole crops, thereby promoting soil conservation through enhanced belowground productivity and root litter input [160] Additionally, intercropping systems have been shown to improve soil fertility and microbial activity essential for sustainable agriculture [161,162].

Crop rotation, which involves changing the types of crops grown in an area each season, disrupts pest and disease cycles and improves soil quality. This practice helps avoid the buildup of pathogens and pests while enhancing soil structure and fertility by alternating deep-rooted and shallow-rooted plants [163]. Furthermore, crop rotations with legumes, such as alfalfa and clover, significantly enhance soil organic carbon sequestration and soil physicochemical properties, contributing to long-term soil sustainability [164]. Implementing these practices can help maintain soil productivity, promote soil improvement, and ensure sustainable agricultural productivity, ultimately contributing to ecological and economic benefits [165].

Education and awareness programs are essential for promoting sustainable land management practices among farmers, especially in the face of climate change. While RSC is traditionally used by indigenous groups and has been inherently conservative, there is a growing need to ensure these practices are sustainable in the long term. These programs can equip farmers with the knowledge and tools necessary to adopt more sustainable techniques, thereby enhancing the benefits of traditional methods and addressing climate resilience. Engaging with indigenous communities in these programs ensures that traditional knowledge is respected and integrated with sustainable practices. This collaborative approach fosters a sense of ownership and empowerment among local farmers, promoting practices that benefit both the environment and local livelihoods. By aligning traditional methods with modern sustainable practices and climate action strategies, education and awareness programs can play a pivotal role in achieving long-term sustainability and resilience in agricultural landscapes.

6. Future Research Directions

Future research in sustainable land management should focus on several key areas to enhance our understanding and implementation of effective practices. A crucial avenue is the development of technology and methodologies to investigate the interaction of fire, soil characteristics, and soil microorganisms. Large-scale analysis using spatial techniques such as remote sensing, unmanned aerial vehicles (UAVs), geographic information systems (GIS), and field measurements can provide valuable insights into the interplay between fire occurrences, land management, and soil degradation across extensive geographical areas [166–168].

Monitoring and understanding soil properties in RSC can be significantly enhanced with real-time information provided by soil sensor technology [169,170]. These sensors can offer crucial insights into how fire and other land management methods affect soil properties, facilitating informed decision-making for sustainable land management. In-depth studies on the dynamics of soil microorganisms and nutrient cycles can be advanced using techniques such as stable isotope probing (SIP) and high-throughput DNA sequencing. SIP allows researchers to trace nutrient flow through microbial communities by incorporating isotopically labeled compounds into the DNA or RNA of active microorganisms [171]. High-throughput DNA sequencing enables comprehensive analysis of microbial diversity and function by rapidly sequencing large volumes of genetic material from soil samples [141]. By integrating these advanced technologies and methodologies, researchers can develop more effective and sustainable land management practices that address the complex interactions between fire, soil characteristics, and soil microorganisms.

Last but not least, the socio-economic aspects of fire management in RSC should focus on understanding the intricate balance between traditional practices, community livelihoods, and sustainable land use. Investigations should delve into the socio-economic drivers behind the use of fire in shifting cultivation, assessing how these practices affect local economies, food security, and cultural heritage. Furthermore, research should explore the effectiveness of community-based fire management strategies and their potential to enhance resilience against environmental and economic pressures. By integrating socio-economic analysis with ecological data, researchers can develop holistic fire management policies that support sustainable development, improve the well-being of local communities, and preserve essential ecosystem services. Furthermore, the economic trade-offs and challenges faced by communities in adopting sustainable fire management practices should be thoroughly investigated. These multi-disciplinary approaches will provide valuable insights into the long-term viability of RSC in the face of changing environmental and socio-economic conditions.

7. Conclusions

This review underscores the intricate effects of low to medium-severity burning, including slash-and-burn and prescribed burning, on soil properties and microbial communities in RSC systems. These fire practices have dual impacts: while they initially cause nutrient losses through volatilization and combustion, they also release essential nutrients such as nitrogen, phosphorus, and potassium from burned vegetation, enhancing their availability to soil microorganisms and plants. The breakdown of organic matter post-fire further increases nutrient availability. Although high surface temperatures during burning may affect soil carbon content, charred residues contribute to long-term soil carbon pools. Recovery of soil properties and microbial communities post-fire is influenced by fire intensity, soil characteristics, and environmental conditions. Fire affects soil microbial diversity and activity, with low-severity burns generally causing minimal short-term changes due to rapid vegetative recovery, while severe burns can lead to long-lasting alterations in microbial community structure. For instance, some microbial species, such as fire-adapted fungi and nitrogen-fixing bacteria, may show resilience or even increased numbers post-fire, whereas others may decline. Understanding these dynamics is crucial for ecosystem restoration and function.

Effective management strategies, including controlled burns, proper land zoning, and sustainable practices such as agroforestry, cover cropping, and crop rotation, are essential for mitigating negative impacts on soil health and microbial communities. Future research should prioritize advancements in technology, such as remote sensing, UAVs, GIS, and soil sensors, to better understand fire interactions with soil and microbial dynamics. Techniques like stable isotope probing and high-throughput DNA sequencing can provide deeper insights into microbial diversity and nutrient cycles. Additionally, exploring the socio-economic dimensions of fire management can help balance traditional practices with sustainable land use, enhancing community resilience and preserving ecosystem services. This holistic approach is vital for achieving long-term sustainability and resilience in fire-affected agricultural systems.

Author Contributions: Conceptualization, N.A. and P.K.; methodology, N.A.; data collection and formal analysis, N.A., P.K., L.W.S. and S.S.; writing—original draft preparation, N.A., P.K., L.W.S. and S.S.; writing—review and editing, N.A., P.K. and S.S.; funding acquisition, N.A. All authors have read and agreed to the published version of the manuscript.

Funding: This research project is supported by Mahidol University (Strategic Research Fund: 2024) (Contract Number: MU-SRF-RS-21 B/67).

Institutional Review Board Statement: Not applicable.

Data Availability Statement: All data generated or analyzed during this study are included in this published article.

Acknowledgments: We would like to thank Mahidol University (Strategic Research Fund: 2024) (Contract Number: MU-SRF-RS-21 B/67) for supporting this study.

Conflicts of Interest: The authors declare no conflicts of interest.

References

1. Fukushima, M.; Kanzaki, M.; Hara, M.; Ohkubo, T.; Preechapanya, P.; Choocharoen, C. Secondary forest succession after the cessation of swidden cultivation in the montane forest area in Northern Thailand. *For. Ecol. Manag.* **2008**, *255*, 1994–2006. [CrossRef]
2. Brussaard, L.; Pulleman, M.M.; Ouédraogo, É.; Mando, A.; Six, J. Soil fauna and soil function in the fabric of the food web. *Pedobiologia* **2007**, *50*, 447–462. [CrossRef]
3. Strydom, T.; Smit, I.P.J.; van Tol, J.J. Short and long-term fire effects on soil C and N in an African savanna. *Geoderma Reg.* **2024**, *37*, e00802. [CrossRef]
4. Muqaddas, B.; Lewis, T.; Esfandbod, M.; Chen, C. Responses of labile soil organic carbon and nitrogen pools to long-term prescribed burning regimes in a wet sclerophyll forest of southeast Queensland, Australia. *Sci. Total Environ.* **2019**, *647*, 110–120. [CrossRef]

5. Roth, H.K.; McKenna, A.M.; Simpson, M.J.; Chen, H.; Srikanthan, N.; Fegel, T.S.; Nelson, A.R.; Rhoades, C.C.; Wilkins, M.J.; Borch, T. Effects of burn severity on organic nitrogen and carbon chemistry in high-elevation forest soils. *Soil. Environ. Health* **2023**, *1*, 100023. [CrossRef]
6. Lucas-Borja, M.E.; de las Heras, J.; Moya Navarro, D.; González-Romero, J.; Peña-Molina, E.; Navidi, M.; Fajardo-Cantos, Á.; Miralles Mellado, I.; Plaza-Alvarez, P.A.; Gianmarco Carrà, B.; et al. Short-term effects of prescribed fires with different severity on rainsplash erosion and physico-chemical properties of surface soil in Mediterranean forests. *J. Environ. Manag.* **2022**, *322*, 116143. [CrossRef]
7. Fox, D.M.; Darboux, F.; Carrega, P. Effects of fire-induced water repellency on soil aggregate stability, splash erosion, and saturated hydraulic conductivity for different size fractions. *Hydrol. Process.* **2007**, *21*, 2377–2384. [CrossRef]
8. Alcañiz, M.; Outeiro, L.; Francos, M.; Farguell, J.; Úbeda, X. Long-term dynamics of soil chemical properties after a prescribed fire in a Mediterranean forest (Montgrí Massif, Catalonia, Spain). *Sci. Total Environ.* **2016**, *572*, 1329–1335. [CrossRef]
9. Grogan, P.; Lalnunmawia, F.; Tripathi, S.K. Shifting cultivation in steeply sloped regions: A review of management options and research priorities for Mizoram state, Northeast India. *Agrofor. Syst.* **2012**, *84*, 163–177. [CrossRef]
10. Howard, R.J. Cultural control of plant diseases: A historical perspective. *Can. J. Plant Pathol.* **1996**, *18*, 145–150. [CrossRef]
11. Obernberger, I.; Biedermann, F.; Widmann, W.; Riedl, R. Concentrations of inorganic elements in biomass fuels and recovery in the different ash fractions. *Biomass Bioenergy* **1997**, *12*, 211–224. [CrossRef]
12. Ketterings, Q.M.; Van Noordwijk, M.; Bigham, J.M. Soil phosphorus availability after slash-and-burn fires of different intensities in rubber agroforests in Sumatra, Indonesia. *Agric. Ecosyst. Environ.* **2002**, *92*, 37–48. [CrossRef]
13. Gay-des-Combes, J.M.; Sanz Carrillo, C.; Robroek, B.J.M.; Jassey, V.E.J.; Mills, R.T.E.; Arif, M.S.; Falquet, L.; Frossard, E.; Buttler, A. Tropical soils degraded by slash-and-burn cultivation can be recultivated when amended with ashes and compost. *Ecol. Evol.* **2017**, *7*, 5378–5388. [CrossRef] [PubMed]
14. Warner, K. *Shifting Cultivators: Local Technical Knowledge and Natural Resource Management in the Humid Tropics*; FAO: Rome, Italy, 2001.
15. Otto, J.S.; Anderson, N.E. Slash-and-Burn Cultivation in the Highlands South: A Problem in Comparative Agricultural History. *Comp. Stud. Soc. Hist.* **1982**, *24*, 131–147. [CrossRef]
16. Schuck, E.C.; Nganje, W.; Yantio, D. The role of land tenure and extension education in the adoption of slash and burn agriculture. *Ecol. Econ.* **2002**, *43*, 61–70. [CrossRef]
17. Vosti, S.A.; Witcover, J. Slash-and-burn agriculture—Household perspectives. *Agric. Ecosyst. Environ.* **1996**, *58*, 23–38. [CrossRef]
18. Mertz, O.; Padoch, C.; Fox, J.; Cramb, R.A.; Leisz, S.J.; Lam, N.T.; Vien, T.D. Swidden change in southeast Asia: Understanding causes and consequences. *Hum. Ecol.* **2009**, *37*, 259–264. [CrossRef]
19. Li, P.; Feng, Z.; Jiang, L.; Liao, C.; Zhang, J. A Review of Swidden Agriculture in Southeast Asia. *Remote Sens.* **2014**, *6*, 1654–1683. [CrossRef]
20. Nakano, K. An Ecological Study or Swidden Agriculture at a Village in Northern Thailand. *Jpn. J. Southeast Asian Stud.* **1978**, *16*, 411–446.
21. Pellegrini, A.F.A.; Hobbie, S.E.; Reich, P.B.; Jumpponen, A.; Brookshire, E.N.J.; Caprio, A.C.; Coetsee, C.; Jackson, R.B. Repeated fire shifts carbon and nitrogen cycling by changing plant inputs and soil decomposition across ecosystems. *Ecol. Monogr.* **2020**, *90*, e01409. [CrossRef]
22. Hauser, S.; Norgrove, L. Slash-and-Burn Agriculture, Effects of. In *Encyclopedia of Biodiversity*, 2nd ed.; Academic Press: Cambridge, MA, USA, 2013; pp. 551–562. [CrossRef]
23. Lal, R. Shifting Cultivation Versus Sustainable Intensification. In *Reference Module in Earth Systems and Environmental Sciences*; Elsevier: New York, NY, USA, 2015; pp. 1–12. [CrossRef]
24. van Vliet, N.; Mertz, O.; Heinimann, A.; Langanke, T.; Pascual, U.; Schmook, B.; Adams, C.; Schmidt-Vogt, D.; Messerli, P.; Leisz, S.; et al. Trends, drivers and impacts of changes in swidden cultivation in tropical forest-agriculture frontiers: A global assessment. *Glob. Environ. Chang.* **2012**, *22*, 418–429. [CrossRef]
25. Schritt, H.; Beusch, C.; Guayasamín, P.R.; Kaupenjohann, M. Transformation of traditional shifting cultivation into permanent cropping systems: A case study in Sarayaku, Ecuador. *Ecol. Soc.* **2020**, *25*, 10. [CrossRef]
26. Tinker, P.B.; Ingram, J.S.I.; Struwe, S. Effects of slash-and-burn agriculture and deforestation on climate change. *Agric. Ecosyst. Environ.* **1996**, *58*, 13–22. [CrossRef]
27. Varma, A. The economics of slash and burn: A case study of the 1997–1998 Indonesian forest fires. *Ecol. Econ.* **2003**, *46*, 159–171. [CrossRef]
28. Brady, N.C. Alternatives to slash-and-burn: A global imperative. *Agric. Ecosyst. Environ.* **1996**, *58*, 3–11. [CrossRef]
29. Kleinman, P.J.A.; Pimentel, D.; Bryant, R.B. The ecological sustainability of slash-and-burn agriculture. *Agric. Ecosyst. Environ.* **1995**, *52*, 235–249. [CrossRef]
30. Ickowitz, A. Shifting Cultivation and Deforestation in Tropical Africa: Critical Reflections. *Dev. Chang.* **2006**, *37*, 599–626. [CrossRef]
31. Kato, M.S.A.; Kato, O.R.; Denich, M.; Vlek, P.L.G. Fire-free alternatives to slash-and-burn for shifting cultivation in the eastern Amazon region: The role of fertilizers. *Field Crops Res.* **1999**, *62*, 225–237. [CrossRef]
32. Hands, M. The search for a sustainable alternative to slash-and-burn agriculture in the World's rain forests: The Guama Model and its implementation. *R. Soc. Open Sci.* **2021**, *8*, 201204. [CrossRef]

33. Arunrat, N.; Sereenonchai, S.; Kongsurakan, P.; Yuttitham, M.; Hatano, R. Variations of soil properties and soil surface loss after fire in rotational shifting cultivation in Northern Thailand. *Front. Environ. Sci.* **2023**, *11*, 1213181. [CrossRef]
34. Laskar, S.Y.; Sileshi, G.W.; Pathak, K.; Debnath, N.; Nath, A.J.; Laskar, K.Y.; Singnar, P.; Das, A.K. Variations in soil organic carbon content with chronosequence, soil depth and aggregate size under shifting cultivation. *Sci. Total Environ.* **2021**, *762*, 143114. [CrossRef]
35. Wang, G.; Zhu, T.; Zhou, J.; Yu, Y.; Petropoulos, E.; Müller, C. Slash-and-burn in karst regions lowers soil gross nitrogen (N) transformation rates and N-turnover. *Geoderma* **2022**, *425*, 116084. [CrossRef]
36. Tomforde, M. The Global in the Local: Contested Resource-use Systems of the Karen and Hmong in Northern Thailand. *J. Southeast Asian Stud.* **2003**, *34*, 347–360. [CrossRef]
37. Moran, J.; NaSuwan, C.; Poocharoen, O.O. The haze problem in Northern Thailand and policies to combat it: A review. *Environ. Sci. Policy* **2019**, *97*, 1–15. [CrossRef]
38. Arunrat, N.; Pumijumnong, N.; Sereenonchai, S. Air-Pollutant Emissions from Agricultural Burning in Mae Chaem Basin, Chiang Mai Province, Thailand. *Atmosphere* **2018**, *9*, 145. [CrossRef]
39. Mostafanezhad, M.; Evrard, O. Chronopolitics of crisis: A historical political ecology of seasonal air pollution in northern Thailand. *Geoforum* **2021**, *124*, 400–408. [CrossRef]
40. Phairuang, W.; Hata, M.; Furuuchi, M. Influence of agricultural activities, forest fires and agro-industries on air quality in Thailand. *J. Environ. Sci.* **2017**, *52*, 85–97. [CrossRef]
41. Arunrat, N.; Sansupa, C.; Sereenonchai, S.; Hatano, R.; Lal, R. Fire-induced changes in soil properties and bacterial communities in rotational shifting cultivation fields in Northern Thailand. *Biology* **2024**, *13*, 383. [CrossRef]
42. Arunrat, N.; Sereenonchai, S.; Hatano, R. Effects of fire on soil organic carbon, soil total nitrogen, and soil properties under rotational shifting cultivation in northern Thailand. *J. Environ. Manag.* **2022**, *302*, 113978. [CrossRef]
43. Arunrat, N.; Sereenonchai, S.; Kongsurakan, P.; Iwai, C.B.; Yuttitham, M.; Hatano, R. Post-fire recovery of soil organic carbon, soil total nitrogen, soil nutrients, and soil erodibility in rotational shifting cultivation in Northern Thailand. *Front. Environ. Sci.* **2023**, *11*, 1117427. [CrossRef]
44. Úbeda, X.; Pereira, P.; Outeiro, L.; Martin, D.A. Effects of fire temperature on the physical and chemical characteristics of the ash from two plots of cork oak (*Quercus suber*). *Land Degrad. Dev.* **2009**, *20*, 589–608. [CrossRef]
45. Certini, G. Effects of fire on properties of forest soils: A review. *Oecologia* **2005**, *143*, 1–10. [CrossRef]
46. Badía, D.; Martí, C. Plant ash and heat intensity effects on chemical and physical properties of two contrasting soils. *Arid Land Res Manag.* **2003**, *17*, 23–41. [CrossRef]
47. Zhao, H.; Tong, D.Q.; Lin, Q.; Lu, X.; Wang, G. Effect of fires on soil organic carbon pool and mineralization in a Northeastern China wetland. *Geoderma* **2012**, *189–190*, 532–539. [CrossRef]
48. Ketterings, Q.M.; Bigham, J.M.; Laperche, V. Changes in Soil Mineralogy and Texture Caused by Slash-and-Burn Fires in Sumatra, Indonesia. *Soil Sci. Soc. Am. J.* **2000**, *64*, 1108–1117. [CrossRef]
49. Memoli, V.; Panico, S.C.; Santorufo, L.; Barile, R.; Di Natale, G.; Di Nunzio, A.; Toscanesi, M.; Trifuoggi, M.; De Marco, A.; Maisto, G. Do Wildfires Cause Changes in Soil Quality in the Short Term? *Int. J. Environ. Res. Public Health* **2020**, *17*, 5343. [CrossRef] [PubMed]
50. Doerr, S.H.; Santín, C.; Mataix-Solera, J. Fire effects on soil. In *Encyclopedia of Biodiversity*, 2nd ed.; Academic Press: Cambridge, MA, USA, 2023; pp. 448–457. [CrossRef]
51. Leal, O.d.A.; Jiménez-Morillo, N.T.; González-Pérez, J.A.; Knicker, H.; de Souza Costa, F.; Jiménez-Morillo, P.N.; de Carvalho Júnior, J.A.; dos Santos, J.C.; Pinheiro Dick, D. Soil Organic Matter Molecular Composition Shifts Driven by Forest Regrowth or Pasture after Slash-and-Burn of Amazon Forest. *Int. J. Environ. Res. Public Health* **2023**, *20*, 3485. [CrossRef]
52. Mataix-Solera, J.; Cerdà, A.; Arcenegui, V.; Jordán, A.; Zavala, L.M. Fire effects on soil aggregation: A review. *Earth-Sci. Rev.* **2011**, *109*, 44–60. [CrossRef]
53. Pellegrini, A.F.A.; Jackson, R.B. The long and short of it: A review of the timescales of how fire affects soils using the pulse-press framework. In *Advances in Ecological Research*; Academic Press: Cambridge, MA, USA, 2020; Volume 62, pp. 147–171, ISBN 9780128211342.
54. Alcañiz, M.; Outeiro, L.; Francos, M.; Úbeda, X. Effects of prescribed fires on soil properties: A review. *Sci. Total Environ.* **2018**, *613–614*, 944–957. [CrossRef]
55. Xue, L.; Li, Q.; Chen, H. Effects of a Wildfire on Selected Physical, Chemical and Biochemical Soil Properties in a Pinus massoniana Forest in South China. *Forests* **2014**, *5*, 2947–2966. [CrossRef]
56. Saplalrinliana, H.; Thakuria, D.; Changkija, S.; Hazarika, S. Impact of shifting cultivation on litter accumulation and properties of Jhum soils of north east India. *J. Indian Soc. Soil Sci.* **2016**, *64*, 402–413. [CrossRef]
57. Mishra, G.; Giri, K.; Jangir, A.; Vasu, D.; Rodrigo-Comino, J. Understanding the effect of shifting cultivation practice (slash-burn-cultivation-abandonment) on soil physicochemical properties in the North-eastern Himalayan region. *Investig. Geogr.* **2021**, *76*, 243–261. [CrossRef]
58. Ekinci, H. Effect of Forest Fire on Some Physical, Chemical and Biological Properties of Soil in Çanakkale, Turkey. *Int. J. Agric. Biol.* **2006**, *8*, 102–106.
59. Li, T.; Jeřábek, J.; Winkler, J.; Vaverková, M.D.; Zumr, D. Effects of prescribed fire on topsoil properties: A small-scale straw burning experiment. *J. Hydrol. Hydromech.* **2022**, *70*, 450–461. [CrossRef]

60. Moreno-Roso, S.; Chávez-Vergara, B.; Solleiro-Rebolledo, E.; Quintero-Gradilla, S.; Merino, A.; Ruiz-Rojas, M. Soil Burn Severities Evaluation Using Micromorphology and Morphometry Traits After a Prescribed Burn in a Managed Forest. *Span. J. Soil Sci.* **2023**, *13*, 11488. [CrossRef]
61. Thomaz, E.L. High fire temperature changes soil aggregate stability in slash-and-burn agricultural systems. *Sci. Agric.* **2017**, *74*, 157–162. [CrossRef]
62. Busse, M.D.; Shestak, C.J.; Hubbert, K.R.; Knapp, E.E. Soil Physical Properties Regulate Lethal Heating during Burning of Woody Residues. *Soil Sci. Soc. Am. J.* **2010**, *74*, 947–955. [CrossRef]
63. Badía, D.; López-García, S.; Martí, C.; Ortíz-Perpiñá, O.; Girona-García, A.; Casanova-Gascón, J. Burn effects on soil properties associated to heat transfer under contrasting moisture content. *Sci. Total Environ.* **2017**, *601–602*, 1119–1128. [CrossRef]
64. Urbanek, E. Why are aggregates destroyed in low intensity fire? *Plant Soil* **2013**, *362*, 33–36. [CrossRef]
65. Ulery, A.L.; Graham, R.C. Forest Fire Effects on Soil Color and Texture. *Soil Sci. Soc. Am. J.* **1993**, *57*, 135–140. [CrossRef]
66. Ying, H.S.; Bin Wasli, M.E.; Perumal, M. Soil characteristics under intensified shifting cultivation for upland rice cultivation in upland Sabal, Sarawak, Malaysia. *Biotropia* **2018**, *25*, 72–83. [CrossRef]
67. Thomaz, E.L.; Antoneli, V.; Doerr, S.H. Effects of fire on the physicochemical properties of soil in a slash-and-burn agriculture. *Catena* **2014**, *122*, 209–215. [CrossRef]
68. Raison, R.J.; Khanna, R.K.; Woods, P.V. Transfer of elements to the atmosphere during low-intensity prescribed fires in three Australian subalpine eucalypt forests. *Can. J. For. Res.* **2011**, *15*, 657–664. [CrossRef]
69. Arocena, J.M.; Opio, C. Prescribed fire-induced changes in properties of sub-boreal forest soils. *Geoderma* **2003**, *113*, 1–16. [CrossRef]
70. Chungu, D.; Ng'andwe, P.; Mubanga, H.; Chileshe, F. Fire alters the availability of soil nutrients and accelerates growth of Eucalyptus grandis in Zambia. *J. For. Res.* **2020**, *31*, 1637–1645. [CrossRef]
71. Escudey, M.; Arancibia-Miranda, N.; Pizarro, C.; Antilén, M. Effect of ash from forest fires on leaching in volcanic soils. *Catena* **2015**, *135*, 383–392. [CrossRef]
72. Ulery, A.L.; Graham, R.C.; Goforth, B.R.; Hubbert, K.R. Fire effects on cation exchange capacity of California forest and woodland soils. *Geoderma* **2017**, *286*, 125–130. [CrossRef]
73. Inbar, A.; Lado, M.; Sternberg, M.; Tenau, H.; Ben-Hur, M. Forest fire effects on soil chemical and physicochemical properties, infiltration, runoff, and erosion in a semiarid Mediterranean region. *Geoderma* **2014**, *221–222*, 131–138. [CrossRef]
74. Sun, W.; Li, Y.; Xu, Z.; Bai, Y.; Bai, S.H. Biochar application for enhancing water and nitrogen use efficiency of understory acacia species in a suburban native forest subjected to nitrogen deposition in Southeast Queensland. *Plant Soil* **2024**. [CrossRef]
75. Shen, J.P.; Chen, C.R.; Lewis, T. Long term repeated fire disturbance alters soil bacterial diversity but not the abundance in an Australian wet sclerophyll forest. *Sci. Rep.* **2016**, *6*, 19639. [CrossRef]
76. Kang, J.W.; Park, Y.D. Effects of deforestation on microbial diversity in a Siberian larch (*Larix sibirica*) stand in Mongolia. *J. For. Res.* **2019**, *30*, 1885–1893. [CrossRef]
77. Rafie, S.A.A.; Blentlinger, L.R.; Putt, A.D.; Williams, D.E.; Joyner, D.C.; Campa, M.F.; Schubert, M.J.; Hoyt, K.P.; Horn, S.P.; Franklin, J.A.; et al. Impact of prescribed fire on soil microbial communities in a Southern Appalachian Forest clear-cut. *Front. Microbiol.* **2024**, *15*, 1322151. [CrossRef] [PubMed]
78. Goberna, M.; García, C.; Insam, H.; Hernández, M.T.; Verdú, M. Burning Fire-Prone Mediterranean Shrublands: Immediate Changes in Soil Microbial Community Structure and Ecosystem Functions. *Microb. Ecol.* **2012**, *64*, 242–255. [CrossRef] [PubMed]
79. Armas-Herrera, C.M.; Martí, C.; Badía, D.; Ortiz-Perpiñá, O.; Girona-García, A.; Porta, J. Immediate effects of prescribed burning in the Central Pyrenees on the amount and stability of topsoil organic matter. *CATENA* **2016**, *147*, 238–244. [CrossRef]
80. Scharenbroch, B.C.; Nix, B.; Jacobs, K.A.; Bowles, M.L. Two decades of low-severity prescribed fire increases soil nutrient availability in a Midwestern, USA oak (Quercus) forest. *Geoderma* **2012**, *183–184*, 80–91. [CrossRef]
81. Pellegrini, A.F.A.; Harden, J.; Georgiou, K.; Hemes, K.S.; Malhotra, A.; Nolan, C.J.; Jackson, R.B. Fire effects on the persistence of soil organic matter and long-term carbon storage. *Nat. Geosci.* **2021**, *15*, 5–13. [CrossRef]
82. Muqaddas, B.; Zhou, X.; Lewis, T.; Wild, C.; Chen, C. Long-term frequent prescribed fire decreases surface soil carbon and nitrogen pools in a wet sclerophyll forest of Southeast Queensland, Australia. *Sci. Total Environ.* **2015**, *536*, 39–47. [CrossRef]
83. Francos, M.; Úbeda, X.; Pereira, P.; Alcañiz, M. Long-term impact of wildfire on soils exposed to different fire severities. A case study in Cadiretes Massif (NE Iberian Peninsula). *Sci. Total Environ.* **2018**, *615*, 664–671. [CrossRef]
84. Francos, M.; Stefanuto, E.B.; Úbeda, X.; Pereira, P. Long-term impact of prescribed fire on soil chemical properties in a wildland-urban interface. Northeastern Iberian Peninsula. *Sci. Total Environ.* **2019**, *689*, 305–311. [CrossRef]
85. Fonseca, F.; de Figueiredo, T.; Nogueira, C.; Queirós, A. Effect of prescribed fire on soil properties and soil erosion in a Mediterranean mountain area. *Geoderma* **2017**, *307*, 172–180. [CrossRef]
86. Díaz-Raviña, M.; Lombao Vázquez, A.; Barreiro Buján, A.I.; Martín Jiménez, A.; Carballas Fernández, T. Medium-term impact of post-fire emergency rehabilitation techniques on a shrubland ecosystem in galicia (NW Spain). *Span. J. Soil Sci.* **2018**, *8*, 322–346. [CrossRef]
87. Wang, Y.; Liu, X.; Yan, Q.; Hu, Y. Impacts of slash burning on soil carbon pools vary with slope position in a pine plantation in subtropical China. *CATENA* **2019**, *183*, 104212. [CrossRef]

88. Moya, D.; Fonturbel, M.T.; Lucas-Borja, M.E.; Peña, E.; Alfaro-Sanchez, R.; Plaza-Álvarez, P.A.; González-Romero, J.; de Las Heras, J. Burning season and vegetation coverage influenced the community-level physiological profile of Mediterranean mixed-mesogean pine forest soils. *J. Environ. Manag.* **2021**, *277*, 111405. [CrossRef] [PubMed]
89. Armas-Herrera, C.M.; Martí, C.; Badía, D.; Ortiz-Perpiñá, O.; Girona-García, A.; Mora, J.L. Short-term and midterm evolution of topsoil organic matter and biological properties after prescribed burning for pasture recovery (Tella, Central Pyrenees, Spain). *L. Degrad. Dev.* **2018**, *29*, 1545–1554. [CrossRef]
90. Fairbanks, D.; Shepard, C.; Murphy, M.; Rasmussen, C.; Chorover, J.; Rich, V.; Gallery, R. Depth and topographic controls on microbial activity in a recently burned sub-alpine catchment. *Soil Biol. Biochem.* **2020**, *148*, 107844. [CrossRef]
91. Hart, B.T.N.; Smith, J.E.; Luoma, D.L.; Hatten, J.A. Recovery of ectomycorrhizal fungus communities fifteen years after fuels reduction treatments in ponderosa pine forests of the Blue Mountains, Oregon. *For. Ecol. Manag.* **2018**, *422*, 11–22. [CrossRef]
92. Kapoor, B.; Onufrak, A.; Klingeman, W.; DeBruyn, J.M.; Cregger, M.A.; Willcox, E.; Trigiano, R.; Hadziabdic, D. Signatures of prescribed fire in the microbial communities of Cornus florida are largely undetectable five months post-fire. *PeerJ* **2023**, *11*, e15822. [CrossRef]
93. Mataix-Solera, J.; Guerrero, C.; García-Orenes, F.; Bárcenas, G.M.; Torres, M.P. Fire effects on soils and restoration strategies. In *Forest Fire Effects on Soil Microbiology*; Cerda, A., Robichaud, P.R., Eds.; CRC Press: Boca Raton, FL, USA, 2009; pp. 149–192.
94. Hamman, S.T.; Burke, I.C.; Stromberger, M.E. Relationships between microbial community structure and soil environmental conditions in a recently burned system. *Soil Biol. Biochem.* **2007**, *39*, 1703–1711. [CrossRef]
95. Smith, N.R.; Kishchuk, B.E.; Mohn, W.W. Effects of wildfire and harvest disturbances on forest soil bacterial communities. *Appl. Environ. Microbiol.* **2008**, *74*, 216–224. [CrossRef]
96. Lombao, A.; Barreiro, A.; Fontúrbel, M.T.; Martín, A.; Carballas, T.; Díaz-Raviña, M. Key factors controlling microbial community responses after a fire: Importance of severity and recurrence. *Sci. Total Environ.* **2020**, *741*, 140363. [CrossRef]
97. Pellegrini, A.F.A.; Caprio, A.C.; Georgiou, K.; Finnegan, C.; Hobbie, S.E.; Hatten, J.A.; Jackson, R.B. Low-intensity frequent fires in coniferous forests transform soil organic matter in ways that may offset ecosystem carbon losses. *Glob. Chang. Biol.* **2021**, *27*, 3810–3823. [CrossRef]
98. Köster, K.; Berninger, F.; Heinonsalo, J.; Lindén, A.; Köster, E.; Ilvesniemi, H.; Pumpanen, J. The long-term impact of low-intensity surface fires on litter decomposition and enzyme activities in boreal coniferous forests. *Int. J. Wildl. Fire* **2016**, *25*, 213. [CrossRef]
99. Miesel, J.R.; Boerner, R.E.J.; Skinner, C.N. Soil nitrogen mineralization and enzymatic activities in fire and fire surrogate treatments in California. *Can. J. Soil Sci.* **2011**, *91*, 935–946. [CrossRef]
100. Moya, D.; Fonturbel, T.; Peña, E.; Alfaro-Sanchez, R.; Plaza-Álvarez, P.A.; González-Romero, J.; Lucas-Borja, M.E.; de Las Heras, J. Fire Damage to the Soil Bacterial Structure and Function Depends on Burn Severity: Experimental Burnings at a Lysimetric Facility (MedForECOtron). *Forests* **2022**, *13*, 1118. [CrossRef]
101. Fioretto, A.; Papa, S.; Pellegrino, A. Effects of fire on soil respiration, ATP content and enzyme activities in Mediterranean maquis. *Appl. Veg. Sci.* **2005**, *8*, 13–20. [CrossRef]
102. Reazin, C.; Morris, S.; Smith, J.E.; Cowan, A.D.; Jumpponen, A. Fires of differing intensities rapidly select distinct soil fungal communities in a Northwest US ponderosa pine forest ecosystem. *For. Ecol. Manag.* **2016**, *377*, 118–127. [CrossRef]
103. Arunrat, N.; Sansupa, C.; Sereenonchai, S.; Hatano, R. Short-term response of soil bacterial and fungal communities to fire in rotational shifting cultivation, northern Thailand. *Appl. Soil Ecol.* **2024**, *196*, 105303. [CrossRef]
104. Fox, S.; Sikes, B.A.; Brown, S.P.; Cripps, C.L.; Glassman, S.I.; Hughes, K.; Semenova-Nelsen, T.; Jumpponen, A. Fire as a driver of fungal diversity—A synthesis of current knowledge. *Mycologia* **2022**, *114*, 215–241. [CrossRef]
105. Cutler, N.A.; Arróniz-Crespo, M.; Street, L.E.; Jones, D.L.; Chaput, D.L.; DeLuca, T.H. Long-Term Recovery of Microbial Communities in the Boreal Bryosphere Following Fire Disturbance. *Microb. Ecol.* **2017**, *73*, 75–90. [CrossRef]
106. Pajares, S.; Bohannan, B.J.M. Ecology of Nitrogen Fixing, Nitrifying, and Denitrifying Microorganisms in Tropical Forest Soils. *Front. Microbiol.* **2016**, *7*, 1045. [CrossRef]
107. Yeager, C.M.; Northup, D.E.; Grow, C.C.; Barns, S.M.; Kuske, C.R. Changes in Nitrogen-Fixing and Ammonia-Oxidizing Bacterial Communities in Soil of a Mixed Conifer Forest after Wildfire. *Appl. Environ. Microbiol.* **2005**, *71*, 2713. [CrossRef]
108. Navarrete, A.A.; Tsai, S.M.; Mendes, L.W.; Faust, K.; De Hollander, M.; Cassman, N.A.; Raes, J.; Van Veen, J.A.; Kuramae, E.E. Soil microbiome responses to the short-term effects of Amazonian deforestation. *Mol. Ecol.* **2015**, *24*, 2433–2448. [CrossRef] [PubMed]
109. Castaño, C.; Hernández-Rodríguez, M.; Geml, J.; Eberhart, J.; Olaizola, J.; Oria-de-Rueda, J.A.; Martín-Pinto, P. Resistance of the soil fungal communities to medium-intensity fire prevention treatments in a Mediterranean scrubland. *For. Ecol. Manag.* **2020**, *472*, 118217. [CrossRef]
110. Cheng, Z.; Wu, S.; Du, J.; Liu, Y.; Sui, X.; Yang, L. Reduced Arbuscular Mycorrhizal Fungi (AMF) Diversity in Light and Moderate Fire Sites in Taiga Forests, Northeast China. *Microorganisms* **2023**, *11*, 1836. [CrossRef]
111. Barraclough, A.D.; Olsson, P.A. Slash-and-Burn Practices Decrease Arbuscular Mycorrhizal Fungi Abundance in Soil and the Roots of *Didierea madagascariensis* in the Dry Tropical Forest of Madagascar. *Fire* **2018**, *1*, 37. [CrossRef]
112. Bastias, B.A.; Anderson, I.C.; Rangel-Castro, J.I.; Parkin, P.I.; Prosser, J.I.; Cairney, J.W.G. Influence of repeated prescribed burning on incorporation of 13C from cellulose by forest soil fungi as determined by RNA stable isotope probing. *Soil Biol. Biochem.* **2009**, *41*, 467–472. [CrossRef]
113. Eivazi, F.; Bayan, M.R. Effects of long-term prescribed burning on the activity of select soil enzymes in an oak–hickory forest. *Can. J. For. Res.* **2011**, *26*, 1799–1804. [CrossRef]

114. Boerner, R.E.J.; Brinkman, J.A. Fire frequency and soil enzyme activity in southern Ohio oak–hickory forests. *Appl. Soil Ecol.* **2003**, *23*, 137–146. [CrossRef]
115. Wang, Y.; Zheng, J.; Liu, X.; Yan, Q.; Hu, Y. Short-term impact of fire-deposited charcoal on soil microbial community abundance and composition in a subtropical plantation in China. *Geoderma* **2020**, *359*, 113992. [CrossRef]
116. Hubbert, K.R.; Wohlgemuth, P.M.; Beyers, J.L.; Narog, M.G.; Gerrard, R. Post-fire soil water repellency, hydrologic response, and sediment yield compared between grass-converted and chaparral watersheds. *Fire Ecol.* **2012**, *8*, 143–162. [CrossRef]
117. Doerr, S.H.; Shakesby, R.A.; Walsh, R.P.D. Soil water repellency: Its causes, characteristics and hydro-geomorphological significance. *Earth-Sci. Rev.* **2000**, *51*, 33–65. [CrossRef]
118. Are, K.S.; Oluwatosin, G.A.; Adeyolanu, O.D.; Oke, A.O. Slash and burn effect on soil quality of an Alfisol: Soil physical properties. *Soil Tillage Res.* **2009**, *103*, 4–10. [CrossRef]
119. Wang, Y.; Hu, X.; Jin, T.; Yang, Y.; Chao, X. Research on the Influence Depth of Soil with Different Burn Severity in the Burned Areas of E'gu Village in Yajiang County. *Earth Sci.* **2019**, *8*, 317–322. [CrossRef]
120. Cawson, J.G.; Nyman, P.; Smith, H.G.; Lane, P.N.J.; Sheridan, G.J. How soil temperatures during prescribed burning affect soil water repellency, infiltration and erosion. *Geoderma* **2016**, *278*, 12–22. [CrossRef]
121. Jiménez-Pinilla, P.; Mataix-Solera, J.; Arcenegui, V.; Delgado, R.; Martín-García, J.M.; Lozano, E.; Martínez-Zavala, L.; Jordán, A. Advances in the knowledge of how heating can affect aggregate stability in Mediterranean soils: A XDR and SEM-EDX approach. *Catena* **2016**, *147*, 315–324. [CrossRef]
122. Ravi, S.; D'Odorico, P.; Herbert, B.E.; Zobeck, T.M.; Over, T.M. Enhancement of wind erosion by fire-induced water repellency. *Water Resour. Res.* **2006**, *42*, W11422. [CrossRef]
123. Whicker, J.J.; Breshears, D.D.; Wasiolek, P.T.; Kirchner, T.B.; Tavani, R.A.; Schoep, D.A.; Rodgers, J.C. Temporal and spatial variation of episodic wind erosion in unburned and burned semiarid shrubland. *J. Environ. Qual.* **2002**, *31*, 599–612. [CrossRef]
124. Eaton, J.M.; Lawrence, D. Loss of carbon sequestration potential after several decades of shifting cultivation in the Southern Yucatán. *For. Ecol. Manag.* **2009**, *258*, 949–958. [CrossRef]
125. Vieira, D.C.S.; Fernández, C.; Vega, J.A.; Keizer, J.J. Does soil burn severity affect the post-fire runoff and interrill erosion response? A review based on meta-analysis of field rainfall simulation data. *J. Hydrol.* **2015**, *523*, 452–464. [CrossRef]
126. Fontúrbel, T.; Carrera, N.; Vega, J.A.; Fernández, C. The effect of repeated prescribed burning on soil properties: A review. *Forests* **2021**, *12*, 767. [CrossRef]
127. Pereira, P.; Martínez-Murillo, J.F.; Francos, M. Environments affected by fire. *Adv. Chem. Pollut. Environ. Manag. Prot.* **2019**, *4*, 119–155. [CrossRef]
128. Creech, M.N.; Katherine Kirkman, L.; Morris, L.A. Alteration and recovery of slash pile burn sites in the restoration of a fire-maintained ecosystem. *Restor. Ecol.* **2012**, *20*, 505–516. [CrossRef]
129. Aboim, M.C.R.; Coutinho, H.L.C.; Peixoto, R.S.; Barbosa, J.C.; Rosado, A.S. Soil bacterial community structure and soil quality in a slash-and-burn cultivation system in Southeastern Brazil. *Appl. Soil Ecol.* **2008**, *38*, 100–108. [CrossRef]
130. López-Poma, R.; Orr, B.J.; Bautista, S. Successional stage after land abandonment modulates fire severity and post-fire recovery in a Mediterranean mountain landscape. *Int. J. Wildl. Fire* **2014**, *23*, 1005–1015. [CrossRef]
131. Fernández, C.; Vega, J.A.; Fonturbel, T.; Pérez-Gorostiaga, P.; Jiménez, E.; Madrigal, J. Effects of wildfire, salvage logging and slash treatments on soil degradation. *Land Degrad. Dev.* **2007**, *18*, 591–607. [CrossRef]
132. Pickup, M.; Wilson, S.; Freudenberger, D.; Nicholls, N.; Gould, L.; Hnatiuk, S.; Delandre, J. Post-fire recovery of revegetated woodland communities in south-eastern Australia. *Austral Ecol.* **2013**, *38*, 300–312. [CrossRef]
133. Kutiel, P.; Shaviv, A. Effect of simulated forest fire on the availability of N and P in mediterranean soils. *Plant Soil* **1989**, *120*, 57–63. [CrossRef]
134. Murdiyarso, D.; Widodo, M.; Suyamto, D. Fire risks in forest carbon projects in Indonesia. *Sci. China (Ser. C)* **2002**, *45*, 65–74.
135. Piché, N.; Kelting, D.L. Recovery of soil productivity with forest succession on abandoned agricultural land. *Restor. Ecol.* **2015**, *23*, 645–654. [CrossRef]
136. Rai, D.; Silveira, M.L.; Strauss, S.L.; Meyer, J.L.; Castellano-Hinojosa, A.; Kohmann, M.M.; Brandani, C.B.; Gerber, S. Short-term prescribed fire-induced changes in soil microbial communities and nutrients in native rangelands of Florida. *Appl. Soil Ecol.* **2023**, *189*, 104914. [CrossRef]
137. Rascio, I.; Curci, M.; Gattullo, C.E.; Lavecchia, A.; Yaghoubi Khanghahi, M.; Terzano, R.; Crecchio, C. Combined Effect of Laboratory-Simulated Fire and Chromium Pollution on Microbial Communities in an Agricultural Soil. *Biol.* **2021**, *10*, 587. [CrossRef] [PubMed]
138. Srikanthasamy, T.; Barot, S.; Koffi, F.K.; Tambosco, K.; Marcangeli, Y.; Carmignac, D.; N'Dri, A.B.; Gervaix, J.; Le Roux, X.; Lata, J.C. Short-term impact of fire on the total soil microbial and nitrifier communities in a wet savanna. *Ecol. Evol.* **2021**, *11*, 9958–9969. [CrossRef] [PubMed]
139. Rietl, A.J.; Jackson, C.R. Effects of the ecological restoration practices of prescribed burning and mechanical thinning on soil microbial enzyme activities and leaf litter decomposition. *Soil Biol. Biochem.* **2012**, *50*, 47–57. [CrossRef]
140. Moura, J.B.; Souza, R.F.; Vieira-Júnior, W.G.; Lucas, L.S.; Santos, J.M.; Silva, S.D.E.; Marín, C. Effects of a megafire on the arbuscular mycorrhizal fungal community and parameters in the Brazilian Cerrado ecosystem. *For. Syst.* **2022**, *31*, e001. [CrossRef]
141. Zhu, L.; Dickson, T.L.; Zhang, Z.; Dere, A.; Xu, J.; Bragg, T.; Tapprich, W.; Lu, G. Effects of burning and mowing on the soil microbiome of restored tallgrass prairie. *Eur. J. Soil Sci.* **2021**, *72*, 385–399. [CrossRef]

142. Garcia-Pausas, J.; Romanyà, J.; Casals, P. Post-fire recovery of soil microbial functions is promoted by plant growth. *Eur. J. Soil Sci.* **2022**, *73*, e13290. [CrossRef]
143. Qiu, D.; Xu, R.; Wu, C.; Mu, X.; Zhao, G.; Gao, P. Vegetation restoration improves soil hydrological properties by regulating soil physicochemical properties in the Loess Plateau, China. *J. Hydrol.* **2022**, *609*, 127730. [CrossRef]
144. Wang, L.; Zhang, J.; Zhao, Y.; Fu, Q.; Li, T. Vegetation restoration and plant roots improve soil infiltration capacity after a severe forest fire in Daxing'anling, northeast China. *J. Soil Water Conserv.* **2022**, *77*, 135–143. [CrossRef]
145. Randriamalala, J.R.; Randriarimalala, J.; Hervé, D.; Carrière, S.M. Slow recovery of endangered xerophytic thickets vegetation after slash-and-burn cultivation in Madagascar. *Biol. Conserv.* **2019**, *233*, 260–267. [CrossRef]
146. Wittenberg, L.; Malkinson, D.; Wittenberg, L.; Malkinson, D. Monitoring Water Repellency Effects on Post-wildfire Infiltration and Runoff. In Proceedings of the EGU General Assembly, Vienna, Austria, 27 April–2 May 2014; p. 6025.
147. Drobyshev, I.; Aleinikov, A.; Lisitsyna, O.; Aleksutin, V.; Vozmitel, F.; Ryzhkova, N. The first annually resolved analysis of slash-and-burn practices in the boreal Eurasia suggests their strong climatic and socio-economic controls. *Veg. Hist. Archaeobot.* **2024**, *33*, 301–312. [CrossRef]
148. Chiroma, A.M.; Alhassan, A.B. A Review of the Impact of Bush Burning on the Environment: Potential Effects on Soil Chemical Attributes. *Int. J. Sci. Environ.* **2023**, *3*, 101–121. [CrossRef]
149. Reang, D.; Nath, A.J.; Sileshi, G.W.; Hazarika, A.; Das, A.K. Post-fire restoration of land under shifting cultivation: A case study of pineapple agroforestry in the Sub-Himalayan region. *J. Environ. Manag.* **2022**, *305*, 114372. [CrossRef]
150. Lintemani, M.G.; Loss, A.; Mendes, C.S.; Fantini, A.C. Long fallows allow soil regeneration in slash-and-burn agriculture. *J. Sci. Food Agric.* **2020**, *100*, 1142–1154. [CrossRef] [PubMed]
151. Condron, L.; Stark, C.; O'Callaghan, M.; Clinton, P.; Huang, Z. The Role of Microbial Communities in the Formation and Decomposition of Soil Organic Matter. In *Soil Microbiology and Sustainable Crop Production*; Springer: Dordrecht, The Netherlands, 2010; pp. 81–118. [CrossRef]
152. Bárcenas-Moreno, G.; García-Orenes, F.; Mataix-Solera, J.; Mataix-Beneyto, J.; Bååth, E. Soil microbial recolonisation after a fire in a Mediterranean forest. *Biol. Fertil. Soils* **2011**, *47*, 261–272. [CrossRef]
153. De Vries, F.T.; Shade, A. Controls on soil microbial community stability under climate change. *Front. Microbiol.* **2013**, *4*, 265. [CrossRef]
154. Fornwalt, P.J.; Rhoades, C.C. Rehabilitating Slash Pile Burn Scars in Upper Montane Forests of the Colorado Front Range. *Nat. Areas J.* **2011**, *31*, 177–182. [CrossRef]
155. Da Silva Neto, E.C.; Pereira, M.G.; Frade, E.F.; Da Silva, S.B.; De Carvalho, J.A.; Dos Santos, J.C. Temporal evaluation of soil chemical attributes after slash-and-burn agriculture in the Western Brazilian Amazon. *Acta Sci. Agron.* **2019**, *41*, e42609. [CrossRef]
156. Szott, L.T.; Palm, C.A.; Buresh, R.J. Ecosystem fertility and fallow function in the humid and subhumid tropics. *Agrofor. Syst.* **1999**, *47*, 163–196. [CrossRef]
157. Lungmuana; Singh, S.B.; Vanthawmliana; Saha, S.; Dutta, S.K.; Rambuatsaiha; Singh, A.R.; Boopathi, T. Impact of secondary forest fallow period on soil microbial biomass carbon and enzyme activity dynamics under shifting cultivation in North Eastern Hill region, India. *CATENA* **2017**, *156*, 10–17. [CrossRef]
158. Chowdhury, F.I.; Barua, I.; Chowdhury, A.I.; Resco de Dios, V.; Alam, M.S. Agroforestry shows higher potential than reforestation for soil restoration after slash-and-burn: A case study from Bangladesh. *Geol. Ecol. Landsc.* **2022**, *6*, 48–54. [CrossRef]
159. Nath, A.J.; Sileshi, G.W.; Laskar, S.Y.; Pathak, K.; Reang, D.; Nath, A.; Das, A.K. Quantifying carbon stocks and sequestration potential in agroforestry systems under divergent management scenarios relevant to India's Nationally Determined Contribution. *J. Clean. Prod.* **2021**, *281*, 124831. [CrossRef]
160. Cong, W.F.; Hoffland, E.; Li, L.; Six, J.; Sun, J.H.; Bao, X.G.; Zhang, F.S.; Van Der Werf, W. Intercropping enhances soil carbon and nitrogen. *Glob. Chang. Biol.* **2015**, *21*, 1715–1726. [CrossRef] [PubMed]
161. Comte, I.; Davidson, R.; Lucotte, M.; de Carvalho, C.J.R.; de Assis Oliveira, F.; da Silva, B.P.; Rousseau, G.X. Physicochemical properties of soils in the Brazilian Amazon following fire-free land preparation and slash-and-burn practices. *Agric. Ecosyst. Environ.* **2012**, *156*, 108–115. [CrossRef]
162. Roa-Fuentes, L.L.; Martínez-Garza, C.; Etchevers, J.; Campo, J. Recovery of Soil C and N in a Tropical Pasture: Passive and Active Restoration. *L. Degrad. Dev.* **2015**, *26*, 201–210. [CrossRef]
163. Mueller, L.; Schindler, U.; Mirschel, W.; Shepherd, T.G.; Ball, B.C.; Helming, K.; Rogasik, J.; Eulenstein, F.; Wiggering, H. Assessing the productivity function of soils. A review. *Agron. Sustain. Dev.* **2010**, *30*, 601–614. [CrossRef]
164. Feiziene, D.; Feiza, V.; Povilaitis, V.; Putramentaite, A.; Janusauskaite, D.; Seibutis, V.; Slepetys, J. Soil sustainability changes in organic crop rotations with diverse crop species and the share of legumes. *Acta Agric. Scand. Sect. B—Soil Plant Sci.* **2016**, *66*, 36–51. [CrossRef]
165. Selim, M.M. A Review of Advantages, Disadvantages and Challenges of Crop Rotations. *Egypt. J. Agron.* **2019**, *41*, 1–10. [CrossRef]
166. Jain, T.B.; Gould, W.A.; Graham, R.T.; Pilliod, D.S.; Lentile, L.B.; Gonzalez, G. A soil burn severity index for understanding soil-fire relations in tropical forests. *AMBIO A J. Hum. Environ.* **2008**, *37*, 563–568. [CrossRef]
167. Padalia, H.; Mondal, P.P. Spatio-Temporal Trends of Fire in Slash and Burn Agriculture Landscape: A Case Study from Nagaland, India. *ISPRS Ann. Photogramm. Remote Sens. Spat. Inf. Sci.* **2014**, *II-8*, 53–59. [CrossRef]
168. Lopresti, A.; Hayden, M.T.; Siegel, K.; Poulter, B.; Stavros, E.N.; Dee, L.E. Remote sensing applications for prescribed burn research. *Int. J. Wildl. Fire* **2024**, *33*, WF23130. [CrossRef]

169. Kayad, A.; Paraforos, D.S.; Marinello, F.; Fountas, S. Latest Advances in Sensor Applications in Agriculture. *Agriculture* **2020**, *10*, 362. [CrossRef]
170. Shakya, A.K.; Ramola, A.; Kandwal, A.; Vidyarthi, A. Soil moisture sensor for agricultural applications inspired from state of art study of surfaces scattering models & semi-empirical soil moisture models. *J. Saudi Soc. Agric. Sci.* **2021**, *20*, 559–572. [CrossRef]
171. Li, Z.; Yao, Q.; Guo, X.; Crits-Christoph, A.; Mayes, M.A.; IV, W.J.H.; Lebeis, S.L.; Banfield, J.F.; Hurst, G.B.; Hettich, R.L.; et al. Genome-Resolved Proteomic Stable Isotope Probing of Soil Microbial Communities Using $^{13}CO_2$ and ^{13}C-Methanol. *Front. Microbiol.* **2019**, *10*, 485423. [CrossRef] [PubMed]

agriculture

MDPI

Review

Soil Health Intensification through Strengthening Soil Structure Improves Soil Carbon Sequestration

Ryusuke Hatano [1,*], Ikabongo Mukumbuta [2] and Mariko Shimizu [3]

1 Soil Science Laboratory, Research Faculty of Agriculture, Hokkaido University, Kita 9 Nishi 9, Kita-ku, Sapporo 060-8589, Japan
2 Golden Valley Agricultural Research Trust, Lusaka P.O. Box RW50834, Zambia; ikabongo1@gmail.com
3 Civil Engineering Research Institute for Cold Region, Hiragishi 1-3-1-34, Toyohira-ku, Sapporo 062-8602, Japan; shimizu-m@ceri.go.jp
* Correspondence: hatano@agr.hokudai.ac.jp

Abstract: Intensifying soil health means managing soils to enable sustainable crop production and improved environmental impact. This paper discusses soil health intensification by reviewing studies on the relationship between soil structure, soil organic matter (SOM), and ecosystem carbon budget. SOM is strongly involved in the development of soil structure, nutrient and water supply power, and acid buffering power, and is the most fundamental parameter for testing soil health. At the same time, SOM can be both a source and a sink for atmospheric carbon. A comparison of the ratio of soil organic carbon to clay content (SOC/Clay) is used as an indicator of soil structure status for soil health, and it has shown significantly lower values in cropland than in grassland and forest soils. This clearly shows that depletion of SOM leads to degradation of soil structure status. On the other hand, improving soil structure can lead to increasing soil carbon sequestration. Promoting soil carbon sequestration means making the net ecosystem carbon balance (NECB) positive. Furthermore, to mitigate climate change, it is necessary to aim for carbon sequestration that can improve the net greenhouse gas balance (NGB) by serving as a sink for greenhouse gases (GHG). The results of a manure application test in four managed grasslands on Andosols in Japan showed that it was necessary to apply more than 2.5 tC ha^{-1} y^{-1} of manure to avoid reduction and loss of SOC in the field. Furthermore, in order to offset the increase in GHG emissions due to N_2O emissions from increased manure nitrogen input, it was necessary to apply more than 3.5 tC $ha^{-1}y^{-1}$ of manure. To intensify soil health, it is increasingly important to consider soil management with organic fertilizers that reduce chemical fertilizers without reducing yields.

Keywords: organic fertilizer; soil carbon sequestration; soil health; soil organic matter; soil structure

1. Introduction

Since the Industrial Revolution, between 1750 and 2011, 555 GtC of CO_2 has been emitted [1]. This includes 375 GtC from fossil fuel combustion and 180 GtC from land use change. Meanwhile, global warming is steadily progressing, with the temperature rise above pre-industrial revolution levels expected to reach 1.5 °C by 2040, making heat waves more likely to occur, and increasing the frequency of droughts and heavy rains that can affect agriculture and natural ecosystems [2]. Due to conventional cultivation, cropland around the world has significantly lower soil carbon content than natural ecosystems [3]. To mitigate this, efforts are underway to add carbon to agricultural soils and return them to the level of natural ecosystems [4]. This has progressed with the proposal of the concept of soil health [5].

The United Nations' Food and Agricultural Organization (FAO) defines soil health as "the ability of the soil to sustain the productivity, diversity, and environmental services of terrestrial ecosystems" [6]. For the assessment of soil health, efforts have been made to identify effective and measurable biological, chemical, and physical indicators that can be controlled through management [7]. Among them, soil organic matter (SOM) is the most widely used [8]. According to Beillouin et al. [9], 15,857 studies on soil organic carbon (SOC) have been reported in more than 150 countries. Of these, 6550 were related to SOC content in land use, land use change, and climate change. The remaining 9307 were studies on the relationship between SOC and soil chemistry, plant productivity, soil biology, greenhouse gases (GHGs), soil physical properties, and water quality. 50% of them were studies on fertility management. By country, the United States and China had the most, followed by Brazil and Canada, Australia, India, and some European countries (UK, Germany, Spain, and Italy), but some regions, such as Africa, have very scarce data available.

In a study on SOC stock, Georgio et al. [3] used data from a total of 1144 soil profiles from 78 papers published since 1960 [3]. They included SOC measurements from managed land (n = 559 profiles) and natural (n = 585 profiles) in all soil types (Alfisols, Andisols, Aridisols, Entisols, Gelisols, Inceptisols, Mollisols, Oxisols, Spodosols, Ultisols, and Vertisols), except Histsols, across a range of mineralogy and vegetation types, with mean temperatures ranging from −2.9 to 29 °C and annual precipitation ranging from 79 to 3806 mm y^{-1}. These data also included data from the Land Use and Coverage Area Frame Survey (LUCAS) [10], which was conducted uniformly by 28 European Union (EU) Member States. LUCAS analyzed the physicochemical properties of the soil at 0–20 cm of the topsoil at a total of 22,000 sites with various land uses (cropland, grassland, forest) in EU countries three times in 2009, 2015, and 2018. The data are freely available from the EUROSTAT website. Mäkipää et al. [11] reported on soil structural degradation using the 2009 LUCAS data.

In an ecosystem, there is an input of organic matter through the net primary production of plants, and an output of organic matter through decomposition, leaching, erosion, and harvesting of organic matter. A positive difference between the input and output causes the accumulation of organic matter in the ecosystem. The accumulation of organic matter in ecosystems is both an effect of removing carbon from the atmosphere and a process of carbon sequestration, which captures carbon in plants and soil. In particular, soil carbon sequestration is enhanced by increasing soil carbon content and soil depth distribution [12].

In ecosystems, capture of carbon from the atmosphere occurs through photosynthesis by plants [13,14]. 80% of the carbon fixed in the ecosystem by plants through photosynthesis is released into the atmosphere through the plants' own autotrophic respiration and the heterotrophic respiration of soil microorganisms, and the rest is retained in plant tissues [15]. In agricultural fields, plant carbon is added to the soil as residue. At the same time, organic carbon can also be added by applying organic fertilizers such as animal manure. Soil carbon loss occurs through heterotrophic respiration of soil microorganisms, which is the microbial decomposition of previously formed carbon contained in SOM [16].

When fresh organic matter is added to soil, it is decomposed by soil microorganisms and becomes microbial biomass, which combines with soil minerals to form organo-mineral complexes and aggregates to stabilize the soil structure [17,18]. Building soil structure increases field water capacity [19]. In the United States, a pedotransfer function has been proposed that regresses available water by soil texture and SOC [20]. Mineral-associated organic carbon (MOC) is difficult for microorganisms to decompose [21], and MOC is essential for improving carbon sequestration in soils. On the other hand, the decomposition process of organic matter produces GHGs such as CO_2, CH_4, and N_2O [22]. If the emissions of these gases into the atmosphere exceed the consumption through plant CO_2 uptake, microbial CH_4 uptake, and microbial N_2O reduction in ecosystems, global warming will be exacerbated. In agroecosystems, harvesting removes organic matter from the field, so there is no chance of it being retained as SOM. Harvesting therefore contributes to global warming. On the other hand, inputting organic fertilizers such as manure or slurry can increase soil organic matter, thereby mitigating global warming.

At the same time, plants need nutrients, especially nitrogen, to fix carbon. In agricultural ecosystems, crops are obtained by inputting nitrogen as fertilizer. Organic agriculture is considered a strategy to increase SOC in agricultural lands. However, a biogeochemical model estimates that organic agriculture without cover crops or manure (no nutrients at all) would reduce carbon inputs to soils worldwide by 40% and reduce SOC by 9% [23]. Cover crops increase SOC and crop yields, and a meta-analysis showed that the greatest increases in yield and SOC were achieved with legume cover crops and nitrogen applications [24]. A long-term experiment on the effects of combined crop rotation and fertilization in Prague-Ruzyně since 1955 showed that SOC was maintained with more rotations and manure applications [25]. Furthermore, a meta-analysis of 13,662 cases in China revealed that increased SOC significantly increased yields of all three crops, maize, wheat, and rice, up to a certain SOC value for each crop. However, the effect of increasing crop yield was only one-fifth of that of nitrogen fertilization [26]. However, excessive fertilizer input increases the outflow of nutrients into the hydrosphere, leading to eutrophication, and increases ammonia volatilization and N_2O emissions to the atmosphere [27]. It has been pointed out that the nitrogen cycle has already exceeded the planetary boundaries that define the "safe area for human activity" [28]. It is desirable to reduce the amount of chemical fertilizer applied to agricultural soils. To address this issue, management techniques are being developed to reduce the amount of chemical fertilizers used by taking into account the amount of nutrients such as nitrogen that are mineralized as organic matter added through cover crops and compost decomposes [29,30]. Especially, soil N_2O is emitted during the processes of nitrification and denitrification, and since denitrification is a reaction that accompanies the decomposition of organic matter, it is possible that organic matter application increases N_2O emissions, and so proper management of nitrogen fertilizer is important.

SOM is central to soil function. Accumulation of SOM improves water retention, permeability, and aeration by building soil structure, and enhances acid buffering, carbon sequestration, and plant nutrient storage and supply. It can be said that it is the most important factor of soil health [8]. SOM is the product of the net ecosystem carbon budget, the difference between carbon inputs (photosynthesis, organic inputs such as manure, and deposition) and carbon outputs (respiration by plants and microorganisms, harvesting, and erosion). Therefore, maintenance of SOM in agricultural land strongly depends on ecosystem carbon and nutrient management practices. It is recommended that to improve soil health, the amount of chemical fertilizers applied is reduced by using nutrients provided by organic matter application.

Supplying organic matter to soil using conservation farming methods promotes the formation of MOC, improves soil structure, achieves soil carbon sequestration, improves soil nutrient and water supply, and directly reduces environmental impact (Figure 1).

Johannes et al. [31] clarified that the SOC/Clay ratio can be an indicator of soil structure status, and soil with an SOC/Clay ratio of 1/13 or less is considered to have a degraded soil structure. Based on this indicator, Mäkipää et al. [11] showed that agricultural soils in the EU were more degraded than grasslands and forests. Soils that have been degraded due to carbon loss need to be restored by applying conservation farming practices. However, the SOC/Clay ratio is an indicator of soil degradation, but does not specifically estimate the increase or decrease in SOC. Therefore, to know the effect of conservation farming practices on the improvement of soil health, carbon balance and GHG balance need to be measured independently.

In this paper, we first discuss recent findings on soil carbon pools and soil carbon accumulation, and the concept of the SOC/Clay ratio as an indicator of soil structure degradation based on the mechanism of soil organic matter accumulation. We then present actual examples of measuring carbon and GHG balances through fertility management with organic matter application and use them to evaluate the soil structure degradation indicator using the SOC/Clay ratio. We also discuss how to mitigate the impact of fertility management with organic matter application on N_2O emissions. From the above, we clarify that intensification of soil health through strengthening soil structure by organic matter application improves soil carbon sequestration and GHG balance.

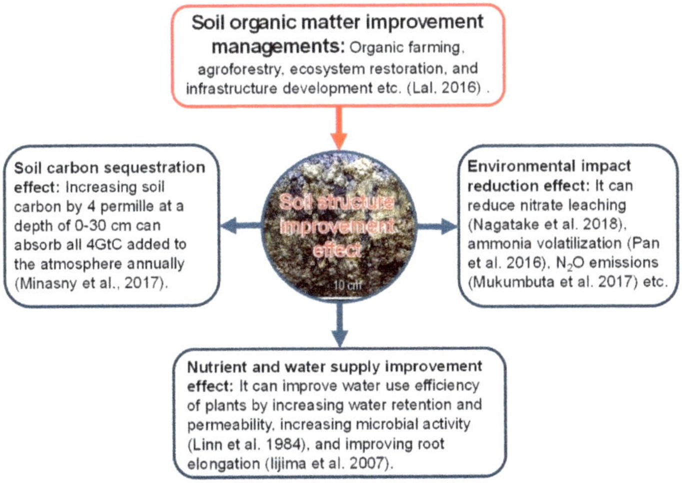

Figure 1. Effects of soil organic matter improvement managements [8] on soil carbon sequestration [4], nutrient and water supply improvement [32,33] and environmental impact reduction [34–36].

2. How Much Organic Matter Can Soil Store?

Ecosystems emit 8.9 GtC y^{-1} annually (fossil fuel combustion 7.8 GtC y^{-1}, land use change 1.1 GtC y^{-1}), and each year 4 GtC y^{-1} is added to the atmosphere as CO_2 [1].

There are two types of SOC, mineral-associated organic carbon (MOC) and particulate organic carbon (POC). POC is non-mineral-associated organic carbon such as peat. MOC is known to be stable and difficult for microorganisms to degrade [21]. Wang and Kuzyakov [37] summarize the mechanism of MOC formation as follows: (i) ligand exchange, (ii) electrostatic attraction, (iii) hydrophobic partition, (iv) cation bridging, (v) co-precipitation, and (vi) physical trapping by microaggregates. Globally, MOC accounts for 34–51% of SOC, but in mineral soils it accounts for ~65% [38]. Georgiou et al. [3] found that soils containing mainly low-activity minerals, such as kaolinite, can adsorb up to 48 gC/kg clay + silt, and soils containing mainly high-activity minerals such as smectite can adsorb up to 86 gC/kg clay + silt. Using these values for the world's soils to determine the MOC content, they showed that most of the world's soils are unsaturated in MOC. According to the report, the SOC at a depth of 1 m in mineral soil, excluding tundra due to insufficient sample sizes, is 1401 GtC, of which the MOC is 899 GtC, and 50% of this exists within the top 30 cm surface layer (Table 1). The maximum amount of MOC (MOCmax), which is calculated from the amount and type of soil minerals, is estimated to be 4596 GtC, and clay minerals can still contribute 3680 GtC. However, if the amount of carbon input is set at the level of the current natural ecosystem, then the SOC content can only accumulate carbon up to the level of the current natural ecosystem, so the MOC that agricultural soils can accumulate is 433 GtC. However, this amount is equivalent to 78% of the ecosystem carbon loss of 555 GtC (375 GtC from fossil fuel combustion and 180 GtC from land use change) since the Industrial Revolution [1]. Furthermore, the deeper soil layers have a large potential for MOC, and it is possible to increase MOC by adding organic matter to the lower layers by deep tillage or cultivating crop varieties with deep root systems. No-tillage increases surface carbon but has been shown to reduce carbon in the subsoil [39]. On the other hand, using deep-rooted oats as a cover crop reduces the hardness of the subsoil, and the roots of the succeeding cash crop also become deeper [40]. The results of 103 carbon addition experiments around the world show that the carbon sequestration rate increases as the degree of MOC saturation (the ratio of MOC to potential) decreases; the rate in soils with only 10% SOC saturation is three times higher than soils with 50% [3].

Table 1. Global carbon stocks (excluding tundra due to insufficient sample sizes). Mineral-associated organic carbon (MOC), soil organic carbon (SOC), mineralogical carbon capacity (MOCmax, which is calculated from the amount and type of soil minerals) and MOC deficit (to the mineralogical capacity and natural land average environmental limit) totaled over all given land-use categories (except tundra) globally [41].

Depth	MOC	SOC	MOC_{max}	MOC Deficit Relative to MOCmax	MOC Deficit Relative to Natural Land Average C Saturation
cm				GtC	
0–30	448	700	1443	990	286
30–100	451	701	3153	2690	147
0–100	899	1401	4596	3680	433

3. Assessment of Soil Structure Based on Soil Organic Matter Accumulation Status

Mäkipää et al. [11] showed that there was no correlation between SOC and clay, silt, or sand content in the top 0–20 cm data of 21,859 European soils (LUCAS soil data, 2009 [10]). In cropland, grassland, and forest, the median SOC content was 14.2, 15.8, and 18.2 g kg^{-1}, respectively, while the median clay content was 210, 150, and 70 g kg^{-1}, respectively. Thus, grassland and forest contain more SOC than cropland, despite having less clay content.

Asano and Wagai [17] investigated the relationship between soil particle size and soil carbon content using allophanic Andosol in Japan. They found that more than 95% of the total C was present as >53 μm aggregates with wet sieving of the aggregates alone, and even after mechanical shaking to break up the water-resistant aggregates, the >53 μm and 2~53 μm fraction contained 37% and 41% of the total C, respectively. Furthermore, when these were completely dispersed by using ultrasound sonication energy, 63% of the total C was present in micro organo-mineral complexes of <2 μm. The C/N ratio of the organic matter was low at 6–10, and microbial decomposition products enriched in stable C and N isotopes (^{13}C and ^{15}N) were predominant. From these findings, it is thought that plant debris supplied to the soil, after being decomposed by microorganisms, preferentially binds to organo-mineral complexes and becomes occluded in aggregates, forming strong aggregates [18].

Beare et al. [42] compared the effects of conventional tillage (CT) and no-tillage (NT) on SOC content and aggregate stability, and found that the SOC content in top 5 cm soil was 18% higher in NT than in CT, and that aggregates larger than 250 μm were significantly more abundant in NT than in CT. In other words, it is thought that tillage increased decomposition of SOC, reduced the production of organo-mineral complexes, and inhibited the growth of aggregates.

Since organo-mineral complex formation is the mechanism of aggregate formation, clay content is an important soil quality. Dexter et al. [43] suggested that clay retains maximum SOC when the SOC/Clay ratio is 1/10. Following this idea, Johannes et al. [31] investigated the relationship between the soil structure condition and the SOC/Clay ratio of 161 Swiss agricultural soils. The soil structure condition was measured using the visual method of Ball et al. [44], and classified the condition as "very good", "good", "adequate" and "bad", using the SOC/Clay ratio of 1/8, 1/10, and 1/13 as thresholds (Table 2). When determining the soil structure condition using the SOC/Clay ratio, grasslands tended to have a good structure of 1/8 or higher. On the other hand, in upland fields, fields with conventional tillage (CT) tended to have a bad level of less than 1/13, while fields with no-tillage (NT) showed a wide variation, averaging around an adequate level of 1/10.

Based on these results, it is reasonable to aim for a SOC/Clay ratio of 1/10 as a rational target for organic carbon for soil management. Using this index, Prout et al. [45] evaluated the soil structure condition from the SOC/Clay ratio in the surface layer 0–15 cm of England and Wales, covering 3809 sites under arable land, grassland and woodland, and found that 38.2, 6.6 and 5.6% of arable, grassland and woodland sites, respectively, were degraded.

Table 2. Expected structure quality as a function of the SOC/Clay ratio [31].

SOC/Clay	Expected Soil Structural Quality
>1/8	Very good
1/10 < SOC/Clay < 1/8	Good
1/13 < SOC/Clay < 1/10	Adequate
<1/13	Bad

Furthermore, Mäkipää et al. [11] evaluated the soil structure condition using the SOC/Clay ratio of 2009 LUCAS soil survey data. As a result, 51.0%, 15.7%, and 4.2% of cropland, grassland, and forest, which account for 44.6%, 21.8%, and 28.6% of the total land cover, have been degraded, respectively, and the degree of degradation was particularly strong in areas with dry summers.

Mäkipää et al. [11] examined the relationship between this deterioration of soil structure and soil carbon loss based on the monitoring data of the SOC stock changes in agricultural soils for the national greenhouse gas inventories [11,46]. However, no significant relationship was found between them, and it was inappropriate to use the deterioration index of soil structure as an index of soil carbon loss. However, the evaluation of soil structure condition by SOC/Clay ratio shows interesting trends when viewed by soil type. The results indicate that despite being rich in highly active clay minerals, Vertisols were the most degraded, which was likely due to cumulative loss of soil carbon due to intensive and long-term agricultural use. Next, Calcisol and Solonchak, found in degraded arid and semi-arid regions, are soils containing high concentrations of calcium and sodium salts, respectively, suggesting the effects of limited productivity in these soils. On the other hand, the podzols under forests usually have sandy soils that are not particularly rich in clay, but the degree of soil structure deterioration was small. This is because Podsols typically have a cool climate, low microbial activity, and a high supply of organic carbon from the forest to the soil. Soil structure deterioration is strongly influenced by SOC dynamics due to climate and land use.

Additionally, allophanic soils (i.e., Andosols) containing large amounts of amorphous minerals can form stable Al and Fe complexes and accumulate SOC [47,48]. Regarding organic matter accumulation, sheet silicates (clays-OH) have the strongest correlation with phenolics, Fe (Fe-O) and Al (Al-O) oxides have a correlation with polysaccharides, and these saccharides were found to capture carbon from microbial necromass, which increased with the addition of organic matter [49]. This indicates that Fe and Al oxides promote organo-organic interactions in soils, providing the basis for large organic matter accumulation in allophanic soils. Furthermore, it has been shown that Fe and Al oxides increase their positive charge as the pH decreases, thereby capturing more negatively charged carboxyl groups [37].

4. Suppressing the Consumption of Soil Organic Matter and Increasing Soil Organic Matter

To prevent SOC from being depleted and to recover and increase SOC from depletion, it is necessary to increase the amount of organic matter input to the soil and to suppress excessive organic matter decomposition.

No-tillage and the application of mulch are effective ways to prevent excessive organic matter decomposition. On the other hand, to increase the amount of organic matter input into the soil, it is effective to apply organic fertilizers and introduce cover crops. Effective methods include applying biochar, restoring degraded land, linking crop cultivation and livestock farming, planting trees, preserving urban ecosystems, and building complex systems for biofuel production. Conservation farming techniques such as organic farming, agroforestry, ecosystem restoration, and infrastructure development have been developed for these practices [8] and these were further important to maintain regional circular economies during global emergencies such as the COVID-19 pandemic which induced heavy disruption of the supply chain of agricultural materials like chemical fertilizers.

As mentioned earlier, the carbon sequestration rate increases as the MOC saturation (the ratio of MOC to the potential) decreases [3], and microbial biomass becomes the MOC through microbial necromass [18,49]. Therefore, inputting fresh organic matter to soils with a low SOC/Clay ratio is more effective in increasing MOC and improving soil structure. Furthermore, considering that the SOC/Clay ratio is lower in the subsoil, cultivation of deep-rooted crops is considered to be effective for increasing MOC.

The application of conservation farming methods has the effect of improving soil structure, sequestering carbon, supplying soil nutrients and water, and reducing environmental burden (Figure 1). By suppressing the depletion of soil organic matter, the SOC/Clay ratio is improved, improving the soil structure status. At the same time, this is nothing but an increase in the amount of soil carbon sequestration. According to Minasny et al. [4], soil carbon content at 30 cm worldwide is estimated to increase by more than 4 per 1000 (0.4%) in the first 20 years following introduction of conservation farming practices [4]. Based on this, they showed that increasing the carbon content by 4 per 1000 could absorb all 4 Gt of carbon that remains in the atmosphere each year. An increase in soil carbon of 4 per 1000 in the top 1 m of agricultural soil will result in SOC sequestration of 2–3 GtC y^{-1}.

Increasing the input of organic fertilizers into the soil and efforts to improve the soil structure will improve the supply of nutrients and water from the soil to plants. Improving water supply improves the efficiency of rainfall use by improving water retention, water permeability, aeration, and rhizome development, as an increase in soil organic matter content improves soil structure [50]. Rooting reaches its maximum when the matric potential is around −10 kPa [32]; when it is wetter than that, there is a lack of air, and when it is drier than that, there is a lack of water and the soil hardness increases, making it impossible for roots to penetrate. Therefore, increased water retention improves rooting. In addition, nitrogen mineralization and nitrification reach their maximum at a water saturation level (WFPS) of 60%, mineralization and nitrification decrease as WFPS increases above this level, and denitrification becomes dominant when WFPS exceeds 80% [33]. Therefore, here too, the development of soil structure leads to an increase in nitrogen supply power. From the above, it is thought that the effects of applying organic fertilizers are more pronounced during droughts [51].

At the same time, the application of organic fertilizers is also expected to have the effect of reducing environmental impact. Nutrient supply from organic fertilizers occurs with mineralization, so the synchronization rate with plant nutrient requirements is higher than chemical fertilizers, and it is effective in suppressing nitrate leaching and N_2O emissions [34]. Because chemical fertilizers are usually applied out of synchronization with plant growth, they easily cause nitrate leaching, ammonia volatilization [35], and N_2O emissions [36]. High concentration of chemical fertilizers can also cause seedling injury [52] and acid disorders of crops [53]. On the other hand, application of organic fertilizer increases CEC, which in turn buffers against acids and prevents a decrease in pH [54].

5. Measurement and Evaluation of Soil Carbon Balance and Greenhouse Gas Balance

According to Mäkipää et al. [11], the SOC/Clay ratio indicates the state of soil structure degradation but does not indicate the dynamic of whether it is improving or degrading. Therefore, to show the effect of soil carbon management on soil health, it is essential to estimate the actual soil carbon balance and GHG balance.

Figure 2 shows a schematic diagram of net ecosystem carbon balance (NECB) and net GHG balance (NGB) regarding GHG emissions and uptakes, harvesting and organic fertilizer application.

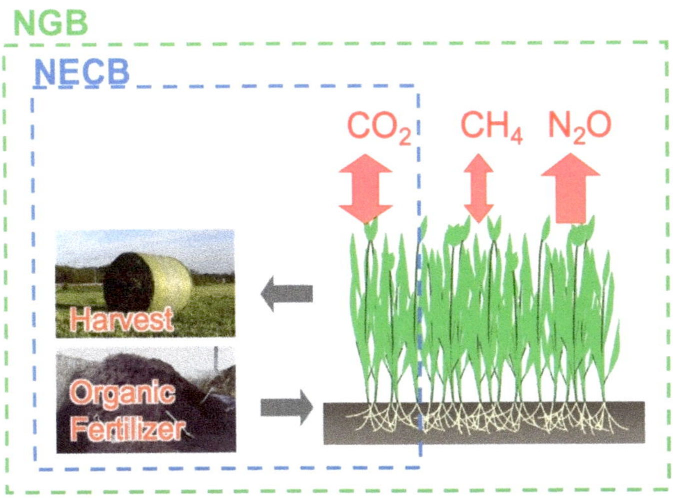

Figure 2. Schematic illustration of net ecosystem carbon balance (NECB) and net greenhouse gas balance (NGB) in agroecosystem.

NECB is a source of SOM and is obtained as net ecosystem production (NEP) [55] and can be written as follows [56]:

$$NECB = NEP + C_{input} - C_{output} \tag{1}$$

where C_{input} is the amount of carbon input to the ecosystem, such as the application of organic fertilizers, and C_{output} is the amount of carbon output through harvesting and erosion. NEP is the difference between net primary production (NPP) of plants and organic matter decomposition by soil microorganisms (Rh), as follows:

$$NEP = NPP - Rh \tag{2}$$

From the above, NECB can be obtained by measuring these parameters in the field. NEP can be estimated using a biometric method that measures plant production by cutting and the amount of SOC decomposed during that period using the chamber methods. At the same time, it can be estimated using micrometeorological methods such as the eddy correlation method [57], which measures and integrates the inflow and outflow of CO_2 between the atmosphere and ecosystem. These can be combined to obtain the net plant root production and plant respiration rate [15].

Furthermore, the CH_4 and N_2O fluxes in the ecosystem are measured using the chamber method and converted into CO_2eq ($GWPCH_4$ and $GWPN_2O$, respectively), and the negative NECB is converted into CO_2eq as the ecosystem CO_2 flux ($GWPCO_2$), and the Global Warming Potential (GWP) is calculated by the sum of these three parameters as follows:

$$GWP = GWPCO_2 + GWPCH_4 + GWPN_2O \tag{3}$$

The coefficients for CO_2eq conversion of CO_2, CH_4, N_2O fluxes are 1 for CO_2, 28 for CH_4, and 265 for N_2O [1].

$$GWPCO_2 = NECB \times (44/12) \times 1$$

$$GWPCH_4 = CH_4 \times (16/12) \times 28$$

$$GWPN_2O = N_2O \times (44/28) \times 265$$

Furthermore, negative values of GWP can be evaluated as the NGB of the ecosystem.

$$NGB = -GWP \tag{4}$$

Since CH_4 is mainly produced under anaerobic conditions as the final step in the anaerobic decomposition of SOM, CH_4 emissions primarily occur in oxygen-limited soils such as wetlands [58]. On the other hand, in well-drained upland soils, CH_4 is normally consumed by oxidation, although it is affected by NH_4 oxidation [59]. N_2O emissions are strongly influenced by the application of nitrogen fertilizers, either chemical, inorganic, or organic fertilizers [60]. N_2O is produced both under aerobic conditions due to nitrification and under anaerobic conditions due to denitrification [61]. How nitrogen fertilization and carbon inputs are managed is key to achieving adequate soil carbon sequestration and environmental conservation and ensuring sustainable crop production, and is detailed in the next section.

6. Proper Application of Organic Fertilizer and Its Effects on Carbon and GHG Balance

Crop growth is limited by the least abundant of the essential elements. Organic fertilizers do not provide an ideal balance of elements for crop growth. Generally, livestock manure has a high potassium content, and the mineralization rate of potassium is higher than that of phosphorus and nitrogen, so if we try to provide the required amount of nitrogen and phosphorus with manure alone, too much potassium will be added. Therefore, as a strict fertilization design for organic fertilizer, the amount of organic fertilizer to be applied should be such that the amount of mineralized nitrogen, phosphorus, and potassium from organic fertilizer is not excessive relative to the element requirements of plants. The missing elements are supplemented with chemical fertilizers [29].

Using this fertilization method, we investigated the effects of three organic fertilizers, manure, slurry, and digestive fluid, on crop growth and GHG emissions in a managed grassland of an Andosol in southern Hokkaido for three years [22]. The results showed that the chemical fertilizer nitrogen application rate in the organic fertilizer treatments was reduced by 10% for manure, 19.7% for slurry, and 29.7% for digestive fluid compared to chemical fertilizer only, but there was no significant difference in grass yield between the fertilizer treatments. Three years of NECB resulted in significantly less carbon loss with organic fertilizer treatment than with chemical fertilizer only. NGB when using organic fertilizers was reduced by 16.5% in slurry, 27.0% in digestive fluid, and 36.2% in manure compared to treatment with chemical fertilizers only. The main effect was due to carbon storage, that is, NECB reduction accounted for more than 90% of the NGB reduction. Manure and digestive fluid reduced N_2O emissions compared to chemical fertilizer only, but slurry increased N_2O. CH_4 was so small that it could be ignored.

This fertilization method was applied to grasslands in four different climates from Hokkaido to Kyushu in Japan, and NECB, N_2O, and CH_4 emissions were measured for three years. As a result, annual NECB and NGB were found to have a significant positive correlation with the amount of carbon applied through manure as shown in Figure 3 [62]. In other words, the NECB was positive when the amount of manure carbon applied was 2.5 tC ha^{-1} y^{-1} or more, and grassland became a net carbon sink. However, the NGB only became positive when the amount of manure carbon applied reached 3.5 tC ha^{-1} y^{-1} or more. This is because applying manure increases the generation of N_2O, so increasing the amount of manure carbon applied by an additional 1 tC ha^{-1} y^{-1} will offset the increase in N_2O and contribute to suppressing global warming.

Figure 4 shows the breakdown of NECB in the four grasslands in Japan shown in Figure 3 for the manure plot and the chemical fertilizer-only plot (i.e., the amount of Manure applied is 0). Net ecosystem production (NEP) was positive in both the chemical fertilizer plot and the manure plot. This indicates that these grassland ecosystems accumulate carbon in their natural state. However, NECB was negative in chemical fertilizer plots. This shows that producing grass using only chemical fertilizers in these grasslands decreases carbon in the ecosystems. As seen earlier, in the manure plot, NECB could be made positive by applying more than 2.5 tC ha^{-1} y^{-1} of manure.

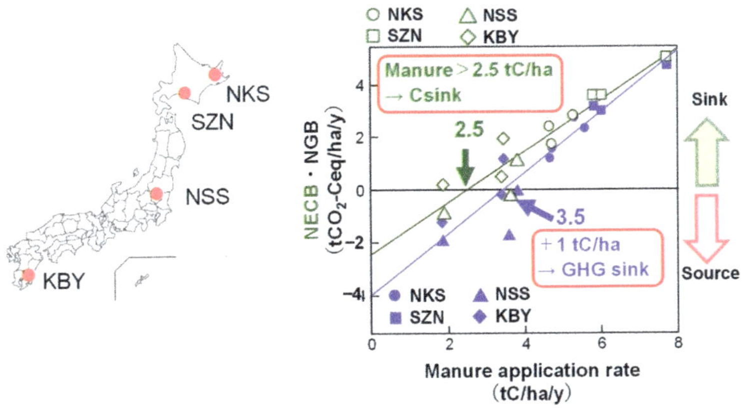

Figure 3. Relationship between manure application rate and net ecosystem carbon balance (NECB) and net greenhouse gas balance (NGB) in Japanese managed grasslands at Nakashibetsu (NKS), Shizunai (SZN), Nas-Shiobara (NSS) and Kobayashi (KBY) (produced from [62]).

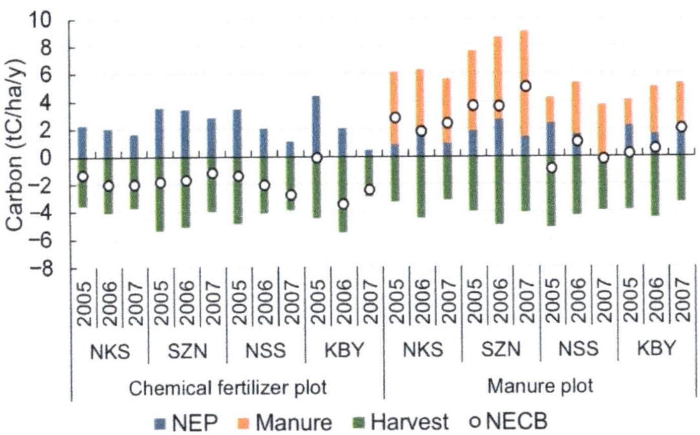

Figure 4. Breakdown of net ecosystem carbon balance (NECB) in four grasslands in Japan: net ecosystem production (NEP), carbon imported through manure application (Manure), carbon exported through harvest (Harvest); Nakashibetsu (NKS), Shizunai (SZN), Nas-Shiobara (NSS) and Kobayashi (KBY) (Produced from [62]).

The SOC/Clay ratio of the cultivated soil in this field was 1/2.3, 1/1.5, 1/5.4, and 1/1.8 for NKS, SZN, NSS, and KBY, respectively. These soils are in a "very good" soil structure condition based on the SOC/Clay ratio index (SOC content was 5.30%, 2.55%, 3.45%, and 5.30%, respectively, and clay content was 12%, 3.7%, 18%, and 9.5%, respectively). However, if grass is produced using only chemical fertilizers and carbon is taken out by harvesting the grass, NECB becomes negative and soil carbon is depleted, decreasing the SOC/Clay ratio.

Table 3 shows the results of estimating how soil carbon would be lost in the four grassland soils if no manure was applied. The number of years that the SOC contained in 0–35 cm of soil will decrease due to NECB without adding the current manure is calculated as "until SOC disappears" and "until the SOC/Clay ratio is 1/8 (good)", "until 1/10 (appropriate)", and "until 1/13 (deterioration)". However, the reduction in SOC was assumed to occur linearly at the current rate of carbon loss (NECB values at chemical fertilizer-only plots). As a result, soil carbon disappears in 55 to 78 years when cultivated without manure. During this period, the soil structure deteriorated, but it took 31 to 65 years for the SOC/Clay ratio to reach a level of 1/10, which is an acceptable level for agricultural soil. It is estimated that it will take 39 to 68 years to reach 1/13 (Table 3). It takes quite a long time for it to deteriorate, but once it starts to deteriorate, it appears to deteriorate rapidly.

Table 3. Estimated number of years until SOC disappears and reduces to the level of SOC/Clay ratio of 1/8, 1/10 and 1/13 by using Clay and SOC contents in 0–35 cm soil depth and net ecosystem carbon balance (NECB) in Japanese managed grasslands at Nakashibetsu (NKS), Shizunai (SZN), Nas-Shiobara (NSS) and Kobayashi (KBY).

				Number of Years until the SOC Reduces to:			
	Clay	SOC	NECB	0	SOC/Clay Ratio		
Site	(0–35 cm)	(0–35 cm)			1/8	1/10	1/13
	t/ha	tC/ha	tC/ha/y	Years			
NKS	272	120	−1.80	67	48	52	55
SZN	126	87	−1.57	55	45	47	49
NSS	720	138	−2.10	66	23	31	39
KBY	282	157	−1.99	79	61	65	68

7. Controlling N_2O Emissions

It is necessary to increase carbon input to soil to offset N_2O emissions, but increasing organic matter application may also exacerbate nutrient imbalances. In this sense, fertilization management that suppresses N_2O emissions is important.

Soil N_2O is produced as a by-product by nitrification processes and as an intermediate product by denitrification processes. Therefore, N_2O emissions increase with increasing N application and N mineralization [63]. Nitrification requires aerobic conditions and NH_4^+, while denitrification requires anaerobic conditions and NO_3^- and organic carbon [64,65]. Through nitrification and denitrification, N_2O is also produced as a by-product and intermediate product. Since nitrification and denitrification are affected by soil moisture [33], N_2O production is also affected by soil moisture [61]. N_2O production starts at 30% of WFPS, peaks at 60% to 70% of WFPS, and decreases when WFPS exceeds 80%, with N_2 predominating.

In nitrification and denitrification, NO is also emitted along with N_2O, but in the nitrification process, NO production is dominant, and the N_2O-N/NO-N ratio is less than 1. In the denitrification process, N_2O production is dominant, and the N_2O-N/NO-N ratio is greater than 100 [66]. In actual fields, investigation of the behavior of NO_3^- is important to understand the production and emission of N_2O. For example, in an onion field of structured clay soil, the above-ground parts and roots are supplied as residue during harvest. It has been reported that when the soil is moist due to rainfall during the harvest season, a large amount of N_2O is emitted, accounting for up to 70% of the annual emissions [67]. At this time, N_2O-N/NO-N exceeded 100, indicating that denitrification was predominant. This indicates that the supply of residual organic matter and rainfall in the onion fields caused denitrification, and the excess NO_3-N left behind is thought to increase N_2O emissions. Similarly, organic fertilizers can significantly increase N_2O emissions if excess NO_3^-N remains. A comparison of chemical fertilizer treatment and cow dung manure application in grassland showed that N_2O emissions increased in the manure application plot [68]. However, there was no significant difference in the relationship of N_2O flux to soil temperature and soil moisture between the chemical fertilizer-only plot and manure plot, and a peak of N_2O flux was observed at 18 °C of soil temperature or above 70–80% of WFPS. Also, since the N_2O-N/NO-N ratio was higher when the N_2O flux was high for both plots of chemical fertilizer only and manure, it was thought that denitrification was the main process for N_2O emission in both plots. Furthermore, the amount of surplus mineral nitrogen (amount of mineral nitrogen applied + amount of mineralized nitrogen—amount of plant nitrogen uptake) showed a significant correlation with N_2O emissions. These results suggest that the cause of N_2O emissions was the residual amount of mineral nitrogen.

Manure releases nitrogen as organic matter decomposes, but in the case of cow manure, the annual decomposition rate of organic nitrogen is less than 20%. However, as mentioned earlier, manure contains potassium with a high release rate. Manure is applied in such an amount that the amount of NPK mineralized does not exceed the amount required by the plants, and the lack of mineralization is supplemented with chemical fertilizers. However, since mineralization occurs from organic fertilizers even after the second year, continuous application of manure can reduce the amount of chemical fertilizer supplements. Jin et al. (2010) reported that chemical nitrogen fertilizer supplements became zero after 4 years, and N_2O emissions were lower in the manure plot than in the chemical fertilizer plot [54]. Additionally, in manure plots, soil pH rarely changes above 5. On the other hand, in the chemically fertilized plots, the soil pH dropped to 4 after three years.

Annual N_2O emissions from grasslands and corn fields were negatively correlated with decreasing mean soil pH when nitrogen was applied to fields, regardless of whether or not manure was applied, when soil pH was less than 6 [36]. Incubation experiments have found that there is no N_2O release at high limestone soil pH above 7 [69]. This is because the optimal pH of incomplete denitrifying bacteria, which causes N_2O release, is lower than pH 6.5, which is the optimal pH of complete denitrifying bacteria. Therefore, to suppress N_2O emissions, it is necessary to raise the soil pH to 6.5 or higher. Maintaining soil pH by applying organic fertilizers is one of the strategies to reduce N_2O emissions and maintain soil health. In addition, crop residues and organic fertilizers with high C/N ratios had significantly lower N_2O emissions per mineralized nitrogen (i.e., emission factors) [22,70–72]. Considering the use of organic fertilizers with high C/N ratios is also important from the perspective of soil carbon sequestration.

As mentioned above, organic matter application maintains SOC and contributes to maintaining good soil structure. The direct effect of improving soil structure is the improvement of nutrient and water supply (Figure 1). This improves the nitrogen balance and suppresses N_2O emissions [68]. In addition, fertilization methods that reduce chemical fertilizers by applying organic matter also suppress N_2O [22].

8. Conclusions

It has become clear that the maintenance effect of soil structure by soil organic matter is essential for soil health. Agricultural soil has clearly lost organic matter due to past cultivation, and the SOC/Clay ratio, which is an index of soil structure condition, is significantly more likely to fall below the deterioration limit of 1/13 in cropland than in grassland or forest. It has become clear that agriculture has been damaging the health of the soil. On the other hand, it was also shown that agricultural soils have the potential capacity to sequester 78% of the 555 GtC released into the atmosphere since the Industrial Revolution as soil organic matter. Soil management that increases rather than decreases soil carbon is necessary. To achieve this, the net ecosystem carbon balance (NECB) must be made positive. In four undegraded managed grasslands from north to south of Japan, it was necessary to apply more than 2.5 tC $ha^{-1}y^{-1}$ of compost carbon to maintain standard yields and achieve positive NECB. Although it is possible to reduce the amount of chemical fertilizer applied by applying compost, N_2O is emitted as a result of fertilization, so it is necessary to achieve carbon sequestration to offset the greenhouse effect caused by this N_2O emission. It was shown that this can be achieved by applying an additional 1 tC $ha^{-1}y^{-1}$ of compost. It has also been recognized that the continuous use of organic fertilizers has a large effect on reducing N_2O emissions. It is increasingly important to consider soil management using organic fertilizers, which reduces the need for chemical fertilizers without reducing yields.

Author Contributions: Conceptualization, R.H.; writing—review and editing, R.H., I.M. and M.S. All authors have read and agreed to the published version of the manuscript.

Funding: This study was partly supported by a Japanese Grant-in-Aid for Science Research from the Ministry of Education, Culture, Sports, Science and Technology (No. 11460028); a research grant provided by the Project entitled 'Establishment of good practices to mitigate Greenhouse Gas emissions from Japanese grasslands' funded by Racing and Livestock Association; and 'Development of Mitigation Technologies to Climate Change in the Agriculture Sector' run by the Ministry of Agriculture, Forestry and Fisheries of Japan.

Institutional Review Board Statement: Not applicable.

Data Availability Statement: All data generated or analyzed during this study are included in this published article.

Acknowledgments: We thank Yo Toma of Hokkaido University for kind suggestions regarding the interpretation of the data.

Conflicts of Interest: The authors declare no conflicts of interest.

References

1. IPCC. *Climate Change 2014. Synthesis Report. Contribution of Working Groups I, II and III to the Fifth Assessment Report of the Intergovernmental Panel on Climate Change*; Core Writing Team, Pachauri, R.K., Meyer, L.A., Eds.; IPCC: Geneva, Switzerland, 2014.
2. IPCC. *Climate Change 2021. The Physical Science Basis. Contribution of Working Group I to the Sixth Assessment Report of the Intergovernmental Panel on Climate Change;* Masson-Delmotte, V., Zhai, P., Pirani, A., Connors, S.L., Péan, C., Berger, S., Caud, N., Chen, Y., Goldfarb, L., Gomis, M.I., et al., Eds.; Cambridge University Press: Cambridge, UK; New York, NY, USA, 2021. [CrossRef]
3. Georgiou, K.; Ahlstrom, A.; Polley, H.W.; Jackson, R.B.; Vindu, O.; Abramoff, R.Z.; Feng, W.; Harden, J.W.; Pellegrini, A.F.A. Global stocks and capacity of mineral-associated soil organic carbon. *Nat. Commun.* **2022**, *13*, 3797. [CrossRef]
4. Minasny, B.; Malone, B.P.; McBratney, A.B.; Angers, D.A.; Arrouays, D.; Chambers, A.; Chaplot, V.; Chen, Z.S.; Cheng, K.; Das, B.S.; et al. Soil carbon 4 per mille. *Geoderma* **2017**, *292*, 59–86. [CrossRef]
5. Lehmann, J.; Bossio, D.A.; Kögel-Knabner, I.; Rillig, M.C. The concept and future prospects of soil health. *Nat. Rev. Earth Environ.* **2020**, *1*, 544–553. [CrossRef] [PubMed]
6. FAO ITPS. Towards a Definition of Soil Health. Soil Letters. 2020. Food and Agriculture Organization of the United Nations (FAO) and Intergovernmental Technical Panel on Soils (ITPS). 2020. Available online: https://openknowledge.fao.org/handle/20.500.14283/cb1110en (accessed on 21 August 2022).
7. Doran, J.W.; Zeiss, M.R. Soil health and sustainability: Managing the biotic component of soil quality. *Appl. Soil Ecol.* **2000**, *15*, 3–11. [CrossRef]
8. Lal, R. Soil health and carbon management. *Food Energy Secur.* **2016**, *5*, 212–222. [CrossRef]
9. Beillouin, D.; Demenois, J.; Cardinael, R.; Berre, D.; Corbeels, M.; Fallot, A.; Boyer, A.; Feder, F. A global database of land management, land-use change and climate change effects on soil organic carbon. *Sci. Data* **2022**, *9*, 228. [CrossRef]
10. Orgiazzi, A.; Ballabio, C.; Panagos, P.; Jones, A.; Fernández-Ugalde, O. LUCAS Soil, the largest expandable soil dataset for Europe: A review. *Eur. J. Soil Sci.* **2018**, *69*, 140–153. [CrossRef]
11. Mäkipää, R.; Menichetti, L.; Martínez-García, E.; Törmänen, T.; Lehtonen, A. Is the organic carbon-to-clay ratio a reliable indicator of soil health? *Geoderma* **2024**, *444*, 116862. [CrossRef]
12. Lal, R. Soil carbon sequestration to mitigate climate change. *Geoderma* **2004**, *123*, 1–22. [CrossRef]
13. Janzen, H.H. Carbon cycling in earth systems—A soil science perspective. *Agric. Ecosyst. Environ.* **2004**, *104*, 399–417. [CrossRef]
14. Trumbore, S. Carbon respired by terrestrial ecosystems–recent progress and challenges. *Glob. Chang. Biol.* **2006**, *2*, 141–153. [CrossRef]
15. Limin, A.; Shimizu, M.; Mano, M.; Ono, K.; Miyata, A.; Wada, H.; Nozaki, H.; Hatano, R. Manure application has an effect on the carbon budget of a managed grassland in southern Hokkaido, Japan. *Soil Sci. Plant Nutr.* **2015**, *61*, 856–872. [CrossRef]
16. Balogh, J.; Papp, M.; Pintér, K.; Fóti, S.; Posta, K.; Eugster, W.; Nagy, Z. Autotrophic component of soil respiration is repressed by drought more than the heterotrophic one in dry grasslands. *Biogeosciences* **2016**, *13*, 5171–5182. [CrossRef]
17. Asano, M.; Wagai, R. Evidence of aggregate hierarchy at micro- to submicron scales in an allophanic Andisol. *Geoderma* **2014**, *216*, 62–74. [CrossRef]
18. Wagai, R.; Kajiura, M.; Asano, M.; Hiradate, S. Nature of soil organo-mineral assemblage examined by sequential density fractionation with and without sonication: Is allophanic soil different? *Geoderma* **2015**, *241–242*, 295–305. [CrossRef]
19. Bagnall, D.K.; Morgan, C.L.S.; Bean, G.M.; Liptzin, D.; Cappellazzi, S.B.; Cope, M.; Greub, K.L.H.; Rieke, E.L.; Norris, C.E.; Tracy, P.W.; et al. Selecting soil hydraulic properties as indicators of soil health: Measurement response to management and site characteristics. *Soil Sci. Soc. Am. J.* **2022**, *86*, 1206–1226. [CrossRef]

20. Bagnall, D.K.; Morgan, C.L.S.; Cope, M.C.; Bean, G.M.; Cappellazzi, S.B.; Greub, K.; Liptzin, D.; Norris, C.L.; Rieke, E.; Tracy, P.; et al. Carbon-sensitive pedotransfer functions for plant available water. *Soil Sci. Soc. Am. J.* **2022**, *86*, 612–629. [CrossRef]
21. Torn, M.S.; Trumbore, S.E.; Chadwick, O.A.; Vitousek, P.M.; Hendricks, D.M. Mineral control of soil organic carbon storage and turnover. *Nature* **1997**, *3603*, 3601–3603. [CrossRef]
22. Kitamura, R.; Sugiyama, C.; Yasuda, K.; Nagatake, A.; Yuan, Y.; Du, J.; Yamaki, N.; Taira, K.; Kawai, M.; Hatano, R. Effects of Three Types of Organic Fertilizers on Greenhouse Gas Emissions in a Grassland on Andosol in Southern Hokkaido, Japan. *Front. Sustain. Food Syst.* **2021**, *5*, 649613. [CrossRef]
23. Gaudaré, U.; Kuhnert, M.; Smith, P.; Martin, M.; Barbieri, P.; Pellerin, S.; Nesme, T. Soil organic carbon stocks potentially at risk of decline with organic farming expansion. *Nat. Clim. Chang.* **2023**, *13*, 719–725. [CrossRef]
24. Vendig, I.; Guzman, A.; De La Cerda, G.; Esquivel, K.; Mayer, A.C.; Ponisio, L.; Bowles, T.M. Quantifying direct yield benefits of soil carbon increases from cover cropping. *Nat. Sustain.* **2023**, *6*, 1125–1134. [CrossRef]
25. Šimon, T.; Madaras, M.; Mayerová, M.; Kunzová, E. Soil Organic Carbon Dynamics in the Long-Term Field Experiments with Contrasting Crop Rotations. *Agriculture* **2024**, *14*, 818. [CrossRef]
26. Ma, Y.; Woolf, D.; Fan, M.; Qiao, L.; Li, R.; Lehmann, J. Global crop production increase by soil organic carbon. *Nat. Geosci.* **2023**, *16*, 1159–1165. [CrossRef]
27. UNEP. *Global Environment Outlook 2000*; UNEP: London, UK; Earthscan: Nairobi, Kenya, 1999; p. 398.
28. Rockström, J.; Steffen, W.; Noone, K.; Persson, A.; Chapin III, F.S.; Lambin, E.F.; Lenton, T.M.; Scheffer, M.; Folke, C.; Schellnhuber, H.J.; et al. A safe operating space for humanity. *Nature* **2009**, *461*, 472–475. [CrossRef] [PubMed]
29. Asaka, D. Fertilizer Recommendations. In *The Soils of Japan*; Hatano, R., Shinjo, H., Takata, Y., Eds.; Springer Nature Singapore Pte Ltd.: Singapore, 2021; pp. 168–169.
30. Karasawa, T. Beneficial effects of cover crops on various soil functions and nutrient supply. *Soil Sci. Plant Nutr.* **2024**, *70*, 237–245. [CrossRef]
31. Johannes, A.; Matter, A.; Schulin, R.; Weisskopf, P.; Baveye, P.C.; Boivin, P. Optimal organic carbon values for soil structure quality of arable soils. Does clay content matter? *Geoderma* **2017**, *302*, 14–21. [CrossRef]
32. Iijima, M.; Kato, J. Combined soil physical stress of soil drying, anaerobiosis and mechanical impedance to seedling root growth of four crop species. *Plant Prod. Sci.* **2007**, *10*, 451–459. [CrossRef]
33. Linn, D.M.; Doran, J.W. Effect of Water Filled Pore Space on Carbon Dioxide and Nitrous Oxide Production in Tilled and Non-Tilled Soils. *Soil Sci. Soc. Am. J.* **1984**, *48*, 1267–1272. [CrossRef]
34. Nagatake, A.; Mukumbuta, I.; Yasuda, K.; Shimizu, M.; Kawai, M.; Hatano, R. Temporal dynamics of nitrous oxide emission and nitrate leaching in renovated grassland with repeated application of manure and/or chemical fertilizer. *Atmosphere* **2018**, *9*, 485. [CrossRef]
35. Pan, B.; Lam, S.K.; Mosier, A.; Luo, Y.; Chen, D. Ammonia volatilization from synthetic fertilizers and its mitigation strategies: A global synthesis. *Agric. Ecosyst. Environ.* **2016**, *232*, 283–289. [CrossRef]
36. Mukumbuta, I.; Shimizu, M.; Jin, T.; Nagatake, A.; Hata, H.; Kondo, S.; Kawai, M.; Hatano, R. Nitrous and nitric oxide emissions from a cornfield and managed grassland: 11 years of continuous measurement with manure and fertilizer applications, and land-use change. *Soil Sci. Plant Nutr.* **2017**, *63*, 185–199. [CrossRef]
37. Wang, C.; Kuzyakov, Y. Soil organic matter priming: The pH effects. *Glob. Chang. Biol.* **2024**, *30*, e17349. [CrossRef]
38. Sokol, N.W.; Whalen, E.D.; Jilling, A.; Kallenbach, C.; Pett-Ridge, J.; Georgiou, K. Global distribution, formation and fate of mineral-associated soil organic matter under a changing climate: A trait-based perspective. *Funct. Ecol.* **2022**, *36*, 1411–1429. [CrossRef]
39. Cai, A.; Han, T.; Ren, T.; Sanderman, J.; Rui, Y.; Wang, B.; Smith, P.; Xu, M.; Li, Y. Declines in soil carbon storage under no tillage can be alleviated in the long run. *Geoderma* **2022**, *425*, 116028. [CrossRef]
40. Nakatsuka, H.; Rakhat, B.; Tamura, K.; Asano, M.; Karasawa, T. Effects of root growth on physicochemical properties of soil profiles and komatsuna productivity in a wild oat cover-crop system. *Pedologist* **2022**, *66*, 3–16.
41. Georgiou, K.; Jackson, R.B.; Vindušková, O.; Abramoff, R.Z.; Ahlström, A.; Feng, W.; Harden, J.W.; Pellegrini, A.F.A.; Polley, H.W.; Soong, J.L.; et al. Supplementary Materials for Global Stocks and Capacity of Mineral-Associated Soil Organic Carbon. 2022. Available online: https://static-content.springer.com/esm/art:10.1038/s41467-022-31540-9/MediaObjects/41467_2022_31540_MOESM1_ESM.pdf (accessed on 4 October 2022).
42. Beare, M.H.; Hendrix, P.F.; Coleman, D.C. Water-stable aggregates and organic matter fractions in conventional and no-tillage soils. *Soil Sci. Soc. Am. J.* **1994**, *58*, 777–786. [CrossRef]
43. Dexter, A.R.; Richard, G.; Arrouays, D.; Czyż, E.A.; Jolivet, C.; Duval, O. Complexed organic matter controls soil physical properties. *Geoderma* **2008**, *144*, 620–627. [CrossRef]
44. Ball, B.C.; Batey, T.; Munkholm, L.J. Field assessment of soil structural quality–a development of the Peerlkamp test. *Soil Use Manag.* **2007**, *23*, 329–337. [CrossRef]
45. Prout, J.M.; Shepherd, K.D.; McGrath, S.P.; Kirk, G.J.D.; Haefele, S.M. What is a good level of soil organic matter? an index based on organic carbon to clay ratio. *Eur. J. Soil Sci.* **2021**, *72*, 2493–2503. [CrossRef]
46. IPCC. 2006 IPCC Guidelines for National Greenhouse Gas Inventories. 2006. Available online: https://www.ipcc-nggip.iges.or.jp/public/2006gl/ (accessed on 14 January 2007).

47. Kleber, M.; Eusterhues, K.; Keiluweit, M.; Mikutta, C.; Mikutta, R.; Nico, P.S. Mineral-organic associations: Formation, properties, and relevance in soil environments. *Adv. Agron.* **2015**, *130*, 1–140.
48. Beare, M.H.; McNeill, S.J.; Curtin, D.; Parfitt, R.L.; Jones, H.S.; Dodd, M.B.; Sharp, J. Estimating the organic carbon stabilisation capacity and saturation deficit of soils: A New Zealand case study. *Biogeochemistry* **2014**, *120*, 71–87. [CrossRef]
49. Kang, J.; Qu, C.; Chen, W.; Cai, P.; Chen, C.; Huang, Q. Organo–organic interactions dominantly drive soil organic carbon accrual. *Glob. Chang. Biol.* **2024**, *30*, e17147. [CrossRef] [PubMed]
50. Wang, X.; Jia, Z.; Liang, L.; Yang, B.; Ding, R.; Nie, J.; Wang, J. Impacts of manure application on soil environment, rainfall use efficiency and crop biomass under dryland farming. *Sci. Rep.* **2016**, *6*, 20994. [CrossRef] [PubMed]
51. Ullah, M.R.; Corneo, P.E.; Dijkstra, F.A. Inter-seasonal Nitrogen Loss with Drought Depends on Fertilizer Management in a Seminatural Australian Grassland. *Ecosystems* **2020**, *23*, 1281–1293. [CrossRef]
52. Stevens, W.B.; Evans, R.G.; Jabro, J.D.; Iversen, W.M. Sugarbeet Productivity as Influenced by Fertilizer Band Depth and Nitrogen Rate in Strip Tillage. *J. Sugar Beet Res.* **2011**, *48*, 137–155. [CrossRef]
53. Fueki, N.; Tani, M.; Higashida, S.; Nakatsu, S. Effect of soil acidity and nitrification of fertilizer introduced by row application on sugar beet growth in several soil types. *Soil Sci. Plant Nutr.* **2004**, *50*, 321–329. [CrossRef]
54. Jin, T.; Shimizu, M.; Marutani, S.; Desyatkin, A.R.; Iizuka, N.; Hata, H.; Hatano, R. Effect of chemical fertilizer and manure application on N_2O emission from reed canary grassland in Hokkaido, Japan. *Soil Sci. Plant Nutr.* **2010**, *56*, 53–65. [CrossRef]
55. Schulze, E.D.; Wirth, C.; Heimann, M. Managing forests after Kyoto. *Science* **2000**, *289*, 2058–2059. [CrossRef]
56. West, T.O.; Marland, G. Net carbon flux from agricultural ecosystems: Methodology for full carbon cycle analyses. *Environ. Pollut.* **2002**, *116*, 439–444. [CrossRef]
57. Wofsy, S.C.; Goulden, M.L.; Munger, J.W.; Fan, S.M.; Bakwin, P.S.; Daube, B.C.; Bassow, S.L.; Bazzaz, F.A. Net exchange of CO_2 in a midlatitude forest. *Science* **1993**, *260*, 1314–1317. [CrossRef]
58. Kimura, M.; Murakami, H.; Wada, H. CO_2, H_2, and CH_4 production in rice rhizosphere. *Soil Sci. Plant Nutr.* **1991**, *37*, 55–60. [CrossRef]
59. Steudler, P.A.; Bowden, R.D.; Melillo, J.M.; Aber, J.D. Influence of nitrogen fertilization on methane uptake in temperate forest soils. *Nature* **1989**, *341*, 314–316. [CrossRef]
60. Bouwman, A.F. Direct emission of nitrous oxide from agricultural soils. *Nutr. Cycl. Agroecosyst.* **1996**, *46*, 53–70. [CrossRef]
61. Bouwman, A.F. Nitrogen oxides and tropical agriculture. *Nature* **1998**, *392*, 866–867. [CrossRef]
62. Hirata, R.; Miyata, A.; Mano, M.; Shimizu, M.; Arita, T.; Kouda, Y.; Matsuura, S.; Niimi, M.; Saigusa, T.; Mori, A.; et al. Carbon dioxide exchange at four intensively managed grassland sites across different climate zones of Japan and the influence of manure application on ecosystem carbon and greenhouse gas budgets. *Agric. For. Meteorol.* **2013**, *177*, 57–68. [CrossRef]
63. Mu, Z.J.; Huang, A.; Kimura, S.D.; Jin, T.; Wei, S.; Hatano, R. Linking N_2O emission to soil mineral N as estimated by CO_2 emission and soil C/N ratio. *Soil Biol. Biochem.* **2009**, *41*, 2593–2597. [CrossRef]
64. Davidson, E.A. Source of nitric oxide and N_2O following wetting of dry soil. *Soil Sci. Soc. Am. J.* **1992**, *56*, 95–102. [CrossRef]
65. Bremner, J.M. Source of nitrous oxide in soils. *Nutr. Cycl. Agroecosyst.* **1997**, *49*, 7–16. [CrossRef]
66. Lipschultz, F.; Zafiriou, O.C.; Wofsy, S.C.; McElroy, M.B.; Valois, F.W.; Watson, S.W. Production of NO and N_2O by soil nitrifying bacteria. *Nature* **1981**, *294*, 641–643. [CrossRef]
67. Kusa, K.; Sawamoto, T.; Hatano, R. Nitrous Oxide Emissions for Six Years from a Gray Lowland Soil Cultivated with Onions in Hokkaido, Japan. *Nutr. Cycl. Agroecosyst.* **2002**, *63*, 239–247. [CrossRef]
68. Shimizu, M.; Marutani, S.; Desyatkin, A.R.; Jin, T.; Nakano, K.; Hata, H.; Hatano, R. Nitrous oxide emissions and nitrogen cycling in managed grassland in Southern Hokkaido, Japan. *Soil Sci. Plant Nutr.* **2010**, *56*, 676–688. [CrossRef]
69. Mukumbuta, I.; Uchida, Y.; Hatano, R. Evaluating the effect of liming on N_2O fluxes from denitrification in an Andosol using the acetylene inhibition and N-15 isotope tracer methods. *Biol. Fertil. Soils* **2018**, *54*, 71–81. [CrossRef]
70. Akiyama, H.; Tsuruta, H. Effect of organic matter application on N_2O, NO, and NO_2 fluxes from an andisol field. *Glob. Biogeochem. Cycles* **2003**, *17*, 1–16. [CrossRef]
71. He, T.; Yuan, J.; Luo, J.; Wang, W.; Fan, J.; Liu, D.; Ding, W. Organic fertilizers have divergent effects on soil N_2O emissions. *Biol. Fertil. Soils* **2019**, *55*, 685–699. [CrossRef]
72. Huang, Y.; Zou, J.; Zheng, X.; Wang, Y.; Xu, X. Nitrous oxide emissions as influenced by amendment of plant residues with different C:N ratios. *Soil Biol. Biochem.* **2004**, *36*, 973–981. [CrossRef]

agriculture

MDPI

Review

Saline–Alkali Soil Reclamation Contributes to Soil Health Improvement in China

Wei Zhu [1], Shiguo Gu [1], Rui Jiang [2,*], Xin Zhang [3] and Ryusuke Hatano [4]

1 College of Civil and Architecture Engineering, Chuzhou University, Chuzhou 239000, China; wzhu@chzu.edu.cn (W.Z.); shiguogu@chzu.edu.cn (S.G.)
2 Research Center for Cultural Landscape Protection and Ecological Restoration, China-Portugal Belt and Road Cooperation Laboratory of Cultural Heritage Conservation Science, Gold Mantis School of Architecture, Soochow University, Suzhou 215006, China
3 State Key Laboratory of Soil and Sustainable Agriculture, Institute of Soil Science, Chinese Academy of Sciences, Nanjing 210008, China; xzhang@issas.ac.cn
4 Research Faculty of Agriculture, Hokkaido University, Sapporo 0608589, Japan; hatano@agr.hokudai.ac.jp
* Correspondence: ruijiang@suda.edu.cn

Abstract: Soil salinization is a significant threat to soil health, especially to the agricultural ecosystem; it reduces vegetation biomass, destroys ecosystem diversity, and limits land use efficiency. This area of investigation has garnered extensive attention in China, especially in the arid and semi-arid areas, totaling 7.66×10^6 ha. A variety of theoretical research and technology developments have contributed to soil water and salt regulation and the screening of salt-tolerant varieties to improve nutrient utilization efficiency and microbial control and reduce ecological problems due to saline-based obstacles. These techniques can be classified into physical treatments, chemical treatments, biological treatments, and combined treatments; these different measures are all aimed at primarily solving saline–alkali stress. In general, the improvement and utilization of saline–alkali soil contribute to soil health improvement, concentrating on high-quality development, food security, ecological security, cultivated land protection, and agricultural upgrading. However, the risks of various technologies in the practical production process should be highlighted; green and healthy measures are still expected to be applied to saline–alkali land.

Keywords: soil health; saline–alkaline land; measures; risk

1. Introduction

The term "saline–alkali soil" describes a range of soil types that contain abnormally high concentrations of soluble salt ions. These ions have a negative impact on the physical, chemical, and biological properties of the soil and plant growth features. Saline soil mainly refers to soil anions from chloride and sulfate, and alkaline soil mainly refers to soil anions from carbonate and bicarbonate [1]. Soil salinization and alkalization have already reduced soil health, and the FAO describes "salinization" as one of the biological chemical indicators of soil health [2], characterized by high pH, soluble salts, and nutrient holding capacity and availability. Furthermore, soil salinization and alkalization reduced functional microbial biomass and activity in farmland soil [3]. Alarmingly, the global extent of saline and sodic soils had expanded to 424 million ha of topsoil (0–30 cm) and 833 million ha of subsoil (30–100 cm) in 2021 [4], having changed from approximately 800 million ha in 2010 [5], indicating a rising trend over the past decade that is anticipated to persist [4]. This has emerged as a pressing global environmental concern, particularly in arid and semi-arid regions [6]. Fortunately, in China, the area of saline–alkali soil shows a decreasing trend, which is a testament to the effective improvement and utilization efforts undertaken in recent decades. According to the second national soil survey, the area of saline–alkali soil is 3.69×10^7 ha, accounting for 4.88% of the total available land [7,8]. The third national soil survey revealed a reduction down to 7.66 × 10^6 ha of saline–alkali soil in China, with this occupying 5.99% of the total available land [9]. Saline–alkali soil is predominantly distributed in three major regions. Firstly, the arid and semi-arid areas of the mid-west, such as Xinjiang, Qinghai, Inner Mongolia, Ningxia, and other areas, account for 96.1% of the total saline–alkali land, which is classified as the chloride–sulfate type [10,11]. Secondly, the northeast regions, including Jilin, Inner Mongolia, Heilongjiang, and other areas, are mainly composed of soda saline–alkali land dominated by carbonates, accounting for 3.2% of the total saline–alkali land [12,13]. Lastly, the eastern coastal region, encompassing Shandong, Jiangsu, and Hebei, which is dominated by chloride-type saline–alkaline soil, accounts for less than 1% of the total saline–alkali land [14,15]. This proves that there is great potential for saline–alkali soil improvement and utilization in China.

Saline–alkali soil improvement and utilization has always been a significant scientific and commercial problem and is of great importance for soil health. Saline–alkali soil is also of great importance to the environmental protection and sustainability of agriculture in China [16]. There is a high population but a limited quantity of land resources; meanwhile, the demand for land resources for economic development is increasing [17], which causes pressure and challenges to sustainable agricultural development. Therefore, salinity and alkalinity reduction play an important role in agricultural production development, land productivity improvement, food security, and cultivated land expansion in China [18]. Meanwhile, saline–alkali soil is an important part of the terrestrial ecosystem [19], and its physical, chemical, and biological properties, as well as the ecosystem material and energy cycling processes, are different from those of other soils [20].

Generally, the improvement and utilization of saline–alkali soil is of great importance in China due to the limited arable land resources and the growing demand for agriculture, and its great significance for improving soil health. In fact, saline–alkali farmland has already been confirmed as one of the main types of low- to medium-yield farmland in China, which is crucial for national food security. In recent years, a large number of studies have been conducted to explore appropriate improvement measures, which is the first step in the reclamation of saline–alkali soil. Further research is required to elucidate the laws of water and salt transport, finally revealing the quality and productivity level of the improved saline–alkali soil. Therefore, in this review, we have summarized the improvement measures for saline–alkali soil, the soil water and salt transport processes, and the fertility and yield of saline–alkali soil. In the last section, we put forward some expectations for the improvement and utilization of saline–alkali soil in China.

Therefore, the summary of the relevant research on saline-alkali land in China is of great significance for guiding sustainable development and utilization. We used Web of Science and China National Knowledge Infrastructure to search peer-reviewed publications between January 2001 and December 2023, including research papers, reviews, and meta-analyses, which are related to saline–alkali land in China. Different combinations of terms were used, such as "saline–alkaline soil", "saline soil", "alkaline soil", "soil salinization", and "salt-affected soil" in the title, abstract, or keywords. We only selected papers reporting field and laboratory studies on the saline–alkaline soil of China, which could contribute the soil health improvement. We mainly discussed improvement measures, the soil water and salt transport process, nutrient change, and yield improvement in saline–alkali soil. Although there are many related research articles and reviews, a comprehensive analysis specifically focused on the improvement measures and their effects on soil water, salt, nutrients, and yields for agriculture is yet to be conducted, especially the different changes in saline–alkali soil caused by different measurements, the water and salt transport process, and the enhancement mechanism of nutrients and yields. Based on the review of research on saline–alkali soil, prospects for saline–alkali soil research in China are proposed.

2. Improvement Measures for Saline–Alkali Soil

In general, the improvement and utilization of saline–alkali soil can be divided into three parts: physical measures, chemical measures, and biological measures; these are important ways to improve soil health, corresponding to soil health indicators to a high degree [2].

2.1. Physical Measures

Physical measures are always used to change the boundary and/or soil profile conditions, as shown in Figure 1, and this is highly related to the saline–alkali soil physical properties, which can reflect the flow patterns of mass and energy in the soil environment, finally determining plant development and microbiological activity [21]. Its essence is to change the upper and lower boundary conditions of the soil, and it is related to soil water-gas parameters. First, the upper boundary conditions, such as agronomic measures (including film mulch, buried layer, and irrigation) and drainage management (e.g., flood irrigation and drip irrigation), are used to change the soil water and heat flux between the soil and air. Second, the cultivation layer conditions in the soil profile, including tillage (e.g., rotary tillage and deep vertically rotary tillage) and agronomic measures (e.g., buried layer), are typically applied to improve the physical soil properties of the topsoil. The bottom boundary condition, such as subsurface drainpipes, shafts, and drainage ditches, controls the groundwater level. New methods for the physical control of salinization have emerged, such as degradable liquid films, biomass materials, and porous adsorption materials, which have developed alongside technological progress.

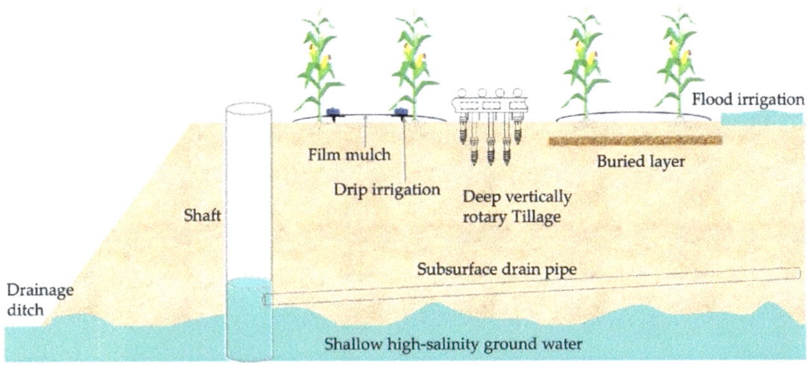

Figure 1. Some physical measures in the saline–alkali soil.

Based on the upper boundary conditions, there are many studies regarding the effect of mulch on soil water and salt. Mulch significantly reduces soil–water evaporation, controls and reduces soil moisture loss through evaporation, limits the upward migration of highly mineralized groundwater under capillary action, and inhibits surface salt accumulation. Mulch also shows a significant effect on soil water storage and salt suppression [22–24]. Plastic film significantly increases soil moisture and decreases salinity in the top 20 cm, especially during the seedling stage [25]. Furthermore, plastic mulching improves plant height and leaf area index, and the yield with mulch is also significantly higher than that without mulch [26]. Mulch types, such as plastic mulch, sand mulch, and straw mulch, have also been introduced for saline soil reclamation, and different mulch types show varying degrees of reduction in electrical conductivity; they can also increase the soil temperature and nutrient availability [27].

Flood irrigation, drip irrigation, saline water, and brackish water irrigation are the most commonly used irrigation methods. Flood irrigation has already been used in the Hetao Irrigation District of Inner Mongolia, China. Autumn irrigation is a traditional salt reduction method in this area, which is not only related to the amount and duration of irrigation water but is also closely linked to the soil freeze–thaw process [28]. Drip irrigation can effectively alleviate soil water shortages [29], and it is used to create a favorable soil water and salinity content in the root zone [30]. Saline water irrigation is recommended in arid and semi-arid areas where freshwater is limited [31], but it requires effective drainage measures; otherwise, it may increase the mineralization of groundwater and cause secondary salinization. Spring irrigation maintained the yield of spring maize when the salinity of irrigation water was <3 g/L, but it was also found that brackish water irrigation may lead to salt accumulation in the soil profile [32].

Plow layer regulation is also important for saline–alkali soil reclamation. Deep vertical rotary tillage management is one of the main tillage techniques widely used in saline soil, and it has the ability to efficiently lower pH, bulk density, and soil solution conductivity, and enhance soil structure [33]. In coastal saline zones, tillage—such as rotational tillage to a depth of 15 cm and deep tillage to a depth of 25–30 cm—is an effective management strategy for enhancing soil qualities, encouraging plant productivity, and raising financial advantages [34]. Straw is one of the most commonly used materials as the buried layer. A buried maize straw layer significantly decreases salt accumulation in the offseason [35], as the buried layer contributes to breaking the continuity of upward capillary movement and salt accumulation [36]. However, we notice that soil water content decreases significantly above the buried layer, so the buried layer must be combined with mulch [37].

Field research in Xinjiang, China, found that subsurface pipe drainage combined with drip irrigation greatly lowered the soil salt content, and the soil salt content decreased as the subsurface pipe spacing decreased after drip irrigation, which was conducted at a soil depth of 0–200 cm [38]. Shafts are always combined with subsurface pipes to collect infiltrated water. Traditional methods of reducing soil salinity include drainage ditches; open ditch drainage treatment was shown to have a stronger desalination impact than subsurface pipe drainage [39], although other research suggested the ability of ditch drainage to reduce soil salinity was only moderate [40].

In general, physical measures require lots of different agricultural machinery and are usually used to intervene in the process of water and salt transport, including soil water infiltration, surface runoff, water–heat exchange between air and soil, etc. Physical measures are the most important methods applied in saline–alkali soil reclamation, which has been widely used, and it is one of the important ways to improve saline–alkali soil health through ameliorating the soil structure, infiltration rate, water holding capacity, etc. Different areas with soil salinization call for different physical measures, one or two measures as the main approach, supplemented by supporting measures, such as water-saving irrigation combined with tillage, film mulch plus buried layer, and or deep tillage.

2.2. Chemical Measures

As shown in Figure 2, there are many soluble ions in saline–alkali soil, and they can be divided into two categories according to the characteristics of salt ions. One is saline soil, which consists of Na^+, K^+, Cl^-, SO_4^{2-}, etc., and the other mainly contains Na^+, HCO_3^-, CO_3^{2-}, etc., and is defined as alkali soil with a high pH value. Soil conditioners are used to improve ion exchange, acid–base neutralization, and ionic balance. These are referred to as chemical measures [41]. The balance of the K^+/Na^+ ratio in soil plays an important role in maintaining crop growth [42]. The most commonly used soil conditioners mainly include calcium-containing compounds such as gypsum, which works more efficiently in alkali soil, helping to decrease pH [43], as well as acidic materials, such as potassium dihydrogen phosphate, ferrous sulfate, and aluminum sulfate, and organic acids such as humic acid and furfural residue. In the saline–sodic soil of China's Songnen Plain, flue gas desulfurization gypsum has been widely used due to its effects on pH, electrical conductivity, and the reduction in exchangeable sodium percentage [44]. The dissolved Ca^{2+} after gypsum application can exchange Na^+ in soil colloids, reducing the exchangeable sodium adsorbed by soil colloids [45]. Many studies have confirmed that gypsum application can improve soil yield [46–48]; however, the excessive use of gypsum can inhibit plant growth and reduce yield [49].

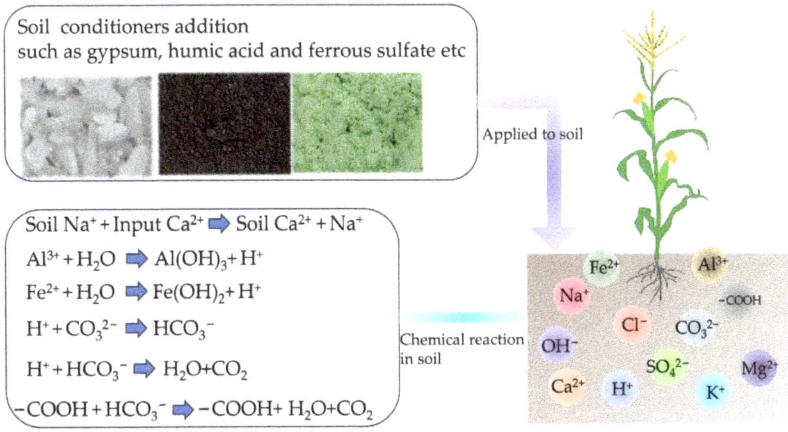

Figure 2. The chemical process in the saline–alkali soil.

Acidic materials are used to improve the neutralization reaction; they mainly affect HCO_3^-, then Na^+ and Cl^-, and the pH is reduced by 10–20% [50,51]. A substantial phosphate precipitate was produced in alkaline soil [52], but the addition of acid phosphate suppressed alkaline stress through the neutralization reaction. Humic acid fertilizer may also have an impact on bacterial and fungal community structures, particularly at the harvest stage when soil nutrient availability and root nutrient absorption are enhanced in saline soil. [53]. The application of humic acid amendments improved the yield and quality of sugar beets in saline–alkali soil, with the yield and sugar production of sugar beets increasing by 11.29% to 32.54% and 13.50% to 38.61%, respectively [54].

Biochar is one of the most used soil conditioners in saline–alkali soil. Quite apart from improving soil structure, biochar addition can replace excess exchangeable sodium in saline–alkali soil by increasing soil organic carbon and cations, thereby reducing the soil's electrical conductivity and salt content [55]. However, biochar with a high pH can increase soil pH [56]. Three different pH levels of corn straw biochar (pH = 8.01), wheat straw biochar (pH = 6.93), and peanut shell biochar (pH = 7.71) were applied in saline–alkali soil. The application of wheat straw biochar with the lowest pH significantly reduced saline soil pH, confirming that the pH difference between biochar and saline–alkali soil may be the main reason for soil pH changes [57]. Furthermore, the carboxyl and other functional groups of biochar alleviate saline–alkali stress on saline–alkali plants [56]. Additionally, biochar altered the C:N ratio in saline–alkali soil [58,59]. The addition of peanut shell biochar in coastal saline–alkali soil reduced the absorption of Na^+ by *Kosteletzkya virginica*, enhanced the absorption of K^+, and improved potassium use efficiency [60]. Soil K^+ content increased by 34.1% and 70.2% when 2% and 2.5% woody biochar was applied to saline–alkali soil, respectively [61]. Biochar also affects physical processes, and a low biochar application rate reduces soil water evaporation, but not at a high biochar application rate. A high application rate clearly demonstrates an increase in the soil's ability to store water while decreasing the surface soil's salinity and sodium adsorption ratio [62].

In addition, some polymer materials are also applied to the improvement and utilization of saline–alkali soil. Polyacrylamide ameliorates saline soil structure, especially the macroaggregate (>0.25 mm) content in coastal areas, improves soil hydraulic properties, and alleviates salt stress [63]. Overall, chemical measures have an obvious effect on soil improvement over a short time, but chemical amendments examine a wider range of potential problems for secondary pollution, and a lifecycle assessment is recommended to evaluate their safety, economy, and long-term effectiveness [41]. A pot experiment confirmed that applying chemical amendments improved soil health, and actually, the combined application of chemical and organic materials was better at improving saline–alkali soil health [64].

Above all, chemical measures aim to change the chemical processes in saline–alkali soil, and chemical amendments would improve ion content and reduce pH, but the environmental effects and influence characteristics should be evaluated and quantified because the exogenous materials may pollute soil, with these affected by the addition of compounds to saline–alkali soil. The implementation of chemical measures depends on the main chemical and nutrient characteristics in the soil, such as calcium-containing compounds usually applied to alkali soil, humic acid, and furfural residue are widely used in saline–alkali soil, which suffers from low soil organic matter.

2.3. Biological Measures

Biological measures refer to salt tolerance improvement and adaptive planting in saline–alkali soil. Root growth and root exudation improve soil physicochemical properties, increase the dry matter accumulation of plants, and remove salt from the soil through crop harvesting. There are three parts of biological measures: plant salt tolerance, soil fertility improvement via plant growth, and desalination via plant harvesting [60]. There are many salt-tolerant plants in the ecological environment, such as *Spartina anglica*, *Phragmites australis*, *Suaeda salsa*, *Salicornia europaea*, *Sesbania cannabina*, *Tamarix chinensis*, *Imperata cylindrica*, *Pennisetum alopecuroides*, *Setaria viridis* and *Cynodon dactylon* [41,65].

These salt-tolerant plants have been used as pioneer plants for the reclamation of saline–alkali soil, especially in heavily saline–alkali soil. In fact, *Tamarix*, *Siberian* white thorn, and sand jujube can reduce soil salinities, pH, and bulk densities. Their root exudates increase soil micro-organisms and enhance the activities of soil cellulase, urease, and dehydrogenase [66]. Planting salt-tolerant trees improves the physical and chemical properties of the soil to varying degrees. The results showed that, after planting, the electrical conductivity was reduced by 70–80%, the content of soil microaggregates (0.25–0.053 mm) was improved by 5.0–5.9%, the bare soil particle size was less than 0.053 mm, and electrical conductivity was 1.65 dS/m [67]. The dry salt discharge and plant salt accumulation theories were proposed to alleviate the stress of saline–alkali soil. This involves retaining low-lying areas near saline–alkali farmland for excess irrigation water and high-salinity groundwater accumulation, allowing the salt to accumulate in low-lying areas or be absorbed by plants [68,69]. In dry salt discharge systems, halophytes promote soil water and salt transport through transpiration, dispersing soil salt in open spaces and effectively improving the efficiency of salt accumulation in saline–alkali wasteland [70,71]. The results showed that *Salicornia* europaea and *Salicornia* salsa can remove 4.7 Mg/ha and 5.2 Mg/ha of salt from heavy saline soil, respectively. The accumulation efficiency of *Salicornia europaea* plants for Na^+ and Cl^- is 2.2 to 2.3 times that of *Salicornia salsa* [72]. Plant salt accumulation contributed to the biological adaptability improvement of saline soil, but further observations of long-term effectiveness are needed due to the frequent salt exchange between soil and groundwater in saline soil areas.

A review concluded that saline–alkali soil mostly inhibited plant growth and development through pH and ionic osmosis [73]. Alkaline pH stress limited root growth by reducing the ethylene and auxin content [74]. Salt-tolerant plants' root and/or leaf cells have the ability to regulate osmotic pressure; some plants' root exudation could isolate the plants from the salt stress [42], and the metabolites, such as sucrose, amino acids, alkaloids, flavonoids, and carotenoids, could also help to deal with the stress of saline–alkali soil [75]. According to metabolome research, the halophytes *Suaeda salsa* and *Salicornia europaea* have significant quantities of branched-chain amino acids, which may help them withstand high saline–alkali stress [76]. In fact, the transgenic technique has already been applied to improve the salt tolerance of plants; for instance, the OsLOL5 gene increases transgenic rice's tolerance to saline–alkali soil by upregulating OsAPX2, OsCAT, Os-Cu/Zn-SOD, OsRGRC2, and ROS detoxification [77]. The decreasing cost and increasing efficiency of genomics have revolutionized the understanding of the mechanism of biological processes and brought new possibilities and options to the salt tolerance of plants and high-yielding crops [42]. The *P. tenuiflora* genome improves the salinity and drought tolerance of cereals [78]. The manufacture of secondary metabolites, the hormone signal transduction system, and antioxidant enzymes may play a role in the tolerance of saline–alkaline soil, which is associated with differentially expressed genes [79].

Soil micro-organisms are the main participants in soil nutrient cycling; moreover, bacteria are one of the most important micro-organisms in saline soil, which affects soil health [80] because the micro-organisms contribute to soil ecological function stability. The reclamation of saline–alkali soil using micro-organisms has been confirmed as a useful method [81]. In the fertilizer market, microbial fertilizers are mainly divided into three categories: agricultural microbial agents, composite microbial fertilizers, and bio-organic fertilizers [81]. Agricultural microbial agents are fertilizer complexes with special species and porous substances such as peat [82]. Composite microbial fertilizers mainly consist of two or more effective bacterial strains [83]. Bio-organic fertilizers refer to the fermentation products of effective bacterial strains and decomposed organic matter, such as manure from poultry and livestock. Microbial fertilizers significantly improve the excretion of Na^+ and uptake of K^+ by plants, thereby increasing the K^+/Na^+ ratio in soil and improving plant nutrient absorption [84]. For instance, microbial fertilizers secrete extracellular polysaccharides on the surface of plant roots, which form bacterial biofilms that protect the roots from sodium chloride stress [85]. Volatile organic compounds released by microbial fertilizers cause a number of physiological changes in plants and increase their resistance to salt and alkali [86]. Moreover, inoculation of indigenous microalgae has been recommended as an eco-friendly and sustainable method [87].

In conclusion, biological measures contribute to salt removal, plant growth, and the integrated effects of micro-organisms, which are closely related to soil ecosystems, and it is important for keeping and/or improving saline–alkali soil health, but there are limited mature and promotable biological measures, for they are governed by biological adaptability and environmental friendliness; moreover, the efficiency of biological measures requires co-operation with other measures.

2.4. Advantages and Disadvantages of Different Measures

Based on the above content, we have summarized the advantages and disadvantages of different measures. As shown in Table 1, physical measures are widely used in various areas, and their principles and methods are simple and mature; however, they have highly relied on agricultural machinery, so their promotion and application are still limited. Chemical measures are more effective, especially in moderately and severely saline soils with poor soil structure and nutrient depletion; however, there are problems such as having only a single effect, short duration, and their use may result in secondary pollution. Theoretically, biological measures have been recommended as eco-friendly and sustainable methods as the plants and soil micro-organisms are important parts of the ecosystem; however, different species of salt-tolerant plants and/or micro-organisms have different climatic and environmental requirements, and inappropriate operation may reduce their improvement effect, meaning their efficiency requires co-operation with other measures. In general, film mulch and irrigation widely used in the arid and semi-arid areas of central and western China; gypsum added in alkali soil of northeast China; subsurface pipe drainage applied in the eastern coastal saline–alkali soil, which is determined by their saline–alkali characteristics (Table S1).

Table 1. The advantages and disadvantages of different measures.

Measures	Advantages	Disadvantages
Physical measures	Strong operability, simple method, Strong applicability, etc.	Water source demand, large engineering quantity, costly, etc.
Chemical measures	Material variety, producing effects quickly, etc.	Short duration, secondary pollution, etc.
Biological measures	Eco-friendly, sustainable, etc.	Time-consuming, regional specificity, etc.

3. Soil Water and Salt Changes in Saline–Alkali Soil

Water and salt transport in saline–alkali soil is an essential issue during improvement and utilization (Figure 3), and soil water and salt are fundamental indicators of soil health. However, high soil water and salt stress deteriorate soil quality and inhibit the growth of plants and soil biological properties [88,89], so the regulation of soil water and salt is included in the management of soil health. Based on the decision regarding water and salt database management, soil water and salt monitoring and simulation proceed from the initial data analyses to any final decision-making. There is a common Chinese saying: "Salt comes with water, and salt goes with water", which indicates that water is the driving force of salt. Numerous elements, including tillage, soil conditioner, irrigation, soil texture, groundwater, and climate, all have an impact on it. Ion exchange between soil particles and soil solution, diffusion in soil solution, and movement with bound water and pore water in soil are the four methods through which salt is transported with water in soil [90]. The potential energy theory is applied to study water and solute transport problems in saturated and unsaturated soils. Based on the water–salt balance theory, a simple soil–water–salt dynamic prediction model is proposed, which is characterized by avoiding the details of unsaturated water movement as much as possible when studying the salt movement process in unsaturated zones [91]. Furthermore, saline–alkali soil reclamation involves the control and management of water and salt movement [90]. Hence, the central issue of saline–alkali soil is soil water and salt transport under various factors. Below, we summarize some of the aspects involved in soil water and salt transport, such as models, irrigation, buried layers, freezing, and thawing, which are highly discussed in China.

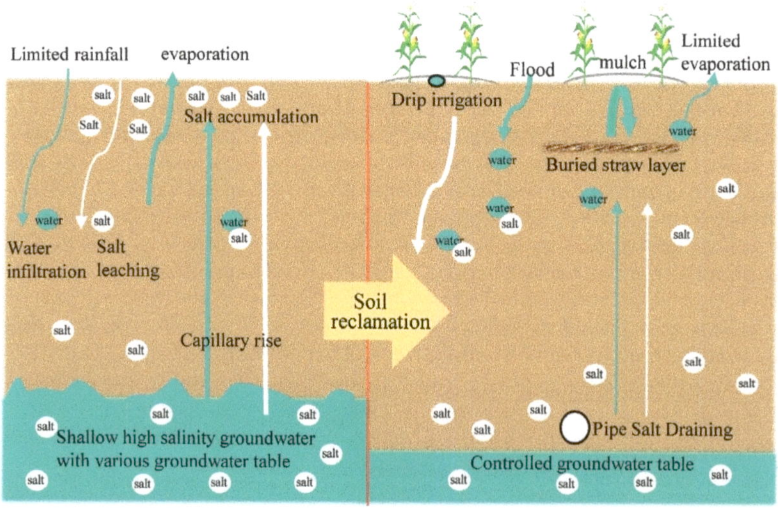

Figure 3. Aspects of soil water and salt transport in saline–alkali soil.

3.1. Soil Water and Salt Simulation and Monitoring

In order to gain a better understanding of the continuous changes in water and salt, model simulation is an effective method, such as HYDRUS, SWAT (Soil and Water Assessment Tool), DRAINMOD (A Hydrological Model for Poorly Drained Soils), SHAW (Simultaneous Heat and Water Model), and COMSOL (COMSOL Multiphysics). HYDRUS is widely used in the study of saline–alkali soil; it can effectively simulate the transformation of water, multi-component solutes, and heat in both saturated and unsaturated zones [92], such as coastal saline soil under film mulch and buried layers [93], the unsaturated root zone of raised land [94], saline water irrigation under subsurface drainage conditions [95], root water uptake calculations [96], and subsurface pipe drainage in reclaiming coastal areas [97]. There is much research on soil solute transport in agricultural systems, while soil solute transport in other ecosystems is limited. Moreover, the root uptake of soil water and nutrients is a key factor affecting the transport of solutes in saline–alkali soil, and the coupling relationship between crop growth and the absorption of water, nutrients (solutes), and solute transport by crop roots are rarely considered. Moreover, almost all models need lots of measured soil parameters, which determine the accuracy of the model, and the accuracy and effectiveness of model applications vary depending on the characteristics of climate, hydrological conditions, and topography. In conclusion, improving simulation accuracy based on a coupling relationship is an urgent problem that needs to be solved, especially in the theory of water and solute transport in the future.

Remote sensing and near-earth sensing technology are important methods for the multi-element, multi-scale, and integrated monitoring of soil salinization [41]. Examples include aviation/satellite optical/microwave remote sensing images [98], magnetic earth conductivity meters [99], ground penetrating radar [100], time-domain reflectometers (TDR), and frequency-domain reflectometers (FDRs) [101,102]. For instance, a remote sensing monitoring model analyzed the spatial dynamics and main factors of soil salinization in the agricultural areas of northern Xinjiang [103]. The combination of the SWAP (Soil–Water–Atmosphere–Plant), GIS (Geographic Information System), and RS (Remote Sensing) models was used to explain the distribution of soil water and salt spatiotemporal evolution at the regional scale under water-saving irrigation conditions [104]. A linear regression model was constructed between magnetic- induced earth conductivity and soil salinity, quantitatively evaluating the spatiotemporal evolution of soil salinization in the Yangtze River Estuary region over the past 10 years [105]. Landsat imagery and magnetic conductivity meter (EM38) data were combined for the study of the spatial variability of regional soil salinity [106]. A soil salinization monitoring model was constructed using unmanned aerial vehicle multispectral remote sensing and GF-1 satellite remote sensing data, and a further-improved TsHARP scale conversion method was applied to soil salinization monitoring through unmanned aerial vehicle and satellite remote sensing upscaling [107]. Overall, model simulation, remote sensing, and near-earth sensing technologies are important ways to accurately quantify soil water and salt transport in saline–alkali soil, and future research should strengthen the collection of observation data at different scales and construct universal scale conversion functions, especially under spatiotemporal hydrological processes, changing climates, special topographic and geological conditions, etc.

3.2. Soil Water and Salt Dynamics under Irrigation

Irrigation is the driving force of soil water and salt transport in saline–alkali soil, as shown in Figure 3. In northwest China, 225 and 300 mm of winter irrigation are advised in order to desalinate the soil, promote cotton growth, and save water [108]. Film mulching plus irrigation also contribute to salt reduction in the root zone, especially in heavily saline soil [109]. It was observed that throughout the irrigation period, the soil salt distribution was more consistent below the top of the ridge than it was below the furrow. Subsequent research using saline water and mulched furrow irrigation revealed that the soil salt content of the surface soil layer beneath the top of the ridge was lower during the irrigation period than that below the bottom of the furrow [110]. Micro-sprinkler irrigation was confirmed as a potential method for alleviating soil salinization, especially in coastal saline soil in northern China [111]. A meta-analysis indicated that drip irrigation generally decreased soil salt content in the root zone by 37.7% relative to flooding irrigation, and a flow rate of 2–4 L/h was recommended in drip emitters, which could have a positive effect on salt control. Furthermore, the salinity of irrigation water should be lower than 2 dS/m [18].

Saline water irrigation is very important for agriculture in the arid regions of northwestern China [32]. It was demonstrated that saline irrigation with a salinity below 10.6 dS/m can lessen freshwater shortages during a lengthy 15-year period of saline water irrigation [112]. After 2 years of saline water irrigation under a subsurface drainage system, the soil salinity reduced annually, and there was no salt buildup in the topsoil [113]. The ridge planting system was more effective under drip irrigation with saline water when planting small shrubs and herbaceous plants in the coastal saline soils [114]. Nevertheless, a study also discovered that following saline irrigation, soil salinization did not increase (<1.0 dS/m) in the 40–60 cm soil depth, where a large number of lateral roots also germinated and spread horizontally, but it did significantly accumulate in the topsoil (crust and 0–10 cm soil layers) [115]. Regardless of the salinity of the irrigation water, heavily saline soil changed to a weakly salinized or even non-saline profile at 0–1 m [116]. The average soil salt concentration in the 1.0 m profile was 336% and 547% of the initial level after 3 years of irrigation with moderately salted and highly salted water, respectively [117]. Generally, various irrigation methods were applied to soil desalination and soil water improvement in the root zone, especially in the arid and semi-arid regions of China.

3.3. Soil Water and Salt Dynamics under Freezing and Thawing

An essential process in saline–alkaline soil is freezing and thawing. Water migrates towards the freezing front mostly due to temperature gradients in the water [118]. In the course of cooling, salt and water go toward the cold end [119], and the area with a high salt content will melt first, followed by the low-salt-content area. This phenomenon is caused by the different freezing points, resulting in salt leaching in the surface layer [120]. However, there are different results regarding the thawing process and salt reduction. It was suggested that soil salinization can occur because more salt is collected in the top soil during the freezing phase than salt leached during the warming period [121]. Particularly in coastal saline soil, the initial water content and bulk density of the soil can also have an impact on the water infiltration and desalination of melting salty ice water. As the original soil water contents and bulk densities diminish, the top soil's salt content also decreases [122]. It was also found that meltwater from saline ice can contribute to the successive infiltration of water with different salinity, resulting in the desalination of coastal saline soil [123]. In the topsoil layers, the soil water content in the upper layer is greatly under the effect of salt-free ice compared to saline ice, but at the deeper soil layers, the situation is reversed [124]. Furthermore, the infiltration rate is faster with saline ice meltwater compared to salt-free ice meltwater when infiltrating into saline soils, and this effect increases with ice salinity level and decreases with ice sodium adsorption ratio [125]. An essential component of soil water and salt transfer occurs when ice and salt undergo phase transitions in saline soils. The findings demonstrated that salt-free soils have less liquid water content and a higher concentration of salt solution as a result of the phase transition between ice and water [126].

3.4. Soil Water and Salt Dynamics under Buried Layers

Previous research has mainly focused on the continuous soil layer, and salt accumulation in the surface layer is the main reason for salt stress, which is caused by capillary action. In order to address this problem, a buried layer is an effective method that has been used in recent years to separate the surface and subsoil, which breaks the continuity of capillary movement and changes soil water and salt migration between the upper and lower layers of soil [36]. The buried layer mainly aims to prevent soil water evaporation and salt accumulation [127]. There are many materials that are used as buried layers, such as straw, wood fiber, biochar, and peat, and straw is the most commonly used buried material. Different buried materials, amounts, and depths contribute to the different characteristics of water and salt transport. A buried straw layer can improve soil water storage in the topsoil by retarding the infiltration process [35] due to the overburden weight compaction effect caused by the buried layer [128,129]. Both a buried straw layer and a buried straw layer plus film mulch could significantly decrease the salt salinity of the top 0–40 cm, especially at the sowing stage of sunflower, and a buried straw layer plus film mulch effectively reduced salt content throughout the growth period [35]. They discovered that adding a layer of straw and gypsum from flue gas desulfurization decreased the salinity and alkalinity of the soil [130]. In comparison to the other methods, the combination of straw layer burial (6.0 t/ha) and surface mulch (3.0 t/ha) was demonstrated to be an effective method of returning straw [131]. A buried wood fiber layer plus plastic film mulch was also applied in coastal saline soil, which decreased water stress and increased efficiency in water utilization throughout the growth season, controlling soil salt content to below 2 g/kg [93]. A buried peat layer reduced the infiltration rate by 68.3% compared with the control, and the buried peat layer increased by 11.9% the 0–20 cm soil water content at the end of the infiltration stage [132]. Buried depth is another factor that affects soil salt and water. For buried layer depths of 30 cm and 50 cm, researchers discovered that when the buried layer was 30 cm deep as opposed to 50 cm deep, the amount of soil water in the 0–30 cm depth decreased more quickly. The wood fiber and biochar layers may both prevent surface soil from becoming too salted [36]. A buried layer is a proper method for soil salt reduction; however, the buried layer limits water's upward movement, and it has difficulties in its operations. Finding effective ways to achieve salt inhibition is one of the topics for the future.

4. Nutrients in Saline–Alkali Soil

Nutrients are important indicators of soil health, including the chemical and biological aspects of nutrient holding capacity, chemical availability, the C:N ratio, and the organic matter in the biological aspect, which are defined by FAO [2]. Saline–alkali soil improvement and utilization are limited by poor nutrients. Considerable research has been conducted on soil fertility in salinized land. Soil organic carbon, nitrogen, and phosphorus are important nutrient components in soil. However, studies have shown that soil organic carbon content in the most saline soil is less than 1% [133]. The nitrogen utilization efficiency of urea in saline farmland is 14 to 29%, while the phosphorus utilization efficiency is 10 to 25%, which is lower than that of conventional farmland [134]. Generally, for saline–alkali soil as one of the main types of low-to-medium-yield farmland in China, soil fertility and/or nutrient storage capacity improvement is of great significance for grain yield increase, carbon sequestration, and increasing soil organic matter. However, the primary factors limiting agricultural productivity are soil salinity and alkalinity [135].

Improving soil fertility and nutrient storage capacity is of great significance to increasing grain production and reducing carbon emissions, especially in saline–alkali soil, as it is severely lacking in nutrients. We find that advances in soil salinity, alkalinity, and nutrients will soon follow, especially regarding the inhibitory mechanism of salinization on soil nutrient storage, the mutual feedback mechanism between soil structure adjustment and nutrient storage capacity expansion, the carbon and nitrogen stabilization mechanism of agricultural and livestock waste resource utilization, and the principle of organic carbon regulation and carbon sink capacity enhancement in saline farmland soil. We also notice that limited studies explain whether to reduce salt first or increase soil nutrients first; theoretically, soil nutrients could be improved after salinity reduction, but there are many studies that confirmed that exogenous organic matter addition contributed to salinity reduction. In this study, we think there is a positive interaction between salinity reduction and soil nutrient improvement when one of them first comes into play, along with the other.

4.1. Soil Organic Carbon and Its Improvement in Saline–Alkali Soil

The soil organic carbon (SOC) content in most saline soils is lower than 1% due to high soil salinity [133]. Soil salinization significantly reduces soil organic carbon and microbial biomass carbon by 20.6% and 36.5%, respectively [136]. The mean soil organic carbon density of natural saline–alkaline wetlands is generally lower than that in other wetlands in China [137]. It has been confirmed that saline–alkali soil amelioration can sequester more carbon [45], and micro-organisms and rhizosphere enzyme activity are what propel the conversion of organic and inorganic carbon in saline–alkaline soil [138].

In saline–alkaline soil, fertilization techniques are useful for controlling soil organic carbon [139]. Combining chemical fertilizer with sheep dung, corn straw, fodder grass, and granular corn straw increases the amount of organic waste applied to the soil, particularly in the free light fraction and organic carbon in the occluded fraction [140]. While adding nitrogen boosts aboveground biomass and encourages plant development, it has little effect on soil organic carbon stores [141]. Under organic supplements, such as humic acid plus organic amendments and biofertilizer plus organic amendments, the soil organic carbon and total C content of the 0–40 cm soil layer increase by 9–40% when compared to chemical fertilizer [135]. Furthermore, researchers concluded that soil pH was a decisive factor over the soil organic carbon in saline–alkali grassland [141]. Sludge-based vermicompost and other organically modified soils were achieved by reducing pH and salinity and increasing soil organic carbon content [140]. When there is enough carbon supplied, fungal diversity decreases, resulting in limited CO_2 emission and ensuring the input of carbon [142]. In the salinization areas of Inner Mongolia, China, the application of conditioners, such as those containing marlstone and enzymes, increased the dissolved organic carbon and fractionated organic carbon content, as well as the number of aggregates of size >0.25 mm, when compared to soil planted with Jerusalem artichoke alone [143]. Deep tillage reduced soil organic carbon accumulation, but the addition of vermicompost compensated for this reduction, significantly increasing soil organic carbon content in the saline–alkali soils [144]. Using biochar is an important way to increase soil carbon. Mineral-associated organic carbon and soil organic carbon increased with the duration of biochar application, but the particulate organic carbon content did not [145]. Rice planting contributed to an increase in soil organic matter by enriching soil microbiome diversity [146]. Planting duration also improved soil organic carbon content. It was confirmed that the carbon pool management index and carbon pool index increased with the increase in rice planting duration in the saline–alkali paddy fields in western Jilin, China [147]. However, not all soil amendments could improve soil organic carbon. Gypsum addition significantly reduced dissolved organic carbon by about 36–47%, whereas gypsum and biochar amendments could enhance the stability of soil organic matter in saline–alkaline paddies [43].

There are other factors that can further influence soil organic carbon in saline–alkali soil, such as irrigation, freezing and thawing, mulching, and the number of reclamation years. While salty water irrigation at low concentrations had no effect on soil carbon sequestration, high concentrations of saline water were detrimental to soil carbon sequestration [148]. The features of the saline–alkali soil in wetlands and the distribution of carbon may be impacted by freeze–thaw cycles. With repeated cycles, the amounts of water-soluble organic carbon, microbial biomass carbon, and rapidly oxidized organic carbon increased, decreased, and then eventually achieved a steady state [149]. Mulching with polyethylene decreased soil organic carbon by 16% and 6%, respectively, with and without organic amendment [150]. Mulching needs to be carried out in conjunction with subsurface organic amendment. Although reclamation time did not increase the amount of carbon in the soil, it did convert particle organic matter into organic matter linked with minerals [151].

4.2. Soil Nitrogen in Saline–Alkali Soil

Nitrogen is important in saline–alkali soil. Saline–alkali soil yield and economic benefits are highly related to N application and the soil salt content. When soil salinity is more than 3.5 g/kg, planting is not advised; however, N treatment might boost crop production and economic advantages when soil salinity is between 1.8 and 2.9 g/kg. When the soil salinity is between 2.9 and 3.5 g/kg, the net economic benefit is negative [152]. The nitrogen use efficiency of cotton was 26.1–47.2%, and the residual N in the soil was 7.7–14.9%, which was contributed via ^{15}N-labeled fertilizer addition [153]. The soil nitrogen content in saline–alkali soil is influenced by multiple factors, and salinity, soil amendments, and irrigation are concluded below.

Soil salt mainly affects the migration and transformation process of nitrogen, such as nitrogen leaching and ammonia volatilization, and it restrains the root uptake of nitrogen [154]. Nitrate leaching occurs when the soil matric potential thresholds are higher than −20 kPa; excessive N fertilizer contributes to nitrate leaching [155]. Soil salinity promoted NH_3 volatilization by about 19.50–22.78% of the N input, and nitrate leaching was 32.89–43.84% in the saline soil of the Yellow River Delta, as calculated using Hydrus-2D [156]. One study found that N loss in the top soil is highly determined by anammox and linked to denitrification in the deep layer, especially in coastal saline–alkali soil. Furthermore, anammox took up 40–87.5% of total N within 0–50 cm soil, and it declined at lower than 50 cm [157]. A field experiment showed that the cumulative emissions of N_2O increased with the increase in soil salt content in the Hetao Irrigation District of Inner Mongolia, China [158]. The transformation of soil nitrogen is mostly fueled by soil microorganisms, and there is a considerable positive correlation between urease activity and soil nitrogen [159]. It was found that reducing nitrogen fertilizer application could retain more NH_4^+-N, contributing to limited nitrate leaching, which is very important in reducing environmental pollution [160].

Different soil amendments also affect nitrogen migration and transformation. Biochar application decreased denitrification rates by 23.63–39.60%; polyacrylamide reduced denitrification and ammonia volatilization rates by 9.87–29.08% and 11.39–19.42%, respectively, whereas the ammonia volatilization rates rose by 9.82–25.58% [161]. Biochar considerably raised the NH_4^+-N contents by 80.08% in saline–alkaline soil and encouraged the conversion of NH_4^+-N into NO_3^--N [162]. Biochar could reduce NH_4^+-N leaching by improving crop nutrient absorption and utilization [163]. Moreover, there were increases in total nitrogen (9–198%) and available nitrogen (12–49%) after biochar application [164]. Vermicompost and humic acid fertilizer showed a trend of reducing N_2O emissions in saline–alkali soil, which contributed to increasing *nosZ* gene copy numbers, especially in the macroaggregate microbial community [165]. Nitrogen and straw addition significantly reduced the soil salt content and increased the nutrient contents [166]. Legumes are suitable for saline soil reclamation, as they can increase the soil nutrient content due to nitrogen-fixing bacteria improvements [167]. Applying humic acid and gypsum in tandem at a weight-based ratio of 1:3 did not result in an increase in nitrogen leaching [168]. Earthworms and arbuscular mycorrhizal fungal hyphae contributed to a reduction in soil salinity and an increase in NO_3^--N, NH_4^+-N, microbial biomass nitrogen, and the abundance of *nifH* and AOB *amoA* genes [169].

Irrigation affects soil mineralization rate. The organic nitrogen mineralization rate was lowered by using brackish water irrigation; the nitrogen mineralization rate and net nitrogen mineralization rate were lowered by 41.07% and 11.62%, respectively [170]. A 10-year field experiment concluded that long-term saline water irrigation significantly increased soil NH_4^+-N concentration and inhibited N_2O emission, and it decreased soil NO_3^--N and total nitrogen content due to the effect of saline water irrigation on the denitrifying bacteria community [171].

4.3. Soil Phosphorus

Phosphorus plays a crucial role in plant nutrition and soil fertility and is a key element in photosynthesis and respiration, as well as nucleic acid and membrane synthesis [172,173]. The deficiency of available phosphorus limits plant production and the development of agriculture [174,175], especially in saline–alkali soil with a high pH and low phosphorus solubility, where plants suffer from salt stress and phosphorus deficiency [176].

Saline–alkali soil affects the effectiveness of phosphorus in two ways. Firstly, soil salt reduces the effectiveness of phosphorus by enhancing the adsorption of phosphorus, and this varies depending on the soil texture [177]. There is a chemical reaction between Ca^{2+} or Mg^{2+} and phosphate ions, and they then combine with hydroxyl and oxy groups to change into hydroxyphosphates and fluorophosphates that are not available to plants. Secondly, salt ions affect the absorption of phosphorus via the cell membrane, especially Cl^- [178]. Salt ions also restrict the mineralization and decomposition of organic phosphorus and reduce the effectiveness of organic phosphorus [179]. The exogenous addition of phosphorus is a promising strategy to improve plant salt tolerance. After phosphorus application, the plant K^+/Na^+ ratio in the leaves, stems, and roots of two alfalfa varieties significantly increased, which indicated that phosphorus fertilizer improved the salt tolerance of the plants [180]. However, it was also found that whether phosphorus addition can enhance plant salt tolerance depends on the water use efficiency and salt content [181]. Therefore, the comprehensive effects of salt stress and phosphorus deficiency will ultimately determine the plant's salt tolerance.

Strategies for improving the effectiveness of phosphorus are of great importance in saline soil, and they can also be divided into three methods, according to Section 2. Cultivation measures can improve soil physical structure, reduce the toxicity of salt to plant roots, and promote plant phosphorus absorption. Moreover, physical measures strengthen phosphorus cycling by improving the soil structure of saline soil, enhancing ventilation and permeability, and increasing plant species diversity [182]. Soil amendments also promote phosphorus availability. Phosphogypsum could reduce the pH of the surrounding soil, which is beneficial for the dissolution and release of inorganic phosphorus [183]. The application of mineral phosphorus and organic fertilizer not only significantly increases the available phosphorus content but also improves the yield [184]. However, the chemical fertilizer of calcium superphosphate with organic fertilizer and amendment ($CaSO_4$) did not improve the effective phosphorus content in moderately saline soil due to the formation of calcium phosphate [185]. Reducing urea addition also improves available phosphorus content [186] and improves alkaline phosphatase activity in the 0–10 cm soil layer [187] and reducing 30% of the nitrogen content via fulvic acid addition improves the activation and cycling of phosphorus by reducing Ca_2-P and Ca_8-P [188]. Biochar addition also increases the effective phosphorus content by 40.72–84.8% compared with conventional phosphorus application [189]. The inoculation of marine phosphate-solubilizing bacteria increased soil available phosphorus content by 12.5%, and it increased by 61.2% when marine phosphate-solubilizing bacteria were combined with organic amendments [190]. Phosphorus leaching was reduced, and phosphorus availability increased via the combined use of phosphorus-accumulating and phosphorus-solubilizing bacteria [191]. *Paenibacillus sp. C1*, a saline–alkaline-tolerant bacterial strain, was found to have good characteristics in terms of acid production and phosphorus dissolution, especially in high salinity and alkalinity soil [192].

In general, the application of phosphorus fertilizer is an effective way to improve soil phosphorus availability and crop yield, but the application amount in saline–alkali soil should be adjusted according to the soil's physical and chemical properties and any amendment.

5. Yield Improvement in Saline–Alkali Soil

The ultimate goal of improving and utilizing saline–alkali soil is increased production, and different improvement measures and regions have shown varying results. We summarized the results related to yield under different improvement measures in three main saline–alkali regions; it was clearly shown that crop yields in saline–alkali soil increased under effective improvement measures. Many studies have confirmed that a high yield is contributed via the combined effect of multiple measures, such as mulching, irrigation, tillage, and soil amendments.

Many studies have confirmed that yield is improved by mulch plus irrigation; the details are shown in Table 2. Mulch plus irrigation was the main reason for yield improvement due to salt leaching under irrigation; mulch reduced soil evaporation, which could reduce salt stress and improve soil water content, thus ensuring crop growth. It was found that salt stress limited cotton yield by 5.2% for every unit increase in soil EC when EC was above 7.7 dS/m [193]. Brackish water and/or saline water have been applied in saline–alkali soil due to the limiting fresh water; although the yield was improved by using these methods, there is a high risk of mineralization in groundwater and secondary salinization. Obtaining a sufficient supply of fresh water is an urgent problem in this area; maybe economic brackish water and/or saline water desalination is the proper method. Moreover, increasing soil organic matter and reducing salt accumulation by using cattle manure [194] and a buried wood layer [93] could also contribute to yield improvement.

Table 2. Yield under mulch combined with other measures.

Measures	Yield	Location	Crop	Reference
Brackish water irrigation	6.0–8.0 Mg/ha	37°31′ N, 116°30′ E	Winter wheat	[195]
Brackish water irrigation	3.0–9.0 Mg/ha	37°31′ N, 116°30′ E	Summer maize	[195]
Brackish water irrigation	11.7–15.5 g/plant	41°35′ N, 86°10′ E	Lint yield	[196]
Drip irrigation	5100–6200 kg/ha	40°53′ N, 86°56′ E	Seed cotton	[197]
Drip irrigation with saline water	1250–3100 kg/ha	44°19′ N, 85°59′ E	Seed cotton	[198]
Cattle manure	5124–6197 kg/ha	38°46′ N, 117°13′ E	Maize	[194]
Wood fibre layer	1000–1232.7 kg/ha	37°45′ N, 118°59′ E	Seed cotton	[93]

Table 3 shows yield improvement under tillage plus mulch or irrigation; it indicates that tillage methods, such as traditional tillage, deep tillage, and smash ridge tillage, represent an important way to affect crop yield in saline–alkali soil. These methods are used to loosen soil; deep tillage and smash ridge tillage can break the plow bottom layer, which contributes to plant root development and improves salt leaching. Tillage is indispensable in saline–alkali soil. Due to the uneven distribution of salt ions and rough ground, no-tillage recommended by some studies cannot be adopted in saline–alkali soil reclamation. The details show that traditional tillage with straw mulching could ultimately increase grain yield by 11.3% when compared to no-tillage or mulch [199]. Deep tillage with straw mulch could also contribute more to yield improvement than deep tillage with no mulch [200]. Moreover, it was found that crop yield increases with an increase in smash ridge tillage depth, and when compared to conventional tillage, the yield rose dramatically from 65.24% to 84.14% when using smashing ridge tillage at 60 cm [201]. Ridge planting *Lycium barbarum* L. with drip irrigation in saline soil guarantees a yield close to the local farmland (~900 kg/ha), which was applied in the saline–sodic wasteland of Ningxia [202]. Breaking the plow bottom using a subsoiling technique in alkali soil can effectively reduce salt content and increase crop yield [203].

Table 3. Yield under tillage combined with other measures.

Measures	Yield	Location	Crop	Reference
Traditional tillage and mulch	4655–5331 kg/ha	36°46′ N, 117°13′ E	Maize	[199]
Deep tillage	10,168–12,288 kg/ha (shoot biomass)	41°04′ N, 108°00′ E	Sunflower	[200]
Conventional tillage	2700–5100 kg/ha	79°25′ N, 40°01′ E	Seed cotton	[201]
Tillage with irrigation amount	45–908 kg/ha	38°47′ N, 106°20′ E	*L. barbarum* L.	[202]

Soil amendment has been widely used in saline–alkali soil, which also contributes to yield improvement. This mainly improves soil structure, increases soil organic matter, and exchanges water-soluble ions. Biochar has been widely used in saline–alkali soil, but the addition of different amounts and types of biochar may result in different yields (Table 4). As a porous material, biochar can enhance the porosity and water-holding capacity of saline–alkali soil and reduce soil bulk density due to its large surface area and pore structure. The inorganic components contained in biochar, such as Ca^{2+} and Mg^{2+}, could reduce the content and relative proportion of soluble Na^{+}, thus reducing soil EC and improving yield. The organic carbon and nutrients of biochar can also increase the soil's organic carbon and nutrient content in saline–alkali soil. Gypsum is usually used in alkali soil for a high yield (Table 4). Desulfurized gypsum has been suggested as a useful soil amendment, particularly in northeast China, west of the Songnen Plain [204]. The dissolution of desulfurized gypsum produces Ca^{2+}, which displaces exchangeable Na^{+} in the soil, thereby reducing the pH value and soil alkalinity and improving soil physicochemical properties. Desulfurized gypsum can enhance the ion adsorption capacity of soil. Chemical polymer materials, such as polyacrylamide, could improve the formation of soil aggregates and enhance soil water retention capacity, ultimately increasing crop yield (Table 4).

Table 4. Yield of saline–alkali soil under soil amendments combined with other measures.

Measures	Yield	Location	Crop	Reference
Biochar	19–35 t/ha (aboveground biomass)	33°33′ N, 120°22′ E	Wheat	[205]
Biochar	20–39 t/ha (aboveground biomass)	33°33′ N, 120°22′ E	Maize	[205]
Corn straw biochar	5.0–7.8 t/ha	44°50′ N, 123°35′ E	Maize	[206]
Biochar	11–14 t/ha	46°37′ N, 125°11′ E	Maize	[207]
K-rich biochar	3–3.5 t/ha	37°55′ N, 118°48′ E	Wheat	[208]
K-rich biochar	5.5–7.5 t/ha	37°55′ N, 118°48′ E	Maize	[208]
Gas desulfurization gypsum	776–1428 t/ha	40°15′ N, 110°50′ E	Sunflower	[209]
Polyacrylamide	842–1531 kg/ha	38°19′ N, 117°23′ E	Cotton	[210]

6. Perspectives on the Improvement and Utilization of Saline Alkali Land

Based on the above contents, we also find the following: (1) Technological innovation in this area is in a bottleneck period. The existing technology mainly involves updates and optimizations of previous technologies, such as irrigation, mulching, and subsurface pipe drainage, but this also reflects the reliability and durability of such technologies. (2) The theory and technology of the green reduction of obstacles in saline–alkali land is limited by practical application. More attention is paid to agriculture; however, the ecological value of saline–alkali soil reclamation using ecological restoration methods is underestimated and/or ignored. Ecological restoration methods should be developed alongside agronomic measures. (3) The technology for improving soil fertility and expanding nutrient storage in saline–alkali land is not mature enough, and its long-term effectiveness needs to be strengthened. (4) The development and application of improved and new materials still face a challenge in saline–alkali soil reclamation to achieve increased efficiency, economy, and environmental friendliness.

7. Conclusions

Various soil improvement measures have been developed and applied to saline–alkali soil. These measures are closely related to soil health based on physical, chemical, and biological indicators. Practical measures depend on regional characteristics, such as salinity and alkalinity, soil ion composition, pH, hydrological processes, groundwater levels, soil texture, and regional policies. The technical modes of the three major saline–alkali areas of China are different due to regional differences. However, with the maturity of technology and the spreading of information, compound measures have been widely promoted in different saline–alkali soils.

With the development of saline–alkali soil research, we have also found that nutrient regulation is also very important for saline–alkali soil production. Soil organic carbon enhancement is a key factor in improving soil fertility, and it is confirmed that there is great potential for carbon sequestration in saline–alkali soil. Improvements in soil nitrogen and phosphorus content are highly related to yield, especially in the plow layer. However, the utilization efficiency of nitrogen and phosphorus is lower than that of general farmland soil due to salt and alkali stress. Overall, soil water and salt regulation represent essential steps in improving saline–alkali land and soil nutrient improvement ensures saline–alkali soil production.

Above all, the diverse types, abundant resources, and vast geographical area of saline–alkali soil in China provide unique research conditions for soil scientists and agriculturists. Research is needed to help improve saline–alkali soil health, which is of great importance to soil health in China and in the world. In the future, we look forward to strengthening the following research areas: (1) the efficient and precise control of salt, as well as safe and economical water use theory and technology; (2) reducing the risk of secondary pollution; (3) promoting nutrient storage capacity and improving nutrient efficiency; and (4) comprehensively considering the development prospects of saline–alkali land in agriculture, resources, ecology, and the environment. More work should be conducted on saline–alkali soil improvement and utilization based on the perspective of soil health.

Supplementary Materials: The following supporting information can be downloaded at https://www.mdpi.com/article/10.3390/agriculture14081210/s1, Table S1. The three typical saline-alkali soil and its' characteristics and technical models.

Author Contributions: Conceptualization, W.Z., X.Z., S.G. and R.J.; methodology, W.Z.; data collection and formal analysis, W.Z.; writing—original draft preparation, W.Z.; writing—review and editing, W.Z., S.G., X.Z., R.H. and R.J.; funding acquisition, W.Z., X.Z., S.G. and R.J. All authors have read and agreed to the published version of the manuscript.

Funding: This research was funded by Chuzhou University Research Initiation Fund Project (No. 2023qd44 and 2023qd49); the open foundation of Kunyu Mountain station of national ecological quality comprehensive monitoring center (No. NIES-KYS-202403); the Key Research Project of Natural Science in Colleges and Universities of Anhui Province (No. 2022AH051096); the Young Backbone Teachers to Visit and Study in China (No. JNFX2023061); the Chinese Academy of Sciences (No. NK2022180405).

Institutional Review Board Statement: Not applicable.

Data Availability Statement: All data generated or analyzed during this study are included in this published article.

Conflicts of Interest: The authors declare no conflicts of interest.

References

1. Wang, Z.Q.; Zhu, S.Q.; Yu, R.P. *Salt-Affected Soils of China*; Science Press: Beijing, China, 1993.
2. FAO. Global Soil Health Indicators and Assessment. Available online: http://www.fao.org/soils-portal/soil-degradation-restoration/global-soil-health-indicators-and-assessment/en/ (accessed on 21 February 2016).
3. Wichern, J.; Wichern, F.; Joergensen, R.G. Impact of salinity on soil microbial communities and the decomposition of maize in acidic soils. *Geoderma* **2006**, *137*, 100–108. [CrossRef]
4. FAO. The World Map of Salt Affected Soil [WWW Document]. FOOD Agric. Organ, United Nations. 2021. Available online: https://www.fao.org/global-soil-partnership/gsasmap/en (accessed on 10 December 2010).
5. FAO. FAO Land and Plant Nutrition Management Service. Available online: http://www.fao.org (accessed on 10 December 2010).
6. Hassani, A.; Azapagic, A.; Shokri, N. Global predictions of primary soil salinization under changing climate in the 21st century. *Nat. Commun.* **2021**, *12*, 6663. [CrossRef]
7. Yang, J.S. Development and prospect of the research on salt-affected soils in China. *Acta Pedol. Sin.* **2008**, *45*, 837–845. (In Chinese)
8. Wang, J.L.; Huang, X.J.; Zhong, T.Y.; Chen, Z.G. Review on sustainable utilization of salt-affected land. *Acta Geogr. Sin.* **2011**, *66*, 673–684. (In Chinese)
9. Ministry of Natural Resources, People's Republic of China. Actively promoting the comprehensive transformation and utilization of saline alkali land. *Qiushi* **2023**, *23*, 15–20. (In Chinese)
10. Liao, J.; Wen, X. Dynamic change of saline and alkali land in Hexi Region based on remote sensing. *J. Arid Land Resour. Environ.* **2011**, *25*, 96–101. (In Chinese)

11. Guo, S.S.; Ruan, B.Q.; Guan, X.Y.; Wang, S.L.; Li, Y.P. Analysis on spatial-temporal evolution of soil salinity and its driving factors in Hetao Irrigation district during recent 30 years. *China Rural Water Hydropower* **2016**, *9*, 159–162+167. (In Chinese)
12. Xu, Z.Q.; Xu, X.H. The cause of formation and characteristics of Soda saline–alkaline land of the Songnen plain and the study progress of control Measures. *Soil Water Conserv. China* **2018**, *2*, 54–59+69. (In Chinese)
13. Zhang, H.Q.; Wang, L.X.; Sun, G.Y.; Yang, Y. Evaluation of saline-alkali land resource and development potential in low Songnen Plains. *Chin. J. Agric. Resour. Reg. Plan.* **2013**, *34*, 6–11. (In Chinese)
14. Zhao, Y.; Wang, L.; Zhao, H.L.; Chen, X.B. Research Status and Prospects of Saline-alkali Land Amelioration in the Coastal Region of China. *Chin. Agric. Sci. Bull.* **2022**, *38*, 67–74. (In Chinese)
15. Yao, R.J.; Yang, J.S.; Zhao, X.F.; Han, J.J. Effect of soil salt-water management on Helianthus growth and salinity distribution characteristics of coastal saline soil in north Jiangsu province. *J. Irrig. Drain.* **2013**, *32*, 5–9. (In Chinese)
16. Yao, R.J.; Li, H.Q.; Zhu, W.; Yang, J.S.; Wang, X.P.; Yin, C.Y.; Jing, Y.P.; Chen, Q.; Xie, W.P. Biochar and potassium humate shift the migration, transformation and redistribution of urea-N in salt-affected soil under drip fertigation: Soil column and incubation experiments. *Irrig. Sci.* **2022**, *40*, 267–282. [CrossRef]
17. Guo, Y.Z.; Liu, Y.S. Poverty alleviation through land assetization and its implications for rural revitalization in China. *Land Use Policy* **2021**, *105*, 105418. [CrossRef]
18. Du, Y.Q.; Liu, X.F.; Zhang, L.; Zhou, W. Drip irrigation in agricultural saline-alkali land controls soil salinity and improves crop yield: Evidence from a global meta-analysis. *Sci. Total Environ.* **2023**, *880*, 163226. [CrossRef] [PubMed]
19. Qiu, L.P.; Kong, W.B.; Zhu, H.S.; Zhang, Q.; Banerjee, S.; Ishii, S.; Sadowsky, M.J.; Gao, J.L.; Feng, C.Z.; Wang, J.J.; et al. Halophytes increase rhizosphere microbial diversity, network complexity and function in inland saline ecosystem. *Sci. Total Environ.* **2022**, *831*, 154944. [CrossRef] [PubMed]
20. Li, J.G.; Pu, L.J.; Han, M.F.; Zhu, M.; Zhang, R.S.; Xiang, Y.Z. Soil salinization research in China: Advances and prospects. *J. Geogr. Sci.* **2014**, *24*, 943–960. [CrossRef]
21. Kim, Y.J.; Choo, B.K.; Cho, J.Y. Effect of gypsum and rice straw compost application on improvements of soil quality during desalination of reclaimed coastal tideland soils: Ten years of long-term experiments. *Catena* **2017**, *156*, 131–138. [CrossRef]
22. Zhu, W.; Yang, J.S.; Yao, R.J.; Wang, X.P.; Xie, W.P. Soil Water and Salt Transport in Medium and Heavy Saline Soils of Yellow River Delta. *Soils* **2021**, *53*, 817–825. (In Chinese)
23. Sun, B.; Xie, J.C.; Wang, N.; Zhu, J.W.; Zhang, J.L.; Li, C.J. An experimental study of straw mulching effects in water and salt in saline soils. *Bull. Soil Water Conserv.* **2011**, *31*, 48–51. (In Chinese)
24. Zhang, J.B.; Yang, J.S.; Li, F.R.; Hou, X.J.; Zhao, M.; Yao, R.J.; Yu, S.P.; Wang, X.P. Effects of farmyard manure and mulching on soil water and salinity in sever salinized tide flat soil of north Jiangsu province. *Acta Pedol. Sin.* **2014**, *51*, 184–188. (In Chinese)
25. Zhang, M.M.; Dong, B.D.; Qiao, Y.Z.; Yang, H.; Wang, Y.K.; Liu, M.Y. Effects of sub-soil plastic film mulch on soil water and salt content and water utilization by winter wheat under different soil salinities. *Field Crop. Res.* **2018**, *225*, 130–140. [CrossRef]
26. Gu, X.Y.; Liang, Z.W.; Huang, L.H.; Ma, H.Y.; Wang, M.M.; Yang, H.Y.; Liu, M.; Lv, H.Y.; Lv, B.S. Effects of plastic film mulching and plant density on rice growth and yield in saline-sodic soil of Northeast China. *J. Food Agri. Environ.* **2012**, *10*, 560–564.
27. Liu, T.; Cao, Y.B.; Zhang, Y.T.; Wang, R.S.; Xiao, H.J.; Wang, B.T.; Si, L.Q. Soil environment and growth adaptation strategies of Amorpha fruticose as affected by mulching in a moderately saline wasteland. *Land Degrad. Dev.* **2020**, *31*, 2672–2683. [CrossRef]
28. Shi, H.B.; Yang, S.Q.; Li, R.P.; Li, X.Y.; Li, W.P.; Yan, J.W.; Miao, Q.F.; Li, Z. Water-saving Irrigation and Utilization Efficiency of Water and Fertilizer in Hetao Irrigation District of Inner Mongolia: Prospect for Future Research. *J. Irrig. Drain.* **2020**, *39*, 1–12. (In Chinese)
29. Shi, K.L.; Lu, T.A.; Zheng, W.G.; Zhang, X.; Zhangzhong, L.L. A Review of the Category, Mechanism, and Controlling Methods of Chemical Clogging in Drip Irrigation System. *Agriculture* **2022**, *12*, 202. [CrossRef]
30. Wan, S.Q.; Jiao, Y.P.; Kang, Y.H.; Jiang, S.F.; Tan, J.L.; Liu, W.; Meng, J. Growth and yield of oleic sunflower (*Helianthus annuus* L.) under drip irrigation in very strongly saline soils. *Irrig. Sci.* **2013**, *31*, 943–957. [CrossRef]
31. Kang, Y.H.; Chen, M.; Wang, S.Q. Effects of drip irrigation with saline water on waxy maize (*Zea mays* L. var. *ceratina* Kulesh) in North China Plain. *Agric. Water Manag.* **2010**, *97*, 1303–1309. [CrossRef]
32. Wang, Q.M.; Huo, Z.L.; Zhang, L.D.; Wang, J.H.; Zhao, Y. Impact of saline water irrigation on water use efficiency and soil salt accumulation for spring maize in arid regions of China. *Agric.Water Manag.* **2016**, *163*, 125–138. [CrossRef]
33. Li, Z.J.; Liu, H.G.; Yang, H.C.; Wang, T.A. Effects of Deep Vertical Rotary Tillage Management Methods on Soil Quality in Saline Cotton Fields in Southern Xinjiang. *Agriculture* **2023**, *13*, 1864. [CrossRef]
34. Zhang, J.S.; Bian, Q.Q.; Miao, Q.; Jiang, X.L.; Wang, Y.H.; Wang, H.Y.; Cui, Z.L. Maize productivity response to combined tillage and mulching in coastal saline zones. *Agron. J.* **2022**, *114*, 784–794. [CrossRef]
35. Zhao, Y.G.; Li, Y.Y.; Wang, J.; Pang, H.C.; Li, Y. Buried straw layer plus plastic mulching reduces soil salinity and increases sunflower yield in saline soils. *Soil Tillage Res.* **2016**, *155*, 363–370. [CrossRef]
36. Zhu, W.; Yang, J.S.; Yao, R.J.; Wang, X.P.; Xie, W.P.; Shi, Z.G. Buried layers change soil water flow and solute transport from the Yellow River Delta, China. *J. Soil. Sediment.* **2021**, *21*, 1598–1608. [CrossRef]
37. Huo, L.; Pang, H.C.; Zhao, Y.G.; Wang, J.; Lu, C.; Li, Y.Y. Buried straw layer plus plastic mulching improves soil organic carbon fractions in an arid saline soil from Northwest China. *Soil Tillage Res.* **2017**, *165*, 286–293. [CrossRef]
38. Heng, T.; Liao, R.K.; Wang, Z.H.; Wu, W.Y.; Li, W.H.; Zhang, J.Z. Effects of combined drip irrigation and sub-surface pipe drainage on water and salt transport of saline-alkali soil in Xinjiang, China. *J. Arid Land* **2018**, *10*, 932–945. [CrossRef]

39. Bai, Z.T.; Liu, H.G.; Li, J.; Li, M.S.; Gong, P.; Li, P.F.; Li, L. Eight-year comparison of agroeconomic benefits of open ditch and subsurface pipe drainage in mulched drip irrigated saline-sodic farmland. *Irrig. Sci.* **2023**, *41*, 687–699. [CrossRef]
40. Wang, Z.H.; Heng, T.; Li, W.H.; Zhang, J.Z.; Zhangzhong, L.L. Effects of subsurface pipe drainage on soil salinity in saline-sodic soil under mulched drip irrigation. *Irrig. Drain.* **2020**, *6*, 95–106. [CrossRef]
41. Yang, J.S.; Yao, R.J.; Wang, X.P.; Xie, W.P.; Zhang, X.; Zhu, W.; Zhang, L. Research on salt-affected soils in China: History, Status Quo and Prospect. *Acta Pedol. Sin.* **2022**, *59*, 10–27. (In Chinese)
42. Rao, Y.; Peng, T.; Xue, S.W. Mechanisms of plant saline–alkaline tolerance. *J. Plant Physiol.* **2023**, *281*, 153916. [CrossRef]
43. Wu, L.P.; Zheng, H.N.; Wang, X.J. Effects of soil amendments on fractions and stability of soil organic matter in saline–alkaline paddy. *J. Environ. Manag.* **2021**, *294*, 112993. [CrossRef]
44. Zhang, W.C.; Zhang, W.X.; Zhao, Y.G.; Wang, S.J.; Liu, J.; Li, Y. Water regime enhances the effects of flue gas desulfurization gypsum on the reclamation of highly saline-sodic soil. *Land Degrad. Dev.* **2023**, *34*, 981–991. [CrossRef]
45. Wang, S.J.; Chen, Q.; Li, Y.; Zhuo, Y.Q.; Xu, L.Z. Research on saline-alkali soil amelioration with FGD gypsum. *Resour. Conserv. Recy.* **2017**, *121*, 82–92. [CrossRef]
46. Cao, Y.N.; Gao, Y.M.; Li, J.S.; Tian, Y.Q. Straw composts, gypsum and their mixtures enhance tomato yields under continuous saline water irrigation. *Agric. Water Manag.* **2019**, *223*, 105721. [CrossRef]
47. Nan, J.K.; Chen, X.M.; Wang, X.Y.; Lashari, M.S.; Wang, Y.M.; Guo, Z.C.; Du, Z.J. Effects of applying flue gas desulfurization gypsum and humic acid on soil physicochemical properties and rapeseed yield of a saline-sodic cropland in the eastern coastal area of China. *J. Soil. Sediment.* **2016**, *16*, 38–50. [CrossRef]
48. Wang, J.; Zhao, A.Q.; Liu, J.L.; Xiao, G.J.; Xu, X. Amendment of Saline–alkaline Soil with Flue-Gas Desulfurization Gypsum in the Yinchuan Plain, Northwest China. *Sustainability* **2023**, *15*, 8658. [CrossRef]
49. Mao, Y.M.; Li, X.P. Amelioration of flue gas desulfurization gypsum on saline-sodic soil of tidal flats and its effects on plant growth. *China Environ. Sci.* **2016**, *36*, 225–231. (In Chinese)
50. Tang, X.; Shang, H.; Liu, G.M.; Yao, Y.T.; Zhang, F.H.; Yang, J.S.; Zhou, L.X.; Chu, R. Effects of combined amendment on improvement of salinized soil and plant growth. *Soils* **2021**, *53*, 1033–1039. (In Chinese)
51. Zhang, J.S.; Yu, B.T.; Zhang, J.F.; Liu, Y.M.; Jiang, X.L.; Cui, Z.L. Effects of different amendments on soil physical and chemical properties and wheat growth in a coastal saline soil. *J. Plant Nutr. Fertil.* **2017**, *23*, 704–711. (In Chinese)
52. Gou, M.M.; Qu, Z.Y.; Wang, F.; Gao, X.Y.; Hu, M. Progress in research on biochar affecting soil-water environment and carbon sequestration-mitigating emissions in agricultural fields. *Trans. Chin. Soc. Agric. Mach.* **2018**, *49*, 1–12. (In Chinese)
53. Liu, M.L.; Wang, C.; Wang, F.Y.; Xie, Y.J. Maize (*Zea mays*) growth and nutrient uptake following integrated improvement of vermicompost and humic acid fertilizer on coastal saline soil. *Appl. Soil Ecol.* **2019**, *142*, 147–154. [CrossRef]
54. Han, K.; Huang, C.Y.; Su, W.B.; Guo, X.X.; Li, Z.; Tian, L.; Jian, C.Y.; Ren, H.M.; Wei, Z.G.; Song, J.J. Effects of humic acid conditioners on the growth and development of sugar beet in saline-alkali soil. *Soil Fertil. Sci. China* **2023**, *11*, 189–194. (In Chinese)
55. Yue, Y.; Guo, W.N.; Lin, Q.M.; Li, G.T.; Zhao, X.R. Improving salt leaching in a simulated saline soil column by three biochars derived from rice straw (*Oryza sativa* L.), sunfower straw (*Helianthus annuus*), and cow manure. *J. Soil Water Conserv.* **2016**, *71*, 467–475. [CrossRef]
56. Liu, S.; Meng, J.; Jiang, L.L.; Yang, X.; Lan, Y.; Cheng, X.Y.; Chen, W.F. Rice husk biochar impacts soil phosphorous availability, phosphatase activities and bacterial community characteristics in three different soil types. *Appl. Soil Ecol.* **2017**, *116*, 12–22. [CrossRef]
57. Sun, H.J.; Lu, H.Y.; Chu, L.; Shao, H.B.; Shi, W.M. Biochar applied with appropriate rates can reduce N leaching, keep N retention and not increase NH_3 volatilization in a coastal saline soil. *Sci. Total Environ.* **2017**, *575*, 820–825. [CrossRef] [PubMed]
58. Guo, R.; Zhou, J.; Hao, W.P.; Gong, D.Z.; Zhong, X.L.; Gu, F.X.; Liu, Q.; Xia, X.; Tian, J.; Li, H.R. Germination, growth, photosynthesis and ionic balance in Setaria viridis seedlings subjected to saline and alkaline stress. *Can. J. Plant Sci.* **2012**, *91*, 1077–1088. [CrossRef]
59. Xu, G.; Lv, Y.C.; Sun, J.N.; Shao, H.B.; Wei, L.L. Recent advances in biochar applications in agricultural soils: Benefits and environmental implications. *Clean–Soil Air Water* **2012**, *40*, 1093–1098. [CrossRef]
60. Zheng, H.; Wang, X.; Chen, L.; Wang, Z.Y.; Xia, Y.; Zhang, Y.P.; Wang, H.F.; Luo, X.X.; Xing, B.S. Enhanced growth of halophyte plants in biochar-amended coastal soil: Roles of nutrient availability and rhizosphere microbial modulation. *Plant Cell Environ.* **2018**, *41*, 517–532. [CrossRef]
61. He, K.; He, G.; Wang, C.; Zhang, H.P.; Xu, Y.; Wang, S.M.; Kong, Y.Z.; Zhou, G.K.; Hu, R.B. Biochar amendment ameliorates soil properties and promotes Miscanthus growth in a coastal saline alkali soil. *Appl. Soil Ecol.* **2020**, *155*, 103674. [CrossRef]
62. Yang, R.Y.; Zhou, C.X.; Zhu, J.J.; Pan, Y.H.; Sun, J.N.; Zhang, Z.H. Effects of biochar application on phreatic water evaporation and water-salt distribution in coastal saline soil. *J. Plant Nutr.* **2019**, *42*, 1243–1253. [CrossRef]
63. Fei, Y.H.; She, D.L.; Gao, L.; Xin, P. Micro-CT assessment on the soil structure and hydraulic characteristics of saline/sodic soils subjected to short-term amendment. *Soil Till. Res.* **2019**, *193*, 59–70. [CrossRef]
64. Rezapour, S.; Nouri, A.; Asadzadeh, F.; Barin, M.; Erpul, G.; Jagadamma, S.; Qin, R.J. Combining chemical and organic treatments enhance remediation performance and soil health in saline-sodic soils. *Commun. Earth Environ.* **2023**, *4*, 285. [CrossRef]

65. Jia, X.; Zhu, Y.; Zhang, R.; Zhu, Z.; Zhao, T.; Cheng, L.; Gao, L.; Liu, B.; Zhang, X.; Wang, Y. Ionomic and metabolomic analyses reveal the resistance response mechanism to saline-alkali stress in *Malus halliana* seedlings. *Plant. Physiol. Biochem.* **2020**, *147*, 77–90. [CrossRef] [PubMed]
66. Zeng, Y.B. Effects of Different Salt-Tolerant Plants Planting on Soil Physicochemical and Biological Properties of Salines of Saline Soil. Master's Thesis, Northwest A&F University, Yangling, China, 2015. (In Chinese).
67. Liu, X.; Wu, H.Y.; Yang, S.; Lu, X.; Wang, J.W.; Chen, Q.X. Formation of soil aggregates and distribution of soil nutrients in rhizosphere of salt-tolerant trees in coastal polder reclamation. *Acta Pedol. Sin.* **2020**, *57*, 1270–1279. (In Chinese)
68. Chen, X.B.; Yang, J.S.; Yang, C.H.; Hu, S.J.; Liu, G.M. Irrigation-drainage management and hydro-salinity balance in Weigan River Irrigation District. *Trans. Chin. Soc. Agric. Eng.* **2008**, *24*, 59–65. (In Chinese)
69. Jiao, H.Q. Soil Water and Salt Dynamics under Mulched Drip Irrigation in an Oasis Cotton Field: Model Development and Application. Ph.D. Thesis, China Agricultural University, Beijing, China, 2018. (In Chinese).
70. Yang, X.; Zhu, Q.; Ma, M.C.; Wu, J.W. Research on the evaporation capacity and water-salt variation characteristics of different types of saline wasteland. *China Rural Water Hydropower* **2019**, *9*, 49–53. (In Chinese)
71. Wang, S.L.; Zhou, H.P.; Qu, X.Y.; Guan, X.Y. Study on directional salt transport and surface salt draining under mulched drip irrigation in arid areas. *J. Hydraul. Eng.* **2013**, *44*, 549–555. (In Chinese)
72. Zhao, Z.Y.; Zhang, K.; Wang, L.; Wang, P.; Tian, C.Y. Desalting effect of halophytes on heavily saline soil. *J. Desert Res.* **2013**, *33*, 1420–1425. (In Chinese)
73. Fang, S.M.; Hou, X.; Liang, X.L. Response mechanisms of plants under saline-alkali stress. *Front. Plant Sci.* **2021**, *12*, 667458. [CrossRef]
74. Li, J.; Xu, H.H.; Liu, W.C.; Zhang, X.W.; Lu, Y.T. Ethylene inhibits root elongation during alkaline stress through AUXIN1 and associated changes in auxin accumulation. *Plant Physiol.* **2015**, *168*, 1777–1791. [CrossRef] [PubMed]
75. Lu, X.Y.; Chen, Q.; Cui, X.Y.; Abozeid, A.; Liu, Y.; Liu, J.; Tang, Z.H. Comparative metabolomics of two saline-alkali tolerant plants *Suaeda glauca* and *Puccinellia tenuiflora* based on GC-MS platform. *Nat. Prod. Res.* **2019**, *35*, 499–502. [CrossRef]
76. Wang, X.; Bai, J.; Wang, W.; Zhang, G.; Yin, S.; Wang, D. A comparative metabolomics analysis of the halophyte Suaeda salsa and *Salicornia europaea*. *Environ. Geochem. Health* **2021**, *43*, 1109–1122. [CrossRef]
77. Guan, Q.J.; Ma, H.Y.; Wang, Z.Y.; Bu, Q.Y.; Liu, S.K. A rice LSD1-like-type ZFP gene OsLOL5 enhances saline–alkaline tolerance in transgenic *Arabidopsis thaliana*, yeast and rice. *BMC Genom.* **2016**, *17*, 142. [CrossRef] [PubMed]
78. Guo, R.; Zhao, L.; Zhang, K.; Gao, D.; Yang, C. Genome of extreme halophyte *Puccinellia tenuiflora*. *BMC Genom.* **2020**, *21*, 311. [CrossRef] [PubMed]
79. Yan, F.; Zhu, Y.; Zhao, Y.; Wang, Y.; Li, J.; Wang, Q.; Liu, Y. De novo transcriptome sequencing and analysis of salt-, alkali-, and drought-responsive genes in *Sophora alopecuroides*. *BMC Genom.* **2020**, *21*, 423. [CrossRef] [PubMed]
80. Chookietwattana, K.; Yuwa-amornpitak, T. Data on soil properties and halophilic bacterial densities in the Na Si nuan secondary forest at Kantharawichai district, maha sarakham, Thailand. *Data Brief* **2019**, *27*, 104582. [CrossRef] [PubMed]
81. Yin, F.T.; Zhang, F.H. Reclamation of abandoned saline-alkali soil increased soil microbial diversity and degradation potential. *Plant Soil* **2022**, *477*, 521–538. [CrossRef]
82. Li, F.S.; Ghanizadeh, H.; Cui, G.L.; Liu, J.Y.; Miao, S.; Liu, C.; Song, W.W.; Chen, X.L.; Cheng, M.Z.; Wang, P.W.; et al. Microbiome-based agents can optimize composting of agricultural wastes by modifying microbial communities. *Bioresour. Technol.* **2016**, *374*, 128765. [CrossRef] [PubMed]
83. Fu, C.Y.; Ma, W.R.; Qiang, B.B.; Jin, X.J.; Zhang, Y.X.; Wang, M.X. Effect of chemical fertilizer with compound microbial fertilizer on soil physical properties and soybean yield. *Agronomy* **2023**, *13*, 2488. [CrossRef]
84. Qin, Y.; Druzhinina, I.S.; Pan, X.Y.; Yuan, Z.L. Microbially mediated plant salt tolerance and microbiome-based solutions for saline agriculture. *Biotechnol. Adv.* **2016**, *34*, 1245–1259. [CrossRef]
85. Dodd, I.C.; Pérez-Alfocea, F. Microbial amelioration of crop salinity stress. *J. Exp. Bot.* **2012**, *63*, 3415–3428. [CrossRef]
86. Liu, X.M.; Zhang, H.M. The effects of bacterial volatile emissions on plant abiotic stress tolerance. *Front. Plant Sci.* **2015**, *6*, 774. [CrossRef]
87. Zhang, C.; Yu, X.W.; Laipan, M.; Wei, T.; Guo, J.K. Soil health improvement by inoculation of indigenous microalgae in saline soil. *Environ. Geochem. Health* **2024**, *46*, 23. [CrossRef] [PubMed]
88. Yao, R.J.; Li, H.Q.; Yang, J.S.; Zhu, W.; Yin, C.Y.; Wang, X.P.; Xie, W.P.; Zhang, X. Combined application of biochar and N fertilizer shifted nitrification rate and amoA gene abundance of ammonia-oxidizing microorganisms in salt-affected anthropogenic-alluvial soil. *Appl. Soil Ecol.* **2022**, *171*, 104348. [CrossRef]
89. Baloch, M.Y.J.; Zhang, W.J.; Sultana, T.; Akram, M.; Shoumik, B.A.; Khan, M.Z.; Farooq, M.A. Utilization of sewage sludge to manage saline-alkali soil and increase crop production: Is it safe or not? *Environ. Technol. Inno.* **2023**, *32*, 103266. [CrossRef]
90. Wang, Z.Q. *Chinese Saline Soil*, 1st ed.; Science Press: Beijing, China, 1993.
91. Zhang, W.Z. *Calculation Method for Unstable Groundwater Flow and Evaluation of Groundwater Resources*, 1st ed.; Wuhan University Press: Wuhan, China, 2013.
92. Šimůnek, J.; van Genuchten, M.T.; Šejna, M. Recent developments and applications of the HYDRUS computer software packages. *Vadose Zone J.* **2016**, *15*, 25. [CrossRef]
93. Zhu, W.; Yang, J.S.; Yao, R.J.; Xie, W.P.; Wang, X.P.; Liu, Y.Q. Soil water-salt control and yield improvement under the effect of compound control in saline soil of the Yellow River Delta, China. *Agric. Water Manag.* **2022**, *263*, 107455. [CrossRef]

94. Hu, Q.L.; Zhao, Y.; Hu, X.L.; Qi, J.; Suo, L.Z.; Pan, Y.H.; Song, B.; Chen, X.B. Effect of saline land reclamation by constructing the "Raised Field-Shallow Trench" pattern on agroecosystems in Yellow River Delta. *Agric. Water Manag.* **2022**, *261*, 107345. [CrossRef]
95. Feng, G.X.; Zhu, C.L.; Wu, Q.F.; Wang, C.; Zhang, Z.Y.; Mwiya, R.M.; Zhang, L. Evaluating the impacts of saline water irrigation on soil water-salt and summer maize yield in subsurface drainage condition using coupled HYDRUS and EPIC model. *Sustainability* **2019**, *11*, 6431. [CrossRef]
96. Zeng, W.Z.; Lei, G.Q.; Zha, Y.Y.; Fang, Y.H.; Wu, J.W.; Huang, J.S. Sensitivity and uncertainty analysis of the HYDRUS-1D model for root water uptake in saline soils. *Crop Pasture Sci.* **2018**, *69*, 163–173. [CrossRef]
97. Lu, P.R.; Yang, Y.J.; Luo, W.; Zhang, Y.; Jia, Z.H. Numerical Simulation of Soil Water-Salt Dynamics and Agricultural Production in Reclaiming Coastal Areas Using Subsurface Pipe Drainage. *Agronomy* **2023**, *13*, 588. [CrossRef]
98. He, F.K.; Pu, S.Y.; Xiao, H.X.; Liu, S.B. Review of remote sensing application in soil degradation. *J. Agric. Resour. Environ.* **2021**, *38*, 10–19. (In Chinese)
99. Yao, R.J.; Yang, J.S.; Wu, D.H.; Xie, W.P.; Gao, P.; Wang, X.P. Geostatistical monitoring of soil salinity for precision management using proximally sensed electromagnetic induction (EMI) method. *Environ. Earth Sci.* **2016**, *75*, 1362. [CrossRef]
100. Gong, H.Z.; Shao, Y.; Brisco, B.; Hu, Q.R.; Tian, W. Modeling the dielectric behavior of saline soil at microwave frequencies. *Can. J. Remote Sens.* **2013**, *39*, 17–26. [CrossRef]
101. Zheng, R.M.; Li, Z.Z.; Gong, Y.S. A Coated Helical Transmission Line Time Domain Transmission Sensor for Measuring Water Content in Saline Soils. *Soil Sci. Soc. Am. J.* **2011**, *75*, 397–407. [CrossRef]
102. Liu, B.; Han, W.T.; Weckler, P.; Guo, W.C.; Wang, Y.; Song, K.X. Detection model for effect of soil salinity and temperature on FDR moisture content sensors. *Appl. Eng. Agri.* **2014**, *30*, 573–582.
103. Chen, S.; Xu, B.; Jin, Y.X.; Huang, Y.L.; Zhang, W.B.; Guo, J.; Shen, G.; Yang, X.C. Remote sensing monitoring and spatial-temporal characteristics analysis of soil salinization in agricultural area of Northern Xinjiang. *Sci. Geogr. Sin.* **2015**, *35*, 1607–1615. (In Chinese)
104. Li, Y. SWAP Model Distributed Simulation of Soil Water and Salt in Hetao Irrigation with Water-Saving Irrigation. Master's Thesis, Inner Mongolia Agricultural University, Hohhot, China, 2012. (In Chinese).
105. Xie, W.P.; Yang, J.S.; Yao, R.J.; Wang, X.P. Spatial and temporal variability of soil salinity in the Yangtze River estuary using electromagnetic induction. *Remote Sens.* **2021**, *13*, 1875. [CrossRef]
106. Wu, Y.K.; Yang, J.S.; Liu, G.M. Spatial variability of soil salinity using data from remote sensing and electromagnetic induction instruments. *Trans. Chin. Soc. Agric. Eng.* **2009**, *25*, 148–152. (In Chinese)
107. Chen, J.Y.; Wang, X.T.; Zhang, Z.T.; Han, J.; Yao, Z.H.; Wei, G.F. Soil salinization monitoring method based on UAV-satellite remote sensing scale-up. *Trans. Chin. Soc. Agric. Mach.* **2019**, *50*, 161–169.
108. Chen, W.J.; Li, M.S.; Li, Q.L. The influence of winter irrigation amount on the characteristics of water and salt distribution and WUE in different saline-alkali farmlands in northwest China. *Sustainability* **2023**, *15*, 15428. [CrossRef]
109. Chen, Y.Q.; Zhang, G.X.; Xu, Y.J.; Wu, L.L.; Huang, Z.G. Changes in root zone salt content and rice yield on saline-sodic soils across a salinity gradient under different irrigation schedules. *J. Food Agric. Environ.* **2013**, *11*, 1499–1505.
110. Chen, L.J.; Feng, Q.; Li, F.R.; Li, C.S. Simulation of soil water and salt transfer under mulched furrow irrigation with saline water. *Geoderma* **2015**, *241*, 87–96. [CrossRef]
111. Chu, L.L.; Kang, Y.H.; Wan, S.Q. Influence of microsprinkler irrigation amount on water, soil, and pH profiles in a coastal saline soil. *Sci. World J.* **2014**, *2014*, 279895. [CrossRef] [PubMed]
112. Hamani, A.K.; Xu, C.P.; Dang, H.K.; Cao, C.Y.; Wang, G.S.; Sun, J.S. Impacts of Saline Water Irrigation on Soil Respiration from Cotton Fields in the North China Plain. *Agronomy* **2023**, *13*, 1197. [CrossRef]
113. Gao, H.; Fu, T.G.; Tang, S.P.; Liu, J.T. Effects of saline water irrigation on winter wheat and its safe utilization under a subsurface drainage system in coastal saline-alkali land of Hebei Province, China. *Irrig. Sci.* **2023**, *41*, 251–260. [CrossRef]
114. Li, X.B.; Kang, Y.H.; Wan, S.Q.; Chen, X.L.; Liu, S.P.; Xu, J.C. Effect of ridge planting on reclamation of coastal saline soil using drip-irrigation with saline water. *Catena* **2017**, *150*, 24–31. [CrossRef]
115. Li, C.J.; Lei, J.Q.; Zhao, Y.; Xu, X.W.; Li, S.Y. Effect of saline water irrigation on soil development and plant growth in the Taklimakan Desert Highway shelterbelt. *Soil Till. Res.* **2015**, *146*, 99–107. [CrossRef]
116. Li, X.B. Vegetation establishment in coastal salt-affected wasteland using drip-irrigation with saline water. *Land Degrad. Dev.* **2019**, *30*, 1423–1436. [CrossRef]
117. Chen, W.P.; Hou, Z.N.; Wu, L.S.; Liang, Y.C.; Wei, C.Z. Evaluating salinity distribution in soil irrigated with saline water in arid regions of northwest China. *Agric. Water Manag.* **2010**, *97*, 2001–2008. [CrossRef]
118. Zhang, J.; Lai, Y.M.; Li, S.Y.; Zhang, M.Y.; You, Z.M.; Liang, T. Numerical study on the spatial-temporal distribution of solute and salt accumulation in saturated sulfate saline soil during freezing-thawing processes. Mechanism and feedback. *Adv. Water Resour.* **2023**, *177*, 104461. [CrossRef]
119. Tian, R.Z.; Zhang, Y.; Xu, A.H.; Li, X.M.; Hou, Y.L.; Tai, B.W. Impact of cooling on water and salt migration of high-chlorine saline soils. *Geofluids* **2021**, *2021*, 8612762. [CrossRef]
120. Zhang, Y.; Yang, J.S.; Yao, R.J. Effects of Saline Ice Water Irrigation on Distribution of Moisture and Salt Content in Coastal Saline Soil. *Aata Pedol. Sin.* **2016**, *53*, 388–400. (In Chinese)
121. Wu, D.Y.; Zhou, X.Y.; Jiang, X.Y. Water and Salt Migration with Phase Change in Saline Soil during Freezing and Thawing Processes. *Groundwater* **2018**, *56*, 742–752. [CrossRef] [PubMed]

122. Guo, K.; Liu, X.J. Effect of initial soil water content and bulk density on the infiltration and desalination of melting saline ice water in coastal saline soil. *Eur. J. Soil Sci.* **2019**, *70*, 1249–1266. [CrossRef]
123. Guo, K.; Liu, X.J. The successive infiltration of various saline waters accelerates the infiltration processes and enhances the salt leaching in a coastal saline soil. *Land Degrad. Dev.* **2023**, *34*, 5083–5095. [CrossRef]
124. Li, Z.G.; Liu, X.J.; Zhang, X.M.; Li, W.Q. Infiltration of melting saline ice water in soil columns: Consequences on soil moisture and salt content. *Agric. Water Manag.* **2008**, *95*, 498–502. [CrossRef]
125. Guo, K.; Liu, X.J. Dynamics of meltwater quality and quantity during saline ice melting and its effects on the infiltration and desalinization of coastal saline soils. *Agric. Water Manag.* **2014**, *139*, 1–6. [CrossRef]
126. Wan, X.S.; Liu, E.L.; Qiu, E.X.; Qu, M.F.; Zhao, X.; Nkiegaing, F.J. Study on phase changes of ice and salt in saline soils. *Cold Reg. Sci. Technol.* **2020**, *172*, 102988. [CrossRef]
127. Yang, M.; Yang, R.Y.; Li, Y.N.; Pan, Y.H.; Sun, J.N.; Zhang, Z.H. Effects of different biomass materials as a salt-isolation layer on water and salt migration in coastal saline soil. *PeeJ* **2021**, *9*, 11766. [CrossRef] [PubMed]
128. Zhang, S.; Kong, D.G.; Chang, X.H.; Zhai, L.M. Effect of straw deep application on soil water storage capacity. *J. Northeast Agric. Univ.* **2010**, *41*, 127–129. (In Chinese)
129. Wang, R.L.; Huang, Y.; Wei, F.L.; Bai, X.W.; Li, E.; Zhang, Y.L. Design and testing of plough for deep furrowing and riding of straw amendment fields. *J. Shenyang Agric. Univ.* **2011**, *42*, 231–234. (In Chinese)
130. Zhao, Y.G.; Li, Y.; Wang, S.J.; Wang, J.; Xu, L.Z. Combined application of a straw layer and flue gas desulphurization gypsum to reduce soil salinity and alkalinity. *Pedosphere* **2020**, *30*, 226–235. [CrossRef]
131. Song, X.L.; Sun, R.J.; Chen, W.F.; Wang, M.H. Effects of surface straw mulching and buried straw layer on soil water content and salinity dynamics in saline soils. *Can. J. Soil Sci.* **2020**, *100*, 58–68. [CrossRef]
132. Yang, M.; Yang, R.Y.; Li, Y.N.; Pan, Y.H.; Sun, J.N.; Zhang, Z.H. Effects of different peat application methods on water and salt migration in a coastal saline soil. *J. Soil Sci. Plant Nutr.* **2022**, *22*, 791–800. [CrossRef]
133. Shen, R.F.; Wang, C.; Sun, B. Soil related scientific and technological problems in implementing strategy of "Storing Grain in Land and Technology". *China Acad. J.* **2018**, *33*, 135–144. (In Chinese)
134. Zhu, H.; Yang, J.S.; Yao, R.J.; Gao, S.; Cao, Y.F.; Sun, Y.P. Effects of partial substitution of organic nitrogen for inorganic nitrogen in fertilization on salinity and nitrogen utilization in salinized coastal soil. *Chin. J. Eco-Agric.* **2019**, *27*, 441–450. (In Chinese)
135. Song, J.S.; Zhang, H.Y.; Chang, F.D.; Yu, R.; Zhang, X.Q.; Wang, X.Q.; Wang, W.N.; Liu, J.M.; Zhou, J.; Li, Y.Y. Humic acid plus manure increases the soil carbon pool by inhibiting salinity and alleviating the microbial resource limitation in saline soils. *Catena* **2023**, *233*, 107527. [CrossRef]
136. Li, S.P.; Zhao, L.; Wang, C.; Huang, H.Y.; Zhuang, M.H. Synergistic improvement of carbon sequestration and crop yield by organic material addition in saline soil: A global meta-analysis. *Sci. Total Environ.* **2023**, *891*, 164530. [CrossRef]
137. Bai, J.; Cui, B.; Deng, W.; Yang, Z.; Wang, Q.; Ding, Q. Soil organic carbon contents of two natural inland saline–alkalined wetlands in northeastern China. *J. Soil Water Conserv.* **2007**, *62*, 447–452.
138. Qu, Y.K.; Tang, J.; Liu, B.; Lyu, H.; Duan, Y.C.; Yang, Y.; Wang, S.N.; Li, Z.Y. Rhizosphere enzyme activities and microorganisms drive the transformation of organic and inorganic carbon in saline-alkali soil region. *Sci. Rep.* **2022**, *12*, 1314. [CrossRef]
139. Zhao, X.M.; Zhu, M.L.; Guo, X.X.; Wang, H.B.; Sui, B.; Zhao, L.P. Organic carbon content and humus composition after application aluminum sulfate and rice straw to soda saline–alkaline soil. *Environ. Sci. Pollut. Res.* **2019**, *26*, 13746–13754. [CrossRef]
140. Chen, X.D.; Wu, J.G.; Opoku-Kwanowaa, Y. Effects of organic wastes on soil organic carbon and surface charge properties in primary saline-alkali soil. *Sustainability* **2019**, *11*, 7088. [CrossRef]
141. Xu, T.T.; Xu, M.; Zhang, M.N.; Letnic, M.; Wang, J.Y.; Wang, L. Spatial effects of nitrogen deposition on soil organic carbon stocks in patchy degraded saline–alkaline grassland. *Geoderma* **2023**, *432*, 116408. [CrossRef]
142. Zhang, P.F.; Jiang, Z.W.; Wu, X.D.; Lu, Q.; Lin, Y.; Zhang, Y.Y.; Zhang, X.; Liu, Y.; Wang, S.Y.; Zang, S.Y. Effects of biochar and organic additives on CO_2 emissions and the microbial community at two water saturations in saline–alkaline soil. *Agronomy* **2023**, *13*, 1745. [CrossRef]
143. Zhou, Y.J.; Shao, T.Y.; Men, G.T.; Chen, J.H.; Li, N.; Gao, X.M.; Long, X.H.; Rengel, Z.; Zhu, M. Application of malrstone-based conditioner and plantation of Jerusalem artichoke improved properties of saline–alkaline soil in Inner Mongolia. *J. Environ. Manag.* **2023**, *329*, 117083. [CrossRef]
144. Ding, Z.L.; Kheir, A.M.S.; Ali, O.A.M.; Hafez, E.M.; ElShamey, E.A.; Zhou, Z.X.; Wang, B.Z.; Lin, X.E.; Ge, Y.; Fahmy, A.E.; et al. A vermicompost and deep tillage system to improve saline-sodic soil quality and wheat productivity. *J. Environ. Manag.* **2021**, *277*, 111388. [CrossRef]
145. Zhang, R.X.; Qu, Z.Y.; Liu, L.; Yang, W.; Wang, L.P.; Li, J.J.; Zhang, D.L. Soil respiration and organic carbon response to biochar and their influencing factors. *Atmosphere* **2022**, *13*, 2038. [CrossRef]
146. Xu, Z.K.; Shao, T.Y.; Lv, Z.X.; Yue, Y.; Liu, A.H.; Long, X.H.; Zhou, Z.S.; Gao, X.M.; Rengel, Z. The mechanisms of improving coastal saline soils by planting rice. *Sci. Total Environ.* **2020**, *703*, 135529. [CrossRef] [PubMed]
147. Liu, Q.; Tang, J. Effects of different rice planting duration on organic carbon components and components and carbon pool management index of saline alkaline soil in western Jilin province, China. *Appl. Ecol. Env. Res.* **2021**, *19*, 2213–2226. [CrossRef]
148. Dong, X.L.; Wang, J.T.; Zhang, X.J.; Dang, H.K.; Singh, B.P.; Liu, X.J.; Sun, H.Y. Long-term saline water irrigation decreased soil organic carbon and inorganic carbon contents. *Agric. Water Manag.* **2022**, *270*, 107760. [CrossRef]

149. Liu, Q.; Tang, J.; He, C.S.; Long, Y.; Wu, C.C. Effects of freeze-thaw cycles on soil properties and carbon distribution in saline–alkaline soil of wetland. *Sensor. Mater.* **2021**, *33*, 285–300. [CrossRef]
150. Zhang, H.Y.; Pang, H.C.; Song, J.S.; Chang, F.D.; Wang, J.; Wang, X.; Zhang, Y.T.; Peixoto, L.; Li, Y.Y. Subsurface organic ameliorant plus polyethylene mulching strengthened soil organic carbon by altering saline soil aggregate structure and regulating the fungal community. *Land Degrad. Dev.* **2022**, *33*, 2543–2553. [CrossRef]
151. Li, X.R.; Liu, Z.; Li, J.; Gong, H.R.; Zhang, Y.T.; Sun, Z.G.; Ouyang, Z. Increase in soil carbon pool stability rather than its stock in coastal saline-alkali ditches following reclamation time. *Agronomy* **2023**, *13*, 2843. [CrossRef]
152. Xie, H.Y.; Li, J.; Zhang, Y.T.; Xu, X.B.; Wang, L.Q.; Ouyang, Z. Evaluation of coastal farming under salinization and optimized fertilization strategies in China. *Sci. Total Environ.* **2021**, *797*, 149038. [CrossRef] [PubMed]
153. Heng, T.; He, X.L.; Yang, G.; Tian, L.J.; Li, F.D.; Yang, L.L.; Zhao, L.; Feng, Y.; Xu, X. Growth and nitrogen status of cotton (*Gossypium hirsutum* L.) under salt stress revealed using ^{15}N-labeled fertilizer. *J. Plant Ecol.* **2022**, *15*, 1213–1226. [CrossRef]
154. Li, H.Q.; Yao, R.J.; Yang, J.S.; Wang, X.P.; Zheng, F.L.; Chen, Q.; Xie, W.P.; Zhang, X. Influencing mechanism of soil salinization on nitrogen transformation processes and efficiency improving methods for high efficient utilization of nitrogen in salinized farmland. *Chin. J. Appl. Ecol.* **2020**, *31*, 3915–3924. (In Chinese)
155. Li, Y.; Xu, X.; Hu, M.; Chen, Z.J.; Tan, J.W.; Liu, L.; Xiong, Y.W.; Huang, Q.Z.; Huang, G.H. Modeling water-salt-nitrogen dynamics and crop growth of saline maize farmland in Northwest China: Searching for appropriate irrigation and N fertilization strategies. *Agric. Water Manag.* **2023**, *282*, 108271. [CrossRef]
156. Zhu, W.; Yang, J.S.; Yao, R.J.; Wang, X.P.; Xie, W.P.; Li, P.G. Nitrate leaching and NH_3 volatilization during soil reclamation in the Yellow River Delta, China. *Environ. Pollut.* **2021**, *286*, 117330. [CrossRef] [PubMed]
157. Hou, M.M.; Xu, R.; Lin, Z.Y.; Xi, D.; Wang, Y.; Wen, J.L.; Nie, S.A.; Zhong, F.L. Vertical characteristics of anaerobic oxidation of ammonium (anammox) in a coastal saline-alkali field. *Soil Till. Res.* **2020**, *198*, 104531. [CrossRef]
158. Yang, W.Z.; Jiao, Y.; Yang, M.D.; Wen, H.Y. N_2O emissions from saline–alkaline soil with different saline–alkaline levels in the Hetao Irrigation District of Inner Mongolia, China. *China Environ. Sci.* **2019**, *39*, 948–953. (In Chinese)
159. Yao, R.; Yang, J.S.; Zhu, W.; Li, H.Q.; Yin, C.Y.; Jing, Y.P.; Wang, X.P.; Xie, W.P.; Zhang, X. Impact of crop cultivation, nitrogen and fulvic acid on soil fungal community structure in salt-affected alluvial fluvo-aquic soil. *Plant Soil* **2021**, *462*, 539–558. [CrossRef]
160. Yin, C.Y.; Li, L.J.; Li, L.J.; Zhao, J.; Yang, J.S.; Zhao, H.G. Impacts of returning straw and nitrogen application on the nitrification and mineralization of nitrogen in saline soil. *Water* **2023**, *15*, 564. [CrossRef]
161. Pan, Y.C.; She, D.L.; Shi, Z.Q.; Chen, X.Y.; Xia, Y.Q. Do biochar and polyacrylamide have synergistic effect on net denitrification and ammonia volatilization in saline soils? *Environ. Sci. Pollut. Res.* **2021**, *28*, 59974–59987. [CrossRef] [PubMed]
162. Ding, J.N. Soil nitrogen transformation and functional microbial abundance in an agricultural soil amended with biochar. *Rev. Bras. Cienc. Solo* **2023**, *47*, e0220156. [CrossRef]
163. Zhu, H.; Yang, J.S.; Yao, R.J.; Wang, X.P.; Xie, W.P.; Zhu, W.; Liu, X.Y.; Cao, Y.F.; Tao, J. Interactive effects of soil amendments (biochar and gypsum) and salinity on ammonia volatilization in coastal saline soil. *Catena* **2020**, *190*, 104527. [CrossRef]
164. Yue, Y.; Lin, Q.M.; Li, G.T.; Zhao, X.R.; Chen, H. Biochar amends saline soil and enhances maize growth: Three-year field experiment findings. *Agronomy* **2023**, *13*, 1111. [CrossRef]
165. Liu, M.L.; Wang, C.; Liu, X.L.; Lu, Y.C.; Wang, Y.F. Saline-alkali soil applied with vermicompost and humic acid fertilizer improved macroaggregate microstructure to enhance salt leaching and inhibit nitrogen losses. *Appl. Soil Ecol.* **2020**, *156*, 103705. [CrossRef]
166. Yang, H.J.; Xia, J.B.; Xie, W.J.; Wei, S.C.; Cui, Q.; Shao, P.S.; Sun, J.K.; Dong, K.K.; Qi, X.C. Effects of straw returning and nitrogen addition on soil quality of a coastal saline soil: A field study of four consecutive wheat-maize cycles. *Land Degrad. Dev.* **2023**, *34*, 2061–2072. [CrossRef]
167. Zheng, Y.F.; Cao, X.W.; Zhou, Y.A.; Li, Z.; Yang, Y.Z.; Zhao, D.L.; Li, Y.Q.; Xu, Z.C.; Zhang, C.S. Effect of planting salt-tolerant legumes on coastal saline soil nutrient availability and microbial communities. *J. Environ. Manag.* **2023**, *345*, 118574. [CrossRef] [PubMed]
168. Chen, J.C.; Hu, G.Q.; Wang, H.; Fu, W.Z. Leaching and migration characteristics of nitrogen during coastal saline soil remediation by combining humic acid with gypsum and bentonite. *Ann. Agr. Sci.* **2023**, *68*, 1–11. [CrossRef]
169. Zhang, W.W.; Wang, C.; Liu, M.L.; Yu, Y.C. Integrated reclamation of saline soil nitrogen transformation in the hyphosphere by earthworms and arbuscular mycorrhizal fungus. *Appl. Soil Ecol.* **2019**, *135*, 137–146. [CrossRef]
170. Lu, J.Y.; Qu, Z.; Li, M.J.; Wang, Q.J. Effects of ionized water irrigation on organic nitrogen mineralization in saline-alkali soil in China. *Agronomy* **2023**, *13*, 2285. [CrossRef]
171. Ma, L.J.; Guo, H.J.; Min, W. Nitrous oxide emission and denitrifier bacteria communities in calcareous soil as affected by drip irrigation with saline water. *Appl. Soil Ecol.* **2019**, *143*, 222–235. [CrossRef]
172. Ma, J.C.; He, P.; Xu, X.P.; He, W.T.; Liu, Y.X.; Yang, F.Q.; Chen, F.; Li, S.T.; Tu, S.H.; Jin, J.Y.; et al. Temporal and spatial changes in soil available phosphorus in China (1990–2012). *Field Crops Res.* **2016**, *192*, 13–20. [CrossRef]
173. Huang, C.Y.; Roessner, U.; Eickmeier, I.; Genc, Y.; Callahan, D.L.; Shirley, N.; Langridge, P.; Bacic, A. Metabolite profiling reveals distinct changes in carbon and nitrogen metabolism in phosphate-deficient barley plants (*Hordeum vulgare* L.). *Plant Cell Physiol.* **2008**, *49*, 691–703. [CrossRef]
174. Shen, J.B.; Yuan, L.X.; Zhang, J.L.; Li, H.; Bai, Z.; Chen, X.; Zhang, W.; Zhang, F. Phosphorus dynamics: From soil to plant. *Plant Physiol.* **2011**, *156*, 997–1005. [CrossRef]

175. Chaves, M.M.; Flexas, J.; Pinheiro, C. Photosynthesis under drought and salt stress: Regulation mechanisms from whole plant to cell. *Ann. Bot.* **2009**, *103*, 551–560. [CrossRef] [PubMed]
176. Huang, B.R. *Role of Root Morphological and Physiological Characteristics in Drought Resistance of Plants*; Plant-Environment Interactions; CRC Press: Beijing, China, 2000. (In Chinese)
177. Girsova, M.A.; Golovina, G.F.; Drozdova, I.A.; Polyakova, I.G.; Antropova, T.V. Infrared studies and spectral properties of photochromic high silica glasses. *Opt. Appl.* **2014**, *44*, 337–344.
178. Papadopoulos, I.; Rendig, V.V. Interactive effects of salinity and nitrogen on growth and yield of tomato plants. *Plant Soil* **1983**, *73*, 47–57. [CrossRef]
179. Gao, S.; Yang, J.S.; Yao, R.J.; Cao, Y.F.; Tang, C. Effects of soil amelioration measures mitigating soil salinity and improving crop P uptake in coastal area of north Jiangsu. *Acta Pedol. Sin.* **2020**, *57*, 1219–1229. (In Chinese)
180. Su, R.; Zhang, Z.K.; Chang, C.; Peng, Q.; Cheng, X.; Pang, J.Y.; He, H.H.; Lambers, H. Interactive effects of phosphorus fertilization and salinity on plant growth, phosphorus and sodium status, and tartrate exudation by roots of two alfalfa cultivars. *Ann. Bot.* **2022**, *129*, 53–64. [CrossRef]
181. Hu, Y.C.; Schmidhalter, U. Drought and salinity: A comparison of their effects on mineral nutrition of plants. *J. Plant Nutr. Soil Sci.* **2005**, *168*, 541–549. [CrossRef]
182. Xian, J.T.; Chen, X.B.; Wang, S.; Zhang, X.L.; Xu, G. Phosphorus availability in saline soil: A review. *Soils* **2023**, *55*, 474–486.
183. Yang, L.L.; Li, J.H. Nutrition and fertilizer effect of saline in China. *Chin. J. Eco-Agric.* **2001**, *9*, 79–81. (In Chinese)
184. Ding, Z.L.; Kheir, A.M.S.; Ali, M.G.M.; Ali, O.A.M.; Abdelaal, A.I.N.; Lin, X.E.; Zhou, Z.X.; Wang, B.Z.; Liu, B.B.; He, Z.L. The integrated effect of salinity, organic amendments, phosphorus fertilizers, and deficit irrigation on soil properties, phosphorus fractionation and wheat productivity. *Sci. Rep.* **2020**, *10*, 2736. [CrossRef] [PubMed]
185. Zhang, M.M.; Chen, C.; Liu, G.M.; Yang, J.S.; Yu, S.P. Suitable utilization of fertilizer and soil modifier to ameliorate physicochemical characteristics of saline-alkali soil and increase crop yields. *Trans. Chin. Soc. Agric. Eng.* **2014**, *30*, 91–98. (In Chinese)
186. Liu, X.Y.; Yang, J.S.; Tao, J.Y.; Yao, R.J.; Wang, X.P.; Xie, W.P.; Zhu, H. Elucidating the effect and interaction mechanism of fulvic acid and nitrogen fertilizer application on phosphorus availability in a salt-affected soil. *J. Soil. Sediment.* **2021**, *21*, 2525–2539. [CrossRef]
187. Liu, X.Y.; Yang, J.S.; Tao, J.Y.; Yao, R.J. Integrated application of inorganic fertilizer with fulvic acid for improving soil nutrient supply and nutrient use efficiency of winter wheat in a salt-affected soil. *Appl. Soil Ecol.* **2022**, *170*, 104255. [CrossRef]
188. Liu, X.Y.; Yang, J.S.; Yao, R.J. Synergistic effects of fertilizer reduction and fulvic acid application on decreasing NaCl content and N, P availability of salinized soil. *Plant Nutr. Fertil. Sci.* **2021**, *27*, 1339–1350. (In Chinese)
189. Gao, S.; Yang, J.S.; Yao, R.J.; Cao, Y.F.; Zhu, H.; Sun, Y.P.; Wang, X.P.; Xie, W.P. Effects of different management on phosphorus fractions in coastal saline soil and phosphorus absorption and utilization by crops. *Soils* **2020**, *52*, 691–698. (In Chinese)
190. Li, Z.; Liu, Z.; Wang, Y.; Wang, X.F.; Liu, P.; Han, M.Y.; Zhou, W.Z. Improving soil phosphorus availability in saline areas by marine bacterium *Bacillus paramycoides*. *Environ. Sci. Pollut. Res.* **2023**, *30*, 112385–112396. [CrossRef]
191. Li, Z.; Wang, Y.; Liu, Z.; Han, F.; Chen, S.G.; Zhou, W.Z. Integrated application of phosphorus-accumulating bacteria and phosphorus-solubilizing bacteria to achieve sustainable phosphorus management in saline soils. *Sci. Total Environ.* **2023**, *885*, 163971. [CrossRef] [PubMed]
192. Zhang, Y.; Lang, L.N.; Sun, Z.X.; Li, M.T. Potential application of paenibacillus sp. C1 to the amelioration of Soda saline–alkaline soil. *Geomicrobiol. J.* **2023**, *40*, 172–182. [CrossRef]
193. Maas, E.V.; Hoffman, G.J. Crop salt tolerance: Current assessment. *J. Irrig. Drain. Div.* **1977**, *103*, 115–134. [CrossRef]
194. Zhang, Y.F.; Wang, W.C.; Yuan, W.; Zhang, R.H.; Xi, X.B. Cattle manure application and combined straw mulching enhance maize (*Zea mays* L.) growth and water use for rain-fed cropping system of coastal saline soils. *Agriculture* **2021**, *11*, 745. [CrossRef]
195. Pang, H.C.; Li, Y.Y.; Yang, J.S.; Liang, Y.S. Effect of brackish water irrigation and straw mulching on soil salinity and crop yields under monsoonal climatic conditions. *Agric. Water Manag.* **2010**, *97*, 1971–1977. [CrossRef]
196. Chen, W.L.; Jin, M.G.; Ferré, T.P.A.; Liu, Y.F.; Huang, J.N.; Xian, Y. Soil conditions affect cotton root distribution and cotton yield under mulched drip irrigation. *Field Crop Res.* **2021**, *249*, 107743. [CrossRef]
197. Hou, X.H.; Xiang, Y.Z.; Fan, J.L.; Zhang, F.C.; Hu, W.H.; Yan, F.L.; Xiao, C.; Li, Y.P.; Cheng, H.L.; Li, Z.J. Spatial distribution and variability of soil salinity in film-mulched cotton fields under various drip irrigation regimes in southern Xinjiang of China. *Soil Till. Res.* **2022**, *223*, 105470. [CrossRef]
198. Ren, F.T.; Yang, G.; Li, W.J.; He, X.L.; Gao, Y.L.; Tian, L.J.; Li, F.D.; Wang, Z.L.; Liu, S.H. Yield-compatible salinity level for growing cotton (*Gossypium hirsutum* L.) under mulched drip irrigation using saline water. *Agric. Water Manag.* **2021**, *250*, 106859. [CrossRef]
199. Zhang, Y.F.; Yuan, W.; Han, L.J. Residue mulching alleviates coastal salt accumulation and stimulates post-fallow crop biomass under a fallow-maize (*Zea mays* L.) rotation system. *Agriculture* **2022**, *14*, 509. [CrossRef]
200. Zhao, Y.G.; Pang, H.C.; Wang, J.; Huo, L.; Li, Y.Y. Effects of straw mulch and buried straw on soil moisture and salinity in relation to sunflower growth and yield. *Field Crop. Res.* **2014**, *161*, 16–25. [CrossRef]
201. Bai, Z.T.; Liu, H.G.; Wang, T.A.; Gong, P.; Li, H.Q.; Li, L.; Xue, B.; Cao, M.H.; Feng, J.P.; Xu, Y.B. Effect of smashing ridge tillage depth on soil water, salinity, and yield in saline cotton fields in South Xinjiang, China. *Water* **2021**, *13*, 3592. [CrossRef]

202. Zhang, T.B.; Zhan, X.Y.; Kang, Y.H.; Wan, S.Q.; Feng, H. Improvements of soil salt characteristics and nutrient status in an impermeable saline-sodic soil reclaimed with an improved drip irrigation while ridge planting *Lycium barbarum* L. *J. Soil. Sediment.* **2017**, *17*, 1126–1139. [CrossRef]
203. Chi, C.M.; Zhao, C.W.; Sun, X.J.; Wang, Z.C. Reclamation of saline-sodic soil properties and improvement of rice (*Oriza sativa* L.) growth and yield using desulfurized gypsum in the west of Songnen Plain, northeast China. *Geoderma* **2012**, *187*, 24–30. [CrossRef]
204. Yuan, J.; Feng, W.Z.; Jiang, X.M.; Wang, J.L. Saline-alkali migration in soda saline soil based on sub-soiling technology. *Desalin. Water Treat.* **2019**, *149*, 352–362. [CrossRef]
205. Cui, L.Q.; Liu, Y.M.; Yan, J.L.; Hina, K.; Hussain, Q.; Qiu, T.J.; Zhu, J.Y. Revitalizing coastal saline-alkali soil with biochar application for improved crop growth. *Ecol. Eng.* **2022**, *179*, 106594. [CrossRef]
206. Zhao, W.; Zhou, Q.; Tian, Z.Z.; Cui, Y.T.; Liang, Y.; Wang, H.Y. Apply biochar to ameliorate soda saline-alkali land, improve soil function and increase corn nutrient availability in the Songnen Plain. *Sci. Total Environ.* **2020**, *722*, 137428. [CrossRef]
207. Wang, Z.H.; Yin, D.W.; Wang, H.Y.; Zhao, C.J.; Li, Z.T. Effects of biochar on waterlogging and the associated change in micro-ecological environment of maize rhizosphere soil in saline-alkali land. *Bioresources* **2020**, *15*, 9303–9323. [CrossRef]
208. Xiao, L.; Yuan, G.; Feng, L.; Bi, D.; Wei, J. Soil properties and the growth of wheat (*Tri aestivum* L.) and maize (*Zea mays* L.) in response to reed (*phragmites communis*) biochar use in a salt-affected soil in the Yellow River Delta. *Agric. Ecosyst. Environ.* **2020**, *303*, 107124. [CrossRef]
209. Zhao, Y.G.; Wang, S.J.; Li, Y.; Zhuo, Y.Q.; Liu, J. Effects of straw layer and flue gas desulfurization gypsum treatments on soil salinity and sodicity in relation to sunflower yield. *Geoderma* **2019**, *352*, 13–21. [CrossRef]
210. Wang, X.B.; Zhao, Q.S.; Hu, Y.J.; Zheng, Y.; Wu, X.P.; Wu, H.J.; Zhang, G.X.; Cai, D.X.; Manzur, C.L. An alternative water source and combined agronomic practices for cotton irrigation in coastal saline soils. *Irrig. Sci.* **2012**, *30*, 221–232. [CrossRef]

agriculture

MDPI

Review

Challenges in Sustainable Agriculture—The Role of Organic Amendments

Manuel Matisic, Ivan Dugan and Igor Bogunovic *

Faculty of Agriculture, University of Zagreb, Svetosimunska 25, 10000 Zagreb, Croatia; mmatisic@agr.hr (M.M.); idugan@agr.hr (I.D.)
* Correspondence: ibogunovic@agr.hr

Abstract: Soil degradation threatens global food security and environmental sustainability, necessitating effective soil management strategies. This review comprehensively examines the impact of organic soil amendments on soil quality and productivity across various soil types and climatic conditions. A review of significant research related to organic amendments was performed using encompassed data from online search engines for studies published up until 31 December 2023. Despite their heterogeneity and use of varying methodologies, the data were narratively synthesized, providing a comprehensive understanding of amendment-induced changes in the chemical and physical properties of soil and the effectiveness of restoration on soil degradation. Organic amendments, including compost, vermicompost, biochar, and pomace, are pivotal in enhancing soil quality by increasing soil organic matter content, fostering aggregate formation, and improving soil structure in the short term. They positively influence water retention capacity, pH levels, nutrient availability, and carbon sequestration. In several studies, amendment-induced changes were absent, indicating that the effects of amendments vary depending on soil texture, application rates, and cropping systems, which emphasizes the need for tailored, sustainable soil management practices. This study concludes that organic amendments are a promising option for structure improvement and organic matter accumulation. It further suggests that an approach that integrates various methods is essential in order to meet desirable soil quality and retain agricultural productivity and offers valuable insights and recommendations for policymakers, practitioners, and researchers. Organic amendments can improve soil ecosystem services and contribute to climate change adaptation. In the future, more attention should be directed to tillage management and soil amendment interaction, as well as their effectiveness over specific periods of time.

Keywords: soil amendments; soil texture; organic matter; soil degradation

1. Background

Soil is an essential natural resource that is necessary for sustainable Earth life [1]. Soils provide us with numerous regulating (e.g., air and water purification), provisioning (e.g., food), and cultural (e.g., recreation) ecosystem services, ensuring human well-being and sustainable socioeconomic development [2]. From the perspective of food demand and climate change, soil provisioning (food, water, raw material, medicinal resources, genetic resources, etc.) and regulating (climate regulation, erosion prevention, etc.) ecosystem services are a main focus of the scientific community [3–6].

Rising population growth and global warming are two of the most critical challenges that currently affect food supply security [7]. Decreasing the amount of arable land available per person generates food supply insecurity. This trend exists because population growth is outstripping the expansion rate of the area used for crop production [8]. Solutions such as genetic modifications, agrochemicals, mineral fertilizers, and growth conditioners are tested in order to ensure food security [9,10]. However, these solutions, along with improper land management and excessive soil exploitation, often deteriorate soil quality, leading to a decline in agricultural productivity and soil degradation [11]. The negative impact of human activities on soil has been ongoing since the first agricultural revolution, although the anthropic impact on soil goes back as far as 13,000 years ago [4].

Soil degradation is also a significant global problem. Approximately one-third of the world's cropland is affected by at least one form of degradation, including soil/water pollution, soil water/wind erosion, the loss of soil organic matter (SOM), nutrient imbalances, salinization and acidification, crust formation, and the loss of soil biodiversity [11,12]. Whilst soil continues to degrade at a rate of 5–10 billion hectares annually [13], the responsibility for addressing soil conservation and management falls on all of us, regardless of background, knowledge, or profession. Therefore, on a global scale, significant efforts must be invested to create sustainable measures in order to mitigate or neutralize land degradation. Several policies have been launched to combat land degradation by promoting sustainable agriculture and management. From the United Nations "2030 Agenda for Sustainable Development", which seeks a more sustainable future through a land degradation-neutral world where food production is intensified on existing cropping and pasture lands under sustainable land management practices [14], to the European Commission's "A Soil Deal for Europe", which addresses the Sustainable Development Goals and the Green Deal, together with the European Union Soil Strategy [15]. All policy goals directly or indirectly depend on soil function, land use, and management [16,17], highlighting the need to ensure and preserve soil quality.

Preserving soil quality is essential in implementing sustainable agriculture and safeguarding ecosystem services [18]. Among several other strategies, such as the implementation of cover crops, conservation tillage practices, balanced fertilization, and crop rotation [11,19], soil quality can be improved by applying soil amendments [20–22]. Soil amendments refer to materials obtained from different processes that are used to improve soil productivity and quality [23].

There are two categories of soil amendments: (1) organic materials, such as biochar, straw, pomace, manure, sawdust, and compost, and (2) inorganic materials, including sand, gypsum, vermiculite, zeolite, and lignite [24,25]. Soil amendments, both organic and inorganic, improve the physical and biological properties of soil, increase carbon sequestration, restore saline and contaminated soils, and increase crop yields and fertilizer efficiency [26]. Generally, their impact on soil quality is primarily positive, regardless of the climate conditions and soil type [27].

The use of organic amendments has its economic aspect. The high prices of inorganic fertilizers favor organic amendments as a desirable and acceptable alternative for meeting the demand for nutrients in plant production. In addition, they also have a positive influence on soil properties and reduce dependence on inorganic fertilizers [28,29]. How and to what extent organic amendments will affect soil properties depends on several factors: type of material used, the amount used in a certain area, duration of application, climate conditions, soil type, soil management practices, and cropping systems [30,31].

When using organic amendments in plant production, emphasis is placed on increasing the soil organic matter (SOM) content, which is a crucial factor because a lack of SOM can lead to multiple negative changes in the soil. Reduced water storage capacity and porosity, soil compaction, and low infiltration capacity are just a few examples [32], as well as increased runoff and the loss of the topsoil layer [33]. All of the mentioned elements emphasize the importance of proper soil management, including soil conditioning using organic amendments. Future paragraphs will detail the impact of soil conditioning on soil quality. Several studies highlighted the adverse effects of improper organic amendment addition in this context. For instance, research conducted in the USA [34] reported a decrease in soil nitrogen following biochar application, regardless of application rate. Similarly, compost derived from the wine industry has been shown to elevate soil pH, which is particularly harmful for crop production in the naturally alkaline soils of South Africa [35]. Conversely, detrimental trends associated with animal manure applications were observed in Bangladesh [36], exacerbating soil acidification issues. The presented examples warn farmers that the use of particular organic amendments should align with crop needs, soil characteristics, and environmental conditions.

This paper presents a unique approach, discussing the impacts of organic amendments, namely compost, vermicompost, biochar, pomace, and manure, on the physical and chemical properties of soil. Our discussion is based on the most recent and relevant review and meta-analysis articles. In cases where review and meta-analysis papers were unavailable for a given topic, we turned to the most recent and relevant research papers (Tables 1–4). Importantly, no previous review or meta-analysis articles have synthesized or integrated all the organic amendments named above into a single source, nor have they linked such practices to the emerging topic of sustainable agriculture and restoration strategies. Further, no collective papers were found that considered the contrasting effect of these amendments on such a range of (physical and chemical) soil properties with different textures. This paper fills that gap, providing a comprehensive review of the impacts of organic amendments on soil properties.

For this review, we extensively searched all published reviews, meta-analyses, and research articles on the available databases that had been published up until 31 December 2023. Our search terms included "soil amendments", "organic amendments", "biochar", "compost", "vermicompost", "manure", "dung", and "pomace", combined with "soil", "soil organic matter", "water erosion", "soil quality", "land degradation", "tillage", "soil management", and other keywords related to land degradation and soil restoration strategies. We only included articles written in English and those that included replicated treatments and control treatments without the use of organic amendments for comparison purposes. The statistical results reported in the published articles were used to draw statistical conclusions about the amendments' impacts. Note that a few relevant articles in this review did not provide statistical changes or numerical differences in the improvement of soil properties, only general statements about the impact of the amendments on some qualitative services (i.e., soil quality, structure improvement, soil fertility, etc.). This review focuses on the most relevant amendment-induced changes in soil systems and land degradation reclamation, and discusses how organic amendments affect agricultural production.

Table 1. Soil amendment impact on soil physical properties.

Rn	Texture	Duration	Tillage	Amendment	Base Material	Rate (t ha^{-1})	BD	MWD	TP	WSA	SWC
[31]	L	ST	Con	Ct		0	-				
				Com	Sheep manure	2	NE				
						4	D				
						6	D				
						8	NE				
						10	D				
				Str	Maize	6	D				
[35]	SL	ST	ND	Ct		0					-
				Com	Winery solid waste	5					NE
						10					NE
						20					NE
						40					NE

Table 1. *Cont.*

Rn	Texture	Duration	Tillage	Amendment	Base Material	Rate (t ha^{-1})	BD	MWD	TP	WSA	SWC
[37]	C	ST	ND	Ct		0		-			
				Com	Sheep manure	4.5		I			
						9		I			
						13.5		I			
						18		I			
				Fm	Sheep manure	4.5		I			
						9		I			
						13.5		I			
						18		I			
[38]	SiCL	LT	Con	Com	Poultry manure	8.97		NE			
						8.97		I			
				Bc	Walnut shells	10		NE			
						10		NE			
[39]	S	ND	ND	Dd	Food waste	20	NE			NE	
				Vcom	Digestate food waste	20	NE			NE	
				Com	Sewage sludge, green waste	20	NE			I	
				Vcom	Sewage sludge	20	NE			I	
				Ct			-			-	
	LC			Dd	Food waste	20	NE				
				Vcom	Digestate food waste	20	NE			I	
				Com	Sewage sludge, green waste	20	NE			I	
				Vcom	Sewage sludge	20	NE			I	
				Ct			-			-	
[40]	SL	LT	ND	Ct		0			-	-	-
				FYM	Solid + liquid phase	15 + 4			I	I	I
[41]	SiC	LT	Con	Ct			-				-
				M	Olive leaves	236	D				NE
				Pom	Olive mill waste	270	D				NE
[42]	SiC	LT	Con	Bf		0	-			-	-
				Ct		0	NE			I	NE
				FYM	Cattle manure	38	D			I	NE
	SiL	LT	Con	Ct		0	-			-	-
				FYM	Cattle barn	20	NE			NE	NE
						30	NE			NE	NE
	SL	LT	Con	Ct		0	-			-	-
				FYM	Cattle slurry	25	NE			NE	I
						37.5	D			NE	I
[43]	SiL	LT	ND	Bc	Paper fiber	10				I	
						20				I	
[44]	L	LT	Con	Ct		0				-	
				Com	Grape pomace, poultry droppings, mown grass, and straw	30				I	
						60				I	

Table 1. *Cont.*

Rn	Texture	Duration	Tillage	Amendment	Base Material	Rate (t ha^{-1})	BD	MWD	TP	WSA	SWC
[45]	C	LT	Con	Ct		0	-		-		
				Gm	ND	ND	D		I		
				FYM		35	D		I		
[46]	SiC	LT	Con	Ct		0	-				
			Min	FYM	ND	15	D				NE
			Red			30	D				NE
[47]	SiL	LT	ND	Ct		0	-	-			
				FYM	ND	10	D	I			
[48]	L	LT	Con	Ct		0	-				
				FYM	Cattle manure—composted	10	NE				
[49]	LS	ST	ND	Ct		0	D				I
	SL			Bc	Pine wood mill waste	4	-				I
						20	D				I
	L					100	-				-
						100	-				I
	SiL					100	D				-
	SiCL					100	D				-
						100	D				I
[50]	SiL	ND	Htt	Ct		0		-		-	
				Bc	Rice straw	0; 11.25; 22.5		NE		I	NE
					Maize straw	0; 11.25; 22.5		NE		I	NE
					Wheat straw	0; 11.25; 22.5		NE		I	I
					Rice husk	0; 11.25; 22.5		NE		I	I
					Bamboo	0; 11.25; 22.5		NE		I	I
[51]	CL	LT	Conservation	Ct	ND	0.015 per plant	-		-		-
				Omw		0.8 per plant	NE		NE		NE
				Com	Olive pomace	0.06 per plant	NE		NE		NE
						0.12 per plant	NE		I		NE
[52]	SiL	ST	Con	Ct		0	-			-	
				Com	Olive pomace	4	NE			NE	D

Table 1. *Cont.*

Rn	Texture	Duration	Tillage	Amendment	Base Material	Rate (t ha^{-1})	BD	MWD	TP	WSA	SWC
[53]	SL	ST	ND	Ct		0	-		-		-
				Bc	Poultry litter waste	12.39	D		I		I
						24.78	D		I		I
						37.17	D		I		I
						49.56	D		I		I
						61.95	D		I		I

Abbreviations: Rn—reference number, C—clay, SiC—silty clay, SiCL—silty clay loam, CL—clay loam, SiL—silt loam, L—loam, SL—sandy loam, LS—loamy sand, S—sand, Bc—biochar, Vcom—vermicompost, Dd—dewatered digestate, FYM—farmyard manure, M—mulch, Pom—pomace, Bf—bare fallow, Gm—green manure, Omw—olive mill wastewater, Htt—hand tool tillage, Str—straw, BD—bulk density, MWD—mean weight diameter, TP—total porosity, WSA—water-stable aggregates, SWC—soil water content, ST—short-term, LT—long term, Con—conventional, Red—reduced, Min—minimum, ND—non-defined, NE—no effect, I—increased, D—decreased.

Table 2. Soil amendment impact on soil chemical properties.

Rn	Texture	Duration	Tillage	Amendment	Base Material	Rate (t ha^{-1})	pH	SOM	TN	P	K	C
[31]	L	ST	Con	Ct		0	-	-				
				Com	Sheep manure	2	D	NE				
						4	D	I				
						6	D	I				
						8	NE	I				
						10	D	I				
				Str	Maize	6	NE	I				
[34]	SiL	ST	Con	Ct		0	-	-		-		
				Bc	Douglas fir	11.2	I	I	D	I		
						22.4	I	I	D	I		
						44.8	I	NE	D	I		
[35]	SL	ST	ND	Ct		0	-					
				Com	Winery solid waste	5	NE					
						10	I					
						20	I					
						40	I					
[36]	ND	ST	ND	Ct		0	-					
				FYM	Cow dung	10	D					
					Chicken manure	10	D					
					Cow dung + chicken manure	10	D					

Table 2. *Cont.*

Rn	Texture	Duration	Tillage	Amendment	Base Material	Rate (t ha^{-1})	pH	SOM	TN	P	K	C
[37]	C	ST	ND	Ct		0	-				-	-
				Com	Sheep manure	4.5	NE				NE	NE
						9	NE				NE	NE
						13.5	NE				I	NE
						18	NE				I	NE
				FYM		4.5	NE				NE	NE
						9	NE				NE	NE
						13.5	NE				I	NE
						18	NE				I	NE
[40]	SL	LT	ND	Ct		0		-				
				FYM	Solid + liquid phase	15 + 4		I				
[41]	SiC	LT	Con	Ct			-	-				
				M	Olive leaves	236	NE	I				
				Pom	Olive mill waste	270	NE	I				
[42]	SiC	LT	Con	Bf		0	-	-	-	-	-	
				Ct		0	NE	NE	NE	D	D	
				FYM	Cattle manure with straw	38	NE	I	I	I	I	
	SiL			Ct		0	-	-	-	-	-	
				FYM	Cattle barn	20	NE	NE	NE	NE	NE	
						30	NE	NE	NE	NE	NE	
	SL			Ct		0	-	-	-	-	-	
				FYM	Cattle slurry	25	NE	I	I	I	I	
						37.5	NE	I	I	I	I	
[47]	SiL	LT	ND	Ct		0	-		-	-	-	
				FYM	ND	10	I		I	I	I	
[48]	L	LT	Con	Ct		0	-	-	-	-	-	
				Residues			NE	I	NE	NE	NE	
				FYM	Cattle manure—composted	10	NE	I	NE	NE	NE	
						60	NE		I	I	I	
[52]	SiL	ST	Con	Ct		0	-	-		-	-	
				Com	Olive pomace	4	NE	NE		NE	I	
[53]	SL	ST	ND	Ct		0	-			-	-	-
				Bc	Poultry litter waste	2.02	I			I	I	I
						4.05	I			I	I	I
						6.07	I			I	I	I
						8.1	I			I	I	I
						10.12	I			I	I	I

Table 2. *Cont.*

Rn	Texture	Duration	Tillage	Amendment	Base Material	Rate (t ha^{-1})	pH	SOM	TN	P	K	C
[54]	S	LT	Htt	Ct		0	-		-	-	-	
				FYM	Sheep manure with straw	20	D		NE	NE	NE	
						40	D		I	I	NE	
						60	D		I	I	NE	
				Com	Sewage sludge	20	D		NE	I	NE	
						40	D		I	I	NE	
						60	D		I	I	NE	
					Municipal solid waste	20	D		NE	NE	NE	
						40	NE		NE	NE	NE	

Abbreviations: Rn—reference number, C—clay, SiC—silty clay, SiL—silt loam, L—loam, SL—sandy loam, S—sand, Ct—control, Com—compost, Bc—biochar, FYM—farmyard manure, M—mulch, Pom—pomace, Bf—bare fallow, Htt—hand tool tillage, Str—straw, pH, SOM—soil organic matter, TN—total nitrogen, P—phosphorous, K—potassium, C—carbon, ST—short-term, LT—long-term, Con—conventional, ND—non-defined, NE—no effect, I—increased, D—decreased.

Table 3. Soil amendment impact on soil chemical properties.

Rn	Texture	Duration	Tillage	Amendment	Base Material	Rate (t ha^{-1})	Na	Ca	Mg	S
[34]	SiL	ST	ND	Ct		0				-
				Bc	Douglas fir	11.2				I
						22.4				I
						44.8				I
[37]	C	ST	ND	Ct		0		-	-	
				Com	Sheep manure	4.5		NE	NE	
						9		NE	NE	
						13.5		NE	NE	
						18		NE	NE	
				Fm	Sheep manure	4.5		NE	NE	
						9		NE	NE	
						13.5		NE	NE	
						18		NE	NE	
[47]	SiL	LT	ND	Ct		0				-
				FYM	ND	10				I
[52]	SiL	ST	Con	Ct		0				-
				Com	Olive pomace	4				NE
[53]	SL	ST	ND	Ct		0	-			
				Bc	Poultry litter waste	2.02	D			
						4.05	D			
						6.07	D			
						8.1	D			
						10.12	D			

Table 3. *Cont.*

Rn	Texture	Duration	Tillage	Amendment	Base Material	Rate (t ha^{-1})	Na	Ca	Mg	S
[54]	S	LT	Htt	Ct		0	-	-	-	
				FYM	Sheep manure with straw	20	NE	NE	NE	
						40	NE	NE	I	
						60	NE	NE	I	
				Com	Sewage sludge	20	NE	NE	NE	
						40	NE	NE	NE	
						60	NE	I	I	
					Municipal solid waste	20	NE	NE	NE	
						40	NE	NE	NE	
						60	NE	NE	I	

Abbreviations: Rn—reference number, C—clay, SiL—silt loam, SL—sandy loam, S—sand, Ct—control, Com—compost, Fm—fresh manure, Bc—biochar, FYM—farmyard manure, Htt—hand tool tillage, Na—sodium, Ca—calcium, Mg—magnesium, S—sulfur, ST—short-term, LT—long-term, Con—conventional, ND—non-defined, NE—no effect, I—increased, D—decreased.

Table 4. Soil amendment impact on soil hydrological response.

Rn	Texture	Study Duration	Amendment	Base Material	Rate	IR	Runoff	SL
[52]	SiL	ST	Ct		0		-	-
			Com	Olive pomace	4		NE	NE
[55]	S	ST	Ct		0	-	-	-
			Com	Cattle manure	0.013/1 *	NE	NE	NE
			FYM	Raw cattle manure	0.013/1 *	NE	NE	NE
	C		Ct		0	-	-	-
			Com	Cattle manure	0.013/1 *	D	NE	NE
			FYM	Raw cattle manure	0.013/1 *	D	NE	D

* Soil was mixed with compost or farmyard manure at 0.013/1 (w/w) ratio of dry manure/soil. Abbreviations: Rn—reference number, C—clay, SiL—silt loam, S—sand, Ct—control, Com—compost, FYM—farmyard manure, ST—short-term, IR—infiltration rate, SL—soil loss, NE—no effect, D—decreased.

2. Organic Amendments' Impact on Soil Properties

2.1. Soil Physical Properties

Multiple studies have shown the beneficial effects of organic amendments on different soil types, such as clay [37], loam [31], silty clay loam [38], and sandy loam [38,56]. These amendments enhance SOM content, a crucial soil component that functions as a binding agent, fostering aggregate formation [32,57,58]. When applied, soil structure, pore space, and aggregation improvement are detected within a few months [31,39]. While each organic amendment had its unique impact, collectively, they contributed to enriching soil quality, ensuring better plant growth and sustainability across diverse environments.

In recent studies exploring the impacts of various organic amendments on soil quality, diverse findings have emerged, shedding light on their effects across different soils and durations of application. Rivier et al. [39] demonstrated a positive influence within 30 days of adding compost and vermicompost derived from sewage sludge and organic residue to sandy and clay soils. This study emphasized the beneficial impact of these amendments on soil quality within a relatively brief duration. Conversely, Wang et al. [38] highlighted the significant effects on silty clay loam soil after six years of poultry manure compost application at a rate of 9 t ha^{-1} annually. However, insignificant differences were observed within a shorter duration of 2 years, suggesting a time-dependent influence of compost on soil quality, indicating that significant improvements may require extended application periods. Contrasting with the positive effects observed of other organic amendments, Goldberg et al. [55] reported a short-term negative impact of farmyard manure (FYM) application on soil structure and erosion resistance. This adverse effect was noted only 21 days post-application in clay and sandy soils. These findings underscore the diverse and time-dependent nature of organic amendments' impact on soil quality. The effects of organic amendments on soil quality depend on soil type and the time they are applied. Organic amendments generally increase SOM content and promote the formation of aggregates, but their impacts can differ significantly across different soils and time frames. Short-term studies show that amendments like compost and vermicompost have positive effects, while others, such as FYM, may have initial adverse effects. However, long-term studies suggest that significant improvements in soil quality may require extended application periods. Therefore, it is essential to consider the type of organic amendment and its duration of application in sustainable soil management practices.

Evidence of elevated concentrations of SOM after the application of different amendments have been found around the globe, including after the addition of FYM [40], biochar [34], composted olive pomace [59], or non-composted olive pomace [41,60]. Although the variety of amendments and their properties can vary, their impact on SOM elevation is generally positive [61]. Organic amendments contain cations of polyvalent metals (Ca^{2+}, Mg^{2+}, Fe^{3+}), which act in a similar manner to the inorganic binders that promote the formation of macro-aggregates [62]. Such phenomena are documented in improved soil structural characteristics like aggregate size and stability. However, amendments can reveal different effects on different soils. For example, Fu et al. [42] explored the effects of FYM on silty clay, silt loam, and sandy loam soils. The water-stable aggregate (WSA) percentage was significantly higher only in silty clay soil under the high dose of application (38 t ha^{-1}) condition. On the opposite side, the long-term application of FYM in a dosage of 15 t ha^{-1} on sandy loam soil creates a residual effect such that even after FYM was stopped, the WSA was at a high level, similar to other treatments with a continuous application of FYM. Therefore, whether FYM will increase or decrease the WSA depends on soil texture and the time of application.

Biochar demonstrated positive effects on WSA levels. Juriga et al. [43] found that applying 10 and 20 t ha^{-1} of biochar to silty loam soil significantly increased the content of the stable macro-aggregates by 9% and 14%, respectively. Simultaneously, it reduced the content of the stable micro-aggregates by 23% and 38%, compared to the total content of water stable micro- and macro-aggregates in the control plots. This trend was associated with an increase in SOMs. Composted materials are another aspect of soil improvement. However, it is necessary to highlight that composts differ depending on the feedstock material and their degree of decomposition, making the final soil results variable. However, composted material applications in sandy and clay soils, in general, positively affect the soil's physical properties [63]. Applying 60 t ha^{-1} of grape pomace compost mixed with mowing residues and straw significantly improves WSA in loamy soils [44]. In clay loam soil, Li et al. [56] observed that pig manure compost and vermicompost significantly increased large macro-aggregates while reducing small macro-aggregates after application at 15 t ha^{-1}. In general, soils behaved after applying organic amendments with lower bulk density, whereas aggregate stability, pore volume, and water infiltration were increased. A similar result was noted after applying FYM in works by Mujdeci et al. [45] and Bogunovic et al. [46]. They revealed the positive impact of FYM in reducing soil compaction levels, measured by the bulk density (BD) and penetration resistance (PR) of silty clay [46] after the application of 15 or 30 t ha^{-1} FYM, and of silt loam soil after the application of 10 t ha^{-1} FYM [47]. Moreover, Jensen et al. [40] observed significantly lower BD values after FYM application than control plots in sandy loam soil. Other than FYM, olive pomace has a high organic matter level, reaching 91% [64], which also helps to reduce compaction. Parras-Alcántara et al. [41] reported significant decreases in BD (1.37 to 1.26 g cm^{-3}) following

11 years of continuous application of olive pomace at 270 t ha^{-1} to silty clay and silty clay loam soil. Blanchet et al. [48], in a long-term study with FYM application (10 t ha^{-1} y^{-1}), found no significant changes for BD compared to other treatments that included mineral fertilizers and crop residue application. Such an absence of differences in BD could result from the tillage system. The diverse findings underscore the nuanced impacts of organic amendments, particularly FYM and olive pomace, on soil BD. These insights highlight the significance of utilizing organic amendments in soil management strategies, recognizing their potential to modify soil physical properties in order to enhance soil quality.

Soil structure predominantly influences the hydrological properties of soil, which regulate physical, chemical, and biological processes within the soil. In this regard, organic amendments are frequently applied in agroecosystems in order to maintain a favorable soil structure, and thus, hydrological and other vital properties [31]. Indeed, it has been shown that organic amendments can significantly affect changes in soil hydrological properties. Dong et al. [31] found that sheep manure compost (2–10 t ha^{-1}) can effectively improve the hydrological properties of loam soil. Improvements were closely connected with soil pore system, aggregation size, WSA, and soil differential porosity, leading to a significant increase in soil infiltration.

Organic amendments impact soil water retention capacity and water content. This is especially pronounced in soils enriched by biochar, FYM, and composts [61]. For instance, FYM has shown multiple beneficial effects on the hydrological properties of soil across different environments, soils, and cropping systems. Adding FYM increases the soil water content by increasing water-holding capacity and retention ability [65,66] and water-use efficiency [67,68]. As reported by Jensen et al. [40], FYM had a beneficial effect on plant available water (PAW) in sandy loam-textured soil compared to the control plots. Fu et al. [42] investigated the effects of FYM on silty clay, silty loam, and sandy loam, and found significant PAW improvements in sandy loam soil; while the other soil types showed a positive response as well, the results were not statistically significant. A similar positive effect of FYM on the PAW of sandy loam soil was reported in a study by Blanco-Canqui et al. [65]. The positive effect of manure on PAW was also found in silt loam soil [69] but not in clay soils [70]. Such discrepancies in results indicate that soil type and environment impact FYM efficiency on soil water characteristics. In this context, Ankenbauer and Loheide [71] found that the effect of organic matter on water retention characteristics was more profound on high sand and silt content soil.

Biochar behaves similarly to FYM in terms of impact on soil hydrological properties. Many studies reported a significant rise in soil retention capacity after biochar application [72–74]. This beneficial effect is mainly localized to sandy soils, as documented in a meta-analysis by Rabbi et al. [75], where biochar-enriched soils successfully reduced plant water stress during dry periods [76,77]. Moreover, soils with low silt content are likely to be more hydrologically responsive to biochar application, and changes were more pronounced at higher rates (20 and 100 t ha^{-1}) compared to the lower one (4 t ha^{-1}) [49]. In current climate crisis, due to climate change, such an effect is desirable in order to survive during pronounced drought periods. However, the success of the biochar application on soil water characteristics, besides soil texture, also depends on biochar particle size, as proved by Lim et al. [78], who found a higher decrease of saturated hydraulic conductivity under larger particle sizes of biochar than under smaller ones, proving that the structure of biochar particles and biochar-soil storage pores contributes to water retention [79]. Biochar also affects water holding capacity (WHC) in two ways. Firstly, the highly porous nature of biochar allows it to retain water, which can increase the overall moisture content. Secondly, biochar improves soil porosity and has hydrophilic functional groups on the surface, enhancing the soil's ability to hold water. However, it is essential to note that the amount of biochar used can limit these effects [80]. Applying 22.5 t ha^{-1} of biochar from various cereals and bamboo to silt loam soil significantly increased WHC by 4.1–11.9% [50]. Conversely, Parras-Alcántara et al. [41] observed a significant decrease in AWC following prolonged application (11 years) of olive pomace at 270 t ha^{-1} on silty clay and silty clay loam-textured soils.

Compost application represents another effective method that improves soil hydrological functions. The positive effect of compost on soil hydrology was recorded for different soil types, including sandy and loamy clay soils [39]. Applying compost by incorporating SOM leads to several changes in the physical and hydrological properties, increasing soil WHC [81]. The effects of compost on soil hydrological functions are more pronounced in coarser-textured soils, whereas the effect is smaller or absent for finer-textured soils [82]. Clay soils have a higher matric potential and smaller pore sizes than sandy soils and, therefore, can hold more water by weight [83]. Although compost can increase WHC due to its high organic matter content, its application does not necessarily affect PAW. Moreover, Kranz et al. [84] reported that when sandy and silt loam soils with medium porosity were amended with high levels of compost, in some cases they would show NE on PAW. Also, they showed that a significant increase in PAW can occur after applying compost to sandy clay loam and high-porosity sandy loam soils. Soil porosity, besides soil texture, also plays a vital role in compost effectivity on soil PAW.

2.2. Soil Chemical Properties

Organic amendments modify chemical properties. For example, they can have a positive effect on soil pH through the intake of organic matter, which is usually lowered after frequent fertilization with nitrogen mineral fertilizers [85]. Li et al. [56] noted a rise in soil pH after the application of 15 t ha^{-1} of pig manure compost or vermicompost to clay loam soil. Similar results were found for other soil types [35] and environments, indicating that compost has a liming potential due to its richness in alkaline cations (K^+, Ca^{2+}, Mg^{2+}), which are released with the mineralization of organic matter [83]. Sometimes, compost reduces soil pH [86], which is probably connected to specific conditions during compost production. The use of winery solid waste compost on sandy loam soil, while contributing to increased P and K mineralization, led to a notable rise in soil pH from 7.28 to 8.18 at an application rate of 40 t ha^{-1} [35].

Biochar also exhibits the capability to increase soil pH. This capability stems from its composition of alkaline substances, comprising ash and carbonate (Ca^{2+}, K^+, and Mg^{2+}), its surface properties, and its ability to reduce the exchangeable acid cations (H^+ and Al^{3+}) [87]. Bista et al. [34] reported a higher pH after biochar application of different dosages (11.2, 22.4, and 44.8 t ha^{-1}). However, biochar produced from various sources and at different temperatures showed diverse effects on soil pH [88,89]. Hossain et al. [89] found that the pH reaction of biochar derived from wastewater sludge changes from acidic to alkaline, with an increase in pyrolysis temperature from 300–700 °C. They suggested that it could be used on soils with an alkaline reaction.

In addition to pH, changes in cation exchange capacity (CEC), electrical conductivity, nutrients, and soil organic carbon (SOC) content occur after applying organic amendments. Biochar produced from various sources and at different temperatures showed diverse effects on soil CEC [88,89]. For instance, biochar produced from coffee husk or chicken manure significantly influenced soil CEC more than that produced from eucalyptus sawdust or sugarcane bagasse. Further, they observed that biochar produced at a temperature of 350 °C has a more significant impact on increasing CEC than biochar produced at temperatures of 500 °C and 750 °C [88]. Moreover, biochar application demonstrated positive implications for carbon sequestration, contributing to soil quality and sustainability [90].

The amount of nutrients released during the decomposition of FYM is influenced by various factors: livestock class, age, growth stage, feed and feeding practices, type and amount of bedding materials, and season (climate conditions) [91]. Most nutrients are expected to be released within the initial three years following application. Hence, reapplying the amendment at least every fourth year, with increased application rates, is recommended, while the annual application of smaller quantities is also suggested [92]. The latter is advised as it has been observed that it can significantly enhance crop yields.

Specifically, Oueriemmi et al. [54] noted a more significant effect on barley yield after applying different amounts of FYM in sandy-textured soil during the second year post-application compared to the first year. FYM application (60 t ha^{-1}) significantly increased barley yield by 51% (2.26 t ha^{-1}) in the first year and 77% (6.96 t ha^{-1}) in the second year compared to the control.

As previously mentioned, Fu et al. [42] explored the impact of FYM on various soil types, finding significant increases in SOC and total nitrogen with different application rates on different soil types and crops, and applying 38 t ha^{-1} of FYM on silty clay soil led to a 19% increase in SOC and total nitrogen. On sandy loam soil, there was a significant increase in SOC with application rates of 25 t ha^{-1} and 37.5 t ha^{-1}, but this increase was only present in the grass and spring barley plots, and ranged from 27–37%. On plots with maize and winter wheat, the SOC increase was insignificant. On silty loam soil, the level of SOC increased by a maximum of 48%, but this increase was insignificant. This study also found improvements in the nitrogen, phosphorus, potassium, and magnesium concentrations, but these differences were insignificant for each soil type. In sandy loam soil, there was a significant increase in total nitrogen (38–42%), phosphorus (320–840%), potassium (145–225%), and magnesium (50–120%).

Following the application of biochar in quantities of 11.2, 22.4, and 44.8 t ha^{-1}, there was a positive impact on the amount and availability of P, K, and S. This was attributed to biochar's ability to supply nutrients to the soil and elevate its pH levels, thereby enhancing their accessibility to plants. However, it should be noted that the nitrogen content in the soil decreased with increasing amounts of biochar, as nitrogen tends to bind to biochar [34].

Olive pomace compost also demonstrated positive effects on the availability of micro- and macro-nutrients, including nitrogen, potassium, zinc, magnesium, and copper, essential for optimal plant growth [93]. However, in order to achieve the effect of enriching the soil with organic carbon qualitatively and quantitatively, the amount of compost that will be applied to the soil should be correctly selected [51]. It was shown that higher amounts do not necessarily produce a proportional increase in SOC. Interestingly, a more favorable effect was achieved with a dose of 60 kg per plant compared to a double dose, particularly on soil with a clay loam texture. Additionally, despite its positive impact on soil and plant growth, applying wet olive pomace at a dosage of 70 t ha^{-1} did not result in significant differences in the crop yield of two wheat cultivars compared to the control without application on sandy loam-textured soil. However, the protein content of wheat grain was positively affected by adding olive pomace, indicating the release of nitrogen in the later stages of plant growth and development [94]. Fernández-Hernández et al. [60], after the application of olive mill waste, found a significant increase in nutrients and a 15% higher olive oil content than those treated with inorganic fertilization. Combining grape pomace with FYM fertilization substantially enhanced maize biomass, and increased soil (SOM, N, P, and K) and plant nutrient content. Additionally, a noteworthy improvement in grape yield was observed, marking a substantial increase of 48% [95].

While organic amendments offer substantial benefits, there are concerns regarding animal-based amendments with a high content of monovalent cations like Na^+ and K^+, as well as NH_4^+ from organic waste nitrogen mineralization. The presence of these components can potentially disturb soil structure by inducing soil colloid dispersion. In order to address this concern, analyzing nutrient concentrations in FYM is imperative. Additionally, accounting for variations across the landscape is crucial for determining and ensuring appropriate application rates, thereby mitigating potential adverse environmental impacts [96]. Different organic amendments exhibit diverse effects on soil properties. FYM, biochar, compost, and pomace display varying impacts based on application rates, soil types, and crops. Understanding these influences is crucial for sustainable soil management and agricultural practices.

3. Impact of Organic Amendments on Several Soil Degradation Processes and Restoration Strategies

Soil erosion is the most widespread degradation process occurring naturally but accelerated through human activities, one of which is agriculture [97]. Factors that affect soil erosion in agricultural areas are climatic conditions; soil composition and properties; shape, degree, and length of the slope; soil cover; tillage system; and the overuse of machinery [98–101]. The process occurs through three phases and involves the separation of soil particles in one location and their movement to another under the influence of water or wind energy. The third phase is soil deposition in another location [102,103]. When erosion occurs, it results in the loss of the upper fertile layer, depleting organic matter and nutrients, and affecting soil quality and productivity [104], while decreasing soil water and infiltration capacity [105,106]. When particles are deposited, soil and water can be contaminated with toxic substances, which also seriously impact the environment [107,108].

The pervasive negative effects of soil erosion have spurred global scientific efforts to develop effective mitigation strategies. A range of measures, including crop rotation, cover crops, mulching, conservation tillage, contour cropping, strip cropping, and the application of different amendments, have been investigated [109–112]. However, none of these measures proves universally applicable. Studies indicate that the most favorable outcomes are achieved by measures focused on increasing SOM, complemented by additional measures based on specific conditions [113–116]. In this context, several papers have provided evidence for the beneficial mitigation of erosion using organic amendments, like swine manure [117], olive pomace [41], FYM, and straw [118]. The erosion reductions are mainly attributed to the increase in soil organic matter, vegetation cover, and the protective role of mulch in protecting the soil from disruptive raindrop energy. However, Dugan et al. [52] did not observe a significant reduction in soil losses following the short-term application of olive pomace at a low rate (4 t ha^{-1}). These findings underscore the importance of selecting appropriate application rates and considering repeated applications for effective soil conservation. When conditioning to reduce degradation, specific conditions of each organic amendment should be considered. For example, when utilizing biochar for soil erosion control, careful attention should be paid to biochar particle size and the applied dosage. Li et al. [119] revealed that for silt loam-textured soil, smaller applied biochar doses, precisely at 1% and 3%, result in reduced soil losses, while a dose of 7% increases soil and biochar erosion. Concerning particle size, the most effective outcomes were noticed for 1–2 mm particles. Coarse biochar particles demonstrated lower susceptibility to runoff loss than finer particles, proving more effective in preventing soil loss, as noted by Peng et al. [117].

Soil compaction represents a prevalent form of soil degradation that adversely impacts various soil properties. Compaction implies an increase in the volume of the soil mass under an external force, to the detriment of the air-filled pores. Simultaneously, adverse alterations occur in pore volume, size, distribution, connection, and curvature [120–122]. This disturbance leads to a reduction in total porosity and an increase in both BD and PR [123,124]. Furthermore, compaction directly changes soil structure. When it comes to agriculture production, compacted soils have numerous implications. They are more difficult to till and usually have lower plant germination, poor root development, lower plant growth and development, reduced nutrient adsorption, and lower yields [125–127]. Additionally, compacted soils have impaired hydrological properties, resulting in reduced infiltration rates, increased surface runoff, and soil erosion [128–130]. The drivers of soil compaction can be natural or anthropogenic, but more often occur due to heavy machinery, many passes, inappropriate tire pressure, and soil tillage in wet conditions [131–134].

BD and PR are the most commonly used parameters for measuring soil compaction [135]. Previously, it was mentioned that applying FYM and other organic amendments raises SOM content and reduces BD [123,136]. However, BD reduction can be achieved in several ways. Firstly, SOM can stimulate the formation of aggregates, which create larger or smaller pores depending on aggregate size. Secondly, when organic amendments are incorporated, they have a lower density than the soil, which causes a decrease in BD through a dilution effect. Such an effect commonly occurs in (i) short-term studies where an insufficient amount of time has passed for significant changes in aggregate formation and soil porosity, and (ii) long-term studies, when a steady state is reached and there is no significant increment in the formation of macro-aggregates, but significant BD decreases still occur. Similar to FYM, biochar mitigates soil compaction level and decreases soil BD and PR after application. However, for complete and significant changes in coarse-textured soils [137] and fine-textured soils [138,139], very high doses (>40 t ha^{-1}) often need to be applied. This is likely due to short periods of time from application to measurement, and the aggregate formation is slow. Biochar's resistance to decomposition makes it challenging to promote aggregate formation and stability in the short term. Similarly, high dosages of biochar reduce BD because of the mixing of materials with different densities. When applying lower doses, biochar more effectively reduces BD in coarse-textured soils than in fine-textured soils, with an average decrease of 14.2% and 9.2%, respectively [140]. Biochar generally increases MWD and aggregate stability, although the absence of positive changes found in some studies may be attributed to site-specific conditions such as climate and clay mineralogy.

Soil sealing and crusting commonly occur during high-intensity precipitation on bare soil or lands with sparse vegetation cover. Raindrops disintegrate soil surface aggregates, clog pores, and reduce soil infiltration. Subsequently, water with dispersed soil particles accumulates on the surface, forming an impermeable layer known as soil crust [141,142]. Thus, sealing and crust formation are closely linked to aggregate stability, soil structure, and the factors promoting them, including increases in SOM, MWD, and the proportion of WSA, which enhance aggregate strength, resistance to external forces, and water movement through the soil [143]. This relationship is evident in previous studies on soil erosion, where the application of organic amendments reduced soil loss and particle detachment due to improvements in soil structure in different textured soils and environments [41,52,62,117,118].

Soil acidity is a prevalent soil degradation process in semi-humid and humid regions [144,145]. Precipitation surplus, the improper application of nitrogen and elemental sulfur fertilizers, and legume cultivation accelerate soil acidification [146–148]. Acid soils have poor soil structure and solubilize iron, aluminium, and manganese, potentially toxic to certain crops [149]. Severe acidification reduces cation exchange capacity and the availability of essential nutrients like phosphorus and molybdenum [150–152]. Soil pH can be mitigated through lime or other acid-neutralizing materials, which can strain farm budgets [153,154].

The impact of FYM on soil acidification varies depending on application conditions. Manure application typically elevates soil pH by introducing base cations and organic matter. As organic matter decomposes, it releases alkalinity through decarboxylation, consuming H^+ and raising soil pH [155]. The pH-increasing effect of FYM has been observed across different soil textures, including clay, silt loam, and sandy loam soils [155–158]. However, Roy and Kashem [36] reported a decrease in soil pH following the application of chicken (10 t ha^{-1}) and cow manure (10 t ha^{-1}) on sandy loam-textured soil exhibiting initial acidity. This pH reduction might be attributed to a weak adsorption capacity, which leads to the leaching of basic cations and the suppression of FYM's positive effect on soil pH. Some soils with high buffering capacities may also resist pH changes [159], and in this case, the addition of organic conditioners should be applied in higher dosages or for prolonged durations. Furthermore, the decomposition of organic matter into humic and fulvic acids could contribute to lowered pH values [160].

Similarly, biochar application can elevate soil pH due to its alkaline nature and oxygenated functional groups [161]. For instance, Da Silva Mendes et al. [53] observed a significant increase in soil pH from 5.35 to 5.85 by applying 10.2 t ha^{-1} of biochar to loamy sand-textured soil. Chintala et al. [162] also noted increases in soil pH with different amounts of biochar application to clay-textured soil, with higher doses leading to greater pH elevations. Other organic amendments, like olive pomace, significantly increase pH in acidic sandy loam [163] and loam [164] soils.

In semi-arid and arid regions, soil salinization and alkalization are prevalent, often due to poor agricultural practices [165–168]. These conditions result in the accumulation of salts in the topsoil, particularly sodium, displacing calcium on the soil's adsorption complex, adversely affecting the soil's physical, chemical, and biological properties. High sodium levels exacerbate soil compaction by clay dispersion [169,170]. Excessive salt concentration in the soil's rhizosphere inhibits plant growth, with some plants experiencing toxicity [171–173]. Salinization and alkalization in agricultural lands are primarily caused by using saltwater for irrigation and inadequate drainage, leading to salt accumulation and waterlogging. Mitigation strategies involve the application of gypsum and sulfur, as well as the using organic amendments, followed by leaching. Organic amendments aim to reduce the sodium adsorption ratio (SAR) and exchangeable sodium percentage (ESP), decrease EC, enhance calcium and magnesium uptake in order to counteract sodium's negative effects, decrease soil pH in order to improve nutrient availability, and increase SOM in order to promote aggregate formation [174]. Researchers used several amendments in their works, such as FYM [175], biochar [176], compost [177], pistachio residue [175], and rice straw [178], which mostly showed a positive impact on soil structure, including a decrease in soil EC and SAR. As well as directly adding organic amendments, several other practices are necessary in order to prevent SOM depletion. Most represented are cover cropping, diverse crop rotation, conservation tillage systems, mulching, crop residue management, balanced fertilization, and the promotion of biodiversity. The utilization of the practices mentioned earlier can enhance SOM, improve soil fertility, and promote sustainable agricultural management practices [178–181]. However, in order to determine the most suitable practice, it is essential to know the site-specific conditions that may affect the success of the reclamation. Figures 1 and 2 show the overall and specific improvements of soil properties in all (Figure 1) soils and across different soil textures (Figure 2).

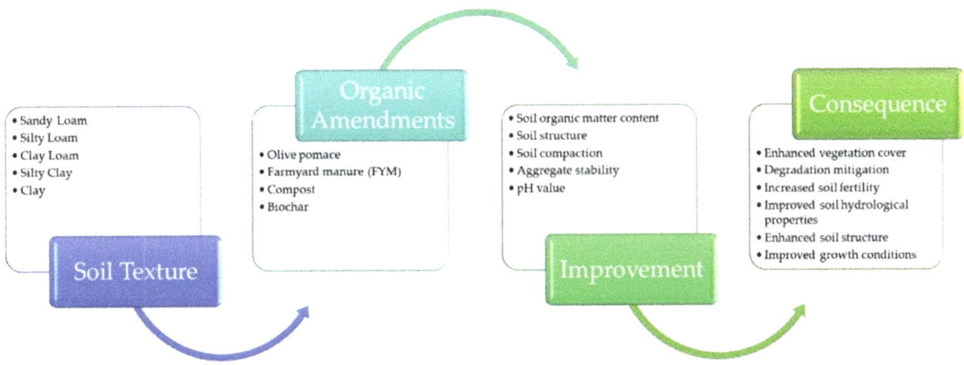

Figure 1. Improved soil properties and their environmental and agricultural impact on all soil textures.

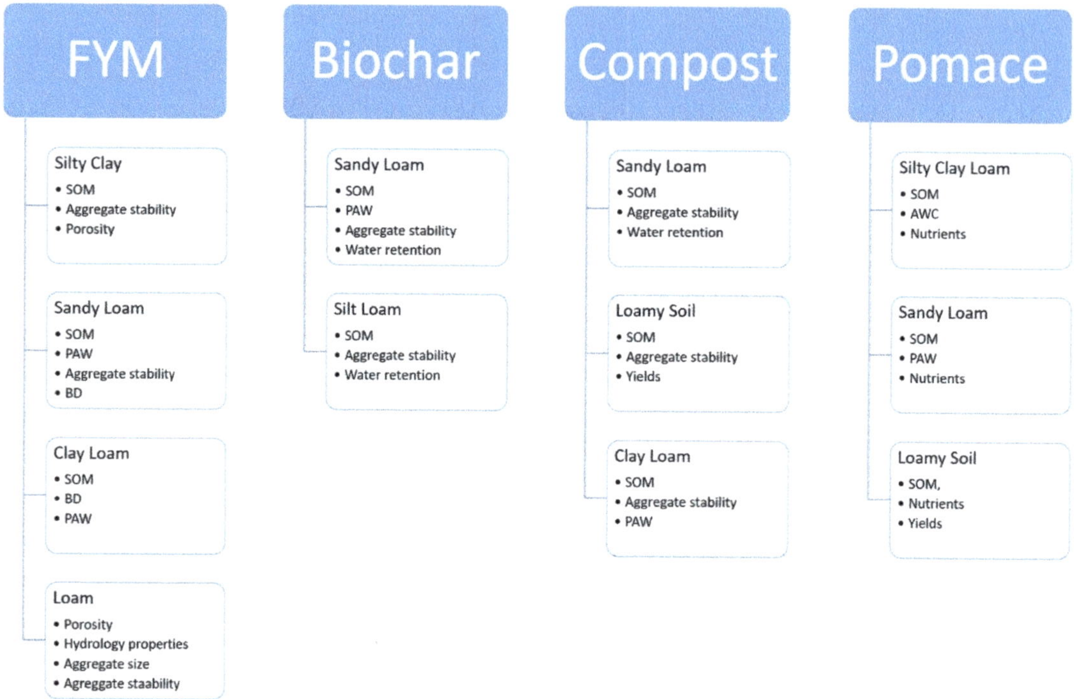

Figure 2. Specific soil property improvements in specific soil textures after several organic amendment applications. Abbreviations: SOM—soil organic matter, PAW—plant available water, BD—bulk density.

4. Guidance for Further Research

In contemporary agricultural research, the strategic integration of soil amendments with diverse tillage systems emerges as a critical pathway for advancing sustainable crop production. Embracing interdisciplinary advancements in technology, soil science, and agronomy, researchers can delve deeper into unlocking novel opportunities in order to enhance soil quality, increase productivity, and mitigate environmental impacts. The future trajectory of soil amendment management hinges on precision agriculture integration, offering researchers a fertile ground for investigation. Leveraging cutting-edge technologies such as remote sensing, GIS, and GPS enables the precise targeting of soil amendments, necessitating further exploration into optimization algorithms and decision-support systems. Real-time monitoring systems and soil sensors present a rich area for research inquiry, particularly in refining their accuracy, reliability, and compatibility with diverse agroecological contexts.

Additionally, exploring the synergistic blending of different soil amendment types holds promise for elevating soil quality and crop productivity, warranting investigation into optimal blends, application rates, and the long-term effects on soil health and ecosystem services. Advancements in biofertilizer technologies offer a frontier for research, particularly in elucidating the mechanisms underlying microbial interactions, optimizing formulations, and assessing their efficacy under varying environmental conditions. Moreover, the paradigm shifts towards data-driven decision-making necessitate interdisciplinary collaborations in order to develop robust predictive models, innovative data analytic techniques, and user-friendly decision support tools tailored to the needs of diverse stakeholders. Future research endeavors should prioritize longitudinal studies, multi-site trials, and meta-analyses to elucidate the long-term impacts of soil amendments on soil health, crop performance, and ecosystem resilience. By continuously refining methodologies, embracing emerging technologies, and fostering collaborative research networks, scientists can unlock the full potential of soil amendments in order to address the evolving challenges of modern agriculture and pave the way towards a more sustainable food system.

5. Conclusions

Organic amendments are indispensable for combating soil degradation processes and restoring soil quality. Most research emphasizes that organic amendments, including pomace, biochar, manure, and compost, offer promising strategies to maintain soil quality and sequester carbon. The effectiveness of these approaches has been duly emphasized. The present study concludes that organic amendments are promising for improving soil structure and carbon sequestration. However, their effectiveness depends on soil texture, climate, and application rates, so tailored approaches are required for optimal results. In several studies, amendment-induced changes were absent, especially for short-term periods, and these may need to be revised in order to contribute to long-term soil resilience to stress and increase soil productivity. This paper further suggests that an approach that integrates different methods is essential for achieving desirable soil quality and maintaining agricultural productivity. It also provides valuable insights and recommendations for policymakers, practitioners, and researchers. Sustainable soil management strategies must effectively combat soil degradation, including comprehensive solutions to combat erosion, compaction, sealing, acidification, and salinization. Organic amendments can improve soil ecosystem services and contribute to climate change adaptation. In the future, more attention should be paid to the interactions between soil management and soil amendments, as well as their effectiveness over time.

Author Contributions: Conceptualization, I.B. and M.M.; methodology, I.B., I.D. and M.M.; validation, I.D., M.M. and I.B.; investigation, I.B., M.M. and I.D.; resources, I.B.; data curation, M.M.; writing—original draft preparation, M.M.; writing—review and editing, I.D., M.M. and I.B.; supervision, I.B.; project administration, I.B.; funding acquisition, I.B. All authors have read and agreed to the published version of the manuscript.

Funding: This work was supported by the Partnership for Research and Innovation in the Mediterranean Area ("the PRIMA Foundation") through the "Soil Health and Agriculture Resilience through an Integrated Geographical Information Systems of Mediterranean Drylands" project (grant agreement number 2211) (SHARInG-MeD).

Institutional Review Board Statement: Not applicable.

Data Availability Statement: The data presented in this study are available on reasonable request from the corresponding author.

Conflicts of Interest: The authors declare no conflicts of interest.

References

1. van Leeuwen, J.P.; Saby, N.P.A.; Jones, A.; Louwagie, G.; Micheli, E.; Rutgers, M.; Schulte, R.P.O.; Spiegel, H.; Toth, G.; Creamer, R.E. Gap assessment in current soil monitoring networks across Europe for measuring soil functions. *Environ. Res. Lett.* **2017**, *12*, 124007. [CrossRef]
2. Ellili-Bargaoui, Y.; Walter, C.; Lemercier, B.; Michot, D. Assessment of six soil ecosystem services by coupling simulation modelling and field measurement of soil properties. *Ecol. Indic.* **2021**, *121*, 107211. [CrossRef]
3. AdhikarI, K.; Hartemink, A.E. Linking soils to ecosystem services—A global review. *Geoderma* **2016**, *262*, 101–111. [CrossRef]
4. Pereira, P.; Bogunovic, I.; Muñoz-Rojas, M.; Brevik, E.C. Soil ecosystem services, sustainability, valuation and management. *Curr. Opin. Environ. Sci. Health* **2018**, *5*, 7–13. [CrossRef]
5. Zurqani, H.A.; Mikhailova, E.A.; Post, C.J.; Schlautman, M.A.; Elhawej, A.R. A review of Libyan soil databases for use within an ecosystem services framework. *Land* **2019**, *8*, 82. [CrossRef]
6. Soto, R.L.; Padilla, M.C.; de Vente, J. Participatory selection of soil quality indicators for monitoring the impacts of regenerative agriculture on ecosystem services. *Ecosyst. Serv.* **2020**, *45*, 101157. [CrossRef]
7. Dimande, P.; Arrobas, M.; Rodrigues, M.Â. Under a Tropical Climate and in Sandy Soils, Bat Guano Mineralises Very Quickly, Behaving More like a Mineral Fertiliser than a Conventional Farmyard Manure. *Agronomy* **2023**, *13*, 1367. [CrossRef]

8. Alexandratos, N.; Bruinsma, J. World Agriculture towards 2030/2050: The 2012 Revision. Available online: https://ageconsearch.umn.edu/record/288998/ (accessed on 8 September 2023).
9. Bamdad, H.; Papari, S.; Lazarovits, G.; Berruti, F. Soil amendments for sustainable agriculture: Microbial organic fertilizers. *Soil Use Manag.* **2022**, *38*, 94–120. [CrossRef]
10. Hlisnikovský, L.; Menšík, L.; Kunzová, E. Development and the Effect of Weather and Mineral Fertilization on Grain Yield and Stability of Winter Wheat following Alfalfa—Analysis of Long-Term Field Trial. *Plants* **2023**, *12*, 1392. [CrossRef]
11. Lal, R. Restoring Soil Quality to Mitigate Soil Degradation. *Sustainability* **2015**, *7*, 5875–5895. [CrossRef]
12. Davis, A.G.; Huggins, D.R.; Reganold, J.P. Linking soil health and ecological resilience to achieve agricultural sustainability. *Front. Ecol. Environ.* **2023**, *21*, 131–139. [CrossRef]
13. Bateman, A.M.; Muñoz-Rojas, M. To whom the burden of soil degradation and management concerns. In *Advances in Chemical Pollution, Environmental Management and Protection*; Pereira, P., Ed.; Elsevier: Amsterdam, The Netherlands, 2019; Volume 4, pp. 1–22. [CrossRef]
14. United Nations. *Transforming our World: The 2030 Agenda for Sustainable Development*; Resolution adopted by the General Assembly on 25 September 2015; United Nations: New York, NY, USA, 2015; 35p. Available online: https://www.un.org/en/development/desa/population/migration/generalassembly/docs/globalcompact/A_RES_70_1_E.pdf (accessed on 15 November 2023).
15. European Commission. *European Missions: A Soil Deal for Europe-100 Living Labs and Lighthouses to Lead the Transition towards Healthy Soils by 2030-Implementation Plan*; European Commission: Brussels, Belgium, 2021; Available online: https://food.ec.europa.eu/system/files/2021-10/f2f_conf_20211015_pres-04.pdf (accessed on 15 November 2023).
16. Bouma, J.; Montanarella, L.; Evanylo, G. The challenge for the soil science community to contribute to the implementation of the UN Sustainable Development Goals. *Soil Use Manag.* **2019**, *35*, 538–546. [CrossRef]
17. Löbmann, M.T.; Maring, L.; Prokop, G.; Brils, J.; Bender, J.; Bispo, A.; Helming, K. Systems knowledge for sustainable soil and land management. *Sci. Total Environ.* **2022**, *822*, 153389. [CrossRef] [PubMed]
18. Pahalvi, H.N.; Rafiya, L.; Rashid, S.; Nisar, B.; Kamili, A.N. Chemical Fertilizers and Their Impact on Soil Health. In *Microbiota and Biofertilizers*; Dar, G.H., Bhat, R.A., Mehmood, M.A., Hakeem, K.R., Eds.; Springer: Cham, Germany, 2021; Volume 2, pp. 1–20. [CrossRef]
19. Farmaha, B.S.; Sekaran, U.; Franzluebbers, A.J. Cover cropping and conservation tillage improve soil health in the southeastern United States. *Agron. J.* **2022**, *114*, 296–316. [CrossRef]
20. Guo, M. The 3R Principles for Applying Biochar to Improve Soil Health. *Soil Syst.* **2020**, *4*, 9. [CrossRef]
21. Urra, J.; Alkorta, I.; Garbisu, C. Potential Benefits and Risks for Soil Health Derived from the Use of Organic Amendments in Agriculture. *Agronomy* **2019**, *9*, 542. [CrossRef]
22. Abdul Halim, N.S.A.; Abdullah, R.; Karsani, S.A.; Osman, N.; Panhwar, Q.A.; Ishak, C.F. Influence of soil amendments on the growth and yield of rice in acidic soil. *Agronomy* **2018**, *8*, 165. [CrossRef]
23. Rakshit, A.; Sarkar, B.; Abhilash, P. In Preface. In *Soil Amendments for Sustainability: Challenges and Perspectives*; Rakshit, A., Sarkar, B., Abhilash, P., Eds.; CRC Press: Boca Raton, FL, USA, 2018; In Preface.
24. Chatzistathis, T.; Papaioannou, E.; Giannakoula, A.; Papadakis, I.E. Zeolite and Vermiculite as Inorganic Soil Amendments Modify Shoot-Root Allocation, Mineral Nutrition, Photosystem II Activity and Gas Exchange Parameters of Chestnut (*Castanea sativa* Mill) Plants. *Agronomy* **2021**, *11*, 109. [CrossRef]
25. Kamali, M.; Sweygers, N.; Al-Salem, S.; Appels, L.; Aminabhavi, T.M.; Dewil, R. Biochar for soil applications-sustainability aspects, challenges and future prospects. *Chem. Eng. J.* **2022**, *428*, 131189. [CrossRef]
26. Głąb, T.; Gondek, K.; Marcińska-Mazur, L.; Jarosz, R.; Mierzwa-Hersztek, M. Effect of organic/inorganic composites as soil amendments on the biomass productivity and root architecture of spring wheat and rapeseed. *J. Environ. Manag.* **2023**, *344*, 118628. [CrossRef]
27. Bogunović, I.; Filipović, V. Mulch as a nature-based solution to halt and reverse land degradation in agricultural areas. *Curr. Opin. Environ. Sci. Health* **2023**, *34*, 100488. [CrossRef]
28. De Corato, U. Agricultural waste recycling in horticultural intensive farming systems by on-farm composting and compost-based tea application improves soil quality and plant health: A review under the perspective of a circular economy. *Sci. Total Environ.* **2020**, *738*, 139840. [CrossRef] [PubMed]
29. Danso, F.; Agyare, W.A.; Bart-Plange, A. Benefits and costs of cultivating rice using biochar-inorganic fertilizer combinations. *J. Sci. Food Agric.* **2023**, *11*, 100491. [CrossRef]
30. Bhogal, A.; Nicholson, F.A.; Rollett, A.; Taylor, M.; Litterick, A.; Whittingham, M.J.; Williams, J.R. Improvements in the quality of agricultural soils following organic material additions depend on both the quantity and quality of the materials applied. *Front. Sustain. Food Syst.* **2018**, *2*, 9. [CrossRef]
31. Dong, L.; Zhang, W.; Xiong, Y.; Zou, J.; Huang, Q.; Xu, X.; Ren, P.; Huang, G. Impact of short-term organic amendments incorporation on soil structure and hydrology in semiarid agricultural lands. *Int. Soil Water Conserv. Res.* **2022**, *10*, 457–469. [CrossRef]
32. Lal, R. Soil organic matter and water retention. *Agron. J.* **2020**, *112*, 3265–3277. [CrossRef]
33. Argaman, E.; Stavi, I. Runoff Mitigation in Croplands: Evaluating the Benefits of Straw Mulching and Polyacrylamide Techniques. *Agronomy* **2023**, *13*, 1935. [CrossRef]

34. Bista, P.; Ghimire, R.; Machado, S.; Pritchett, L. Biochar Effects on Soil Properties and Wheat Biomass vary with Fertility Management. *Agronomy* **2019**, *9*, 623. [CrossRef]
35. Masowa, M.M.; Dlamini, P.; Babalola, O.O.; Mulidzi, A.R.; Kutu, F.R. In-field assessment of soil pH and mineralization of phosphorus and potassium following the application of composted winery solid waste in sandy loam Ferric Luvisol. *Emir. J. Food Agric.* **2023**, *35*, 666–673. [CrossRef]
36. Roy, S.; Kashem, M.A. Effects of organic manures in changes of some soil properties at different incubation periods. *Open J. Soil Sci.* **2014**, *4*, 43613. [CrossRef]
37. de Melo, T.R.; Figueiredo, A.; Machado, W.; Tavares Filho, J. Changes on soil structural stability after in natura and composted chicken manure application. *Int. J. Recycl. Org. Waste Agric.* **2019**, *8*, 333–338. [CrossRef]
38. Wang, D.; Lin, J.Y.; Sayre, J.M.; Schmidt, R.; Fonte, S.J.; Rodrigues, J.L.; Scow, K.M. Compost amendment maintains soil structure and carbon storage by increasing available carbon and microbial biomass in agricultural soil–A six-year field study. *Geoderma* **2022**, *427*, 116117. [CrossRef]
39. Rivier, P.; Jamniczky, D.; Nemes, A.; Makó, A.; Barna, G.; Uzinger, N.; Rékási, M.; Farkas, C. Short-term effects of compost amendments to soil on soil structure, hydraulic properties, and water regime. *J. Hydrol. Hydromech.* **2022**, *70*, 74–88. [CrossRef]
40. Jensen, J.L.; Schjønning, P.; Christensen, B.T.; Munkholm, L.J. Suboptimal fertilisation compromises soil physical properties of a hard-setting sandy loam. *Soil. Res.* **2016**, *55*, 332–340. [CrossRef]
41. Parras-Alcántara, L.; Lozano-García, B.; Keesstra, S.; Cerdà, A.; Brevik, E.C. Long-term effects of soil management on ecosystem services and soil loss estimation in olive grove top soils. *Sci. Total Environ.* **2016**, *571*, 498–506. [CrossRef]
42. Fu, Y.; de Jonge, L.W.; Moldrup, P.; Paradelo, M.; Arthur, E. Improvements in soil physical properties after long-term manure addition depend on soil and crop type. *Geoderma* **2022**, *425*, 116062. [CrossRef]
43. Juriga, M.; Aydın, E.; Horák, J.; Chlpík, J.; Rizhiya, E.Y.; Buchkina, N.P.; Balashov, E.V.; Šimanský, V. The importance of initial application and reapplication of biochar in the context of soil structure improvement. *J. Hydrol. Hydromech.* **2021**, *69*, 87–97. [CrossRef]
44. Novotná, J.; Badalíková, B. The Soil Structure Changes under Varying Compost Dosage. *Agriculture Pol'nohospodárstvo* **2018**, *64*, 143–148. [CrossRef]
45. Mujdeci, M.; Isildar, A.A.; Uygur, V.; Alaboz, P.; Unlu, H.; Senol, H. Cooperative effects of field traffic and organic matter treatments on some compaction-related soil properties. *Solid Earth* **2017**, *8*, 189–198. [CrossRef]
46. Bogunovic, I.; Pereira, P.; Galic, M.; Kisic, I. Tillage system and farmyard manure impact on soil physical properties, CO_2 emissions, and crop yield in an organic farm located in a Mediterranean environment (Croatia). *Environ. Earth Sci.* **2020**, *79*, 70. [CrossRef]
47. Patial, D.; Sankhyan, N.K.; Sharma, R.P.; Dev, P.; Anjali. Assessing Soil Physical and Chemical Properties Under Long Term Fertilization After Forty-Eight Years in North-Western Himalayas. *Commun. Soil Sci. Plant Anal.* **2022**, *53*, 2257–2270. [CrossRef]
48. Blanchet, G.; Gavazov, K.; Bragazza, L.; Sinaj, S. Responses of soil properties and crop yields to different inorganic and organic amendments in a Swiss conventional farming system. *Agric. Ecosyst. Environ.* **2016**, *230*, 116–126. [CrossRef]
49. Peake, L.R.; Reid, B.J.; Tang, X. Quantifying the influence of biochar on the physical and hydrological properties of dissimilar soils. *Geoderma* **2014**, *235*, 182–190. [CrossRef]
50. Yang, C.D.; Lu, S.G. Effects of five different biochars on aggregation, water retention and mechanical properties of paddy soil: A field experiment of three-season crops. *Soil Tillage Res.* **2021**, *205*, 104798. [CrossRef]
51. Vignozzi, N.; Andrenelli, M.C.; Agnelli, A.E.; Fiore, A.; Pellegrini, S. Short-Term Effect of Different Inputs of Organic Amendments from Olive Oil Industry By-Products on Soil Organic Carbon and Physical Properties. *Land* **2023**, *12*, 1628. [CrossRef]
52. Dugan, I.; Pereira, P.; Barcelo, D.; Bogunovic, I. Conservation practices reverse soil degradation in Mediterranean fig orchards. *Geoderma Reg.* **2023**, *36*, e00750. [CrossRef]
53. Da Silva Mendes, J.; Fernandes, J.D.; Chaves, L.H.G.; Guerra, H.O.C.; Tito, G.A.; de Brito Chaves, I. Chemical and physical changes of soil amended with biochar. *Water Air Soil Pollut.* **2021**, *232*, 338. [CrossRef]
54. Oueriemmi, H.; Kidd, P.; Trasar-Cepeda, C.; Rodríguez-Garrido, B.; Zoghlami, R.; Ardhaoui, K.; Prieto-Fernández, Á.; Moussa, M. Evaluation of Composted Organic Wastes and Farmyard Manure for Improving Fertility of Poor Sandy Soils in Arid Regions. *Agriculture* **2021**, *11*, 415. [CrossRef]
55. Goldberg, N.; Nachshon, U.; Argaman, E.; Ben-Hur, M. Short term effects of livestock manures on soil structure stability, runoff and soil erosion in semi-arid soils under simulated rainfall. *Geosciences* **2020**, *10*, 213. [CrossRef]
56. Li, P.; Kong, D.; Zhang, H.; Xu, L.; Li, C.; Wu, M.; Jiao, J.; Li, D.; Xu, L.; Hu, F. Different regulation of soil structure and resource chemistry under animal-and plant-derived organic fertilizers changed soil bacterial communities. *Appl. Soil Ecol.* **2021**, *165*, 104020. [CrossRef]
57. Golchin, A.; Baldock, J.A.; Oades, J.M. A model linking organic matter decomposition, chemistry, and aggregate dynamics. In *Soil Processes and the Carbon Cycle*; Lal, R., Kimble, J.M., Follett, R.F., Stewart, B.A., Eds.; CRC Press: Boca Raton, FL, USA, 2018; pp. 245–266.
58. Hartmann, M.; Six, J. Soil structure and microbiome functions in agroecosystems. *Nat. Rev. Earth Environ.* **2023**, *4*, 4–18. [CrossRef]
59. Innangi, M.; Niro, E.; D'Ascoli, R.; Danise, T.; Proietti, P.; Nasini, L.; Regini, L.; Castaldi, S.; Fioretto, A. Effects of olive pomace amendment on soil enzyme activities. *Appl. Soil Ecol.* **2017**, *119*, 242–249. [CrossRef]

60. Fernández-Hernández, A.; Roig, A.; Serramiá, N.; Civantos, C.G.O.; Sánchez-Monedero, M.A. Application of compost of two-phase olive mill waste on olive grove: Effects on soil, olive fruit and olive oil quality. *Waste Manag.* **2014**, *34*, 1139–1147. [CrossRef] [PubMed]
61. Siedt, M.; Schäffer, A.; Smith, K.E.; Nabel, M.; Roß-Nickoll, M.; van Dongen, J.T. Comparing straw, compost, and biochar regarding their suitability as agricultural soil amendments to affect soil structure, nutrient leaching, microbial communities, and the fate of pesticides. *Sci. Total Environ.* **2021**, *751*, 141607. [CrossRef] [PubMed]
62. Li, Y.; Feng, G.; Tewolde, H.; Yang, M.; Zhang, F. Soil, biochar, and nitrogen loss to runoff from loess soil amended with biochar under simulated rainfall. *J. Hydrol.* **2020**, *591*, 125318. [CrossRef]
63. Amlinger, F.; Peyr, S.; Geszti, J.; Dreher, P.; Weinfurtner, K.; Nortcliff, S. *Evaluierung der Nachhaltig Positiven Wirkung von Kompost auf Die Fruchtbarkeit und Produktivität von Böden*; Bundesministerium für Land-und Forstwirtschaft, Umwelt und Wasserwirtschaft, Lebensministerium: Vienna, Austria, 2006; p. 245.
64. Galic, M.; Bogunovic, I. Use of organic amendment from olive and wine industry in agricultural land: A review. *Agric. Conspec. Sci.* **2018**, *83*, 123–129.
65. Blanco-Canqui, H.; Hergert, G.W.; Nielsen, R.A. Cattle manure application reduces soil compactibility and increases water retention after 71 years. *Soil Sci. Soc. Am. J.* **2015**, *79*, 212–223. [CrossRef]
66. Wang, X.; Wang, L.; Wang, T. Effect of Replacing Mineral Fertilizer with Manure on Soil Water Retention Capacity in a Semi-Arid Region. *Agronomy* **2023**, *13*, 2272. [CrossRef]
67. Wang, X.; Jia, Z.; Liang, L.; Yang, B.; Ding, R.; Nie, J.; Wang, J. Impacts of manure application on soil environment, rainfall use efficiency and crop biomass under dryland farming. *Sci. Rep.* **2016**, *6*, 20994. [CrossRef]
68. Yessoufou, M.W.; Tovihoudji, P.G.; Zakari, S.; Adjogboto, A.; Djenontin, A.J.; Akponikpè, P.I. Hill-placement of manure and fertilizer for improving maize nutrient-and water-use efficiencies in the northern Benin. *Heliyon* **2023**, *9*, E17823. [CrossRef]
69. Liu, C.A.; Li, F.R.; Zhou, L.M.; Zhang, R.H.; Lin, S.L.; Wang, L.J.; Siddique, K.H.M.; Li, F.M. Effect of organic manure and fertilizer on soil water and crop yields in newly-built terraces with loess soils in a semi-arid environment. *Agric. Water Manag.* **2013**, *117*, 123–132. [CrossRef]
70. Miller, J.J.; Beasley, B.W.; Drury, C.F.; Larney, F.J.; Hao, X.; Chanasyk, D.S. Influence of long-term feedlot manure amendments on soil hydraulic conductivity, water-stable aggregates, and soil thermal properties during the growing season. *Can. J. Soil Sci.* **2018**, *98*, 421–435. [CrossRef]
71. Ankenbauer, K.J.; Loheide, S.P. The effects of soil organic matter on soil water retention and plant water use in a meadow of the Sierra Nevada, CA. *Hydrol. Process.* **2017**, *31*, 891–901. [CrossRef]
72. Igalavithana, A.D.; Ok, Y.S.; Niazi, N.K.; Rizwan, M.; Al-Wabel, M.I.; Usman, A.R.; Moon, D.H.; Lee, S.S. Effect of corn residue biochar on the hydraulic properties of sandy loam soil. *Sustainability* **2017**, *9*, 266. [CrossRef]
73. Baiamonte, G.; Crescimanno, G.; Parrino, F.; De Pasquale, C. Effect of biochar on the physical and structural properties of a sandy soil. *Catena* **2019**, *175*, 294–303. [CrossRef]
74. Ni, J.J.; Bordoloi, S.; Shao, W.; Garg, A.; Xu, G.; Sarmah, A.K. Two-year evaluation of hydraulic properties of biochar-amended vegetated soil for application in landfill cover system. *Sci. Total Environ.* **2020**, *712*, 136486. [CrossRef]
75. Rabbi, S.M.; Minasny, B.; Salami, S.T.; McBratney, A.B.; Young, I.M. Greater, but not necessarily better: The influence of biochar on soil hydraulic properties. *Eur. J. Soil Sci.* **2021**, *72*, 2033–2048. [CrossRef]
76. Alkharabsheh, H.M.; Seleiman, M.F.; Battaglia, M.L.; Shami, A.; Jalal, R.S.; Alhammad, B.A.; Almutairi, K.F.; Al-Saif, A.M. Biochar and its broad impacts in soil quality and fertility, nutrient leaching and crop productivity: A review. *Agronomy* **2021**, *11*, 993. [CrossRef]
77. Zheng, J.; Wang, S.; Wang, R.; Chen, Y.; Siddique, K.H.; Xia, G.; Chi, D. Ameliorative roles of biochar-based fertilizer on morpho-physiological traits, nutrient uptake and yield in peanut (*Arachis hypogaea* L.) under water stress. *Agric. Water Manag.* **2021**, *257*, 107129. [CrossRef]
78. Lim, T.J.; Spokas, K.A.; Feyereisen, G.; Novak, J.M. Predicting the impact of biochar additions on soil hydraulic properties. *Chemosphere* **2016**, *142*, 136–144. [CrossRef]
79. Barnes, R.T.; Gallagher, M.E.; Masiello, C.A.; Liu, Z.; Dugan, B. Biochar-induced changes in soil hydraulic conductivity and dissolved nutrient fluxes constrained by laboratory experiments. *PLoS ONE* **2014**, *9*, e108340. [CrossRef]
80. Chang, Y.; Rossi, L.; Zotarelli, L.; Gao, B.; Shahid, M.A.; Sarkhosh, A. Biochar improves soil physical characteristics and strengthens root architecture in Muscadine grape (*Vitis rotundifolia* L.). *Chem. Biol. Technol. Agric.* **2021**, *8*, 7. [CrossRef]
81. Bondì, C.; Castellini, M.; Iovino, M. Compost amendment impact on soil physical quality estimated from hysteretic water retention curve. *Water* **2022**, *14*, 1002. [CrossRef]
82. Brown, S.; Cotton, M. Changes in Soil Properties and Carbon Content Following Compost Application: Results of On-farm Sampling. *Compost Sci. Util.* **2011**, *19*, 88–97. [CrossRef]
83. Adugna, G. A review on impact of compost on soil properties, water use and crop productivity. *Acad. Res. J. Agric. Sci. Res.* **2016**, *4*, 93–104. [CrossRef]
84. Kranz, C.N.; McLaughlin, R.A.; Amoozegar, A.; Heitman, J.L. Influence of compost amendment rate and level of compaction on the hydraulic functioning of soils. *J. Am. Water Resour. Assoc.* **2023**, *59*, 1115–1127. [CrossRef]
85. Assefa, S.; Tadesse, S. The principal role of organic fertilizer on soil properties and agricultural productivity–A review. *Agric. Res. Technol. Open Access* **2019**, *22*, 556192. [CrossRef]

86. Abujabhah, I.S.; Bound, S.A.; Doyle, R.; Bowman, J.P. Effects of biochar and compost amendments on soil physico-chemical properties and the total community within a temperate agricultural soil. *Appl. Soil Ecol.* **2016**, *98*, 243. [CrossRef]
87. Hailegnaw, N.S.; Mercl, F.; Pračke, K.; Száková, J.; Tlustoš, P. Mutual relationships of biochar and soil pH, CEC, and exchangeable base cations in a model laboratory experiment. *J. Soils Sediments* **2019**, *19*, 2405–2416. [CrossRef]
88. Domingues, R.R.; Sánchez-Monedero, M.A.; Spokas, K.A.; Melo, L.C.A.; Trugilho, P.F.; Valenciano, M.N.; Silva, C.A. Enhancing Cation Exchange Capacity of Weathered Soils Using Biochar: Feedstock, Pyrolysis Conditions and Addition Rate. *Agronomy* **2020**, *10*, 824. [CrossRef]
89. Hossain, M.K.; Strezov, V.; Chan, K.Y.; Ziolkowski, A.; Nelson, P.F. Influence of pyrolysis temperature on production and nutrient properties of wastewater sludge biochar. *J. Environ. Manag.* **2011**, *92*, 223–228. [CrossRef] [PubMed]
90. Cooper, J.; Greenberg, I.; Ludwig, B.; Hippich, L.; Fischer, D.; Glaser, B.; Kaiser, M. Effect of biochar and compost on soil properties and organic matter in aggregate size fractions under field conditions. *Agric. Ecosyst. Environ.* **2020**, *295*, 106882. [CrossRef]
91. Spaeth, K.E., Jr. *Soil Health on the Farm, Ranch, and in the Garden*; Springer: Cham, Switzerland, 2020; pp. 227–295. [CrossRef]
92. Barker, A.V. Management of Farm Manures. In *Science and Technology of Organic Farming*, 1st ed.; Barker, A.V., Ed.; CRC Press: Boca Raton, FL, USA, 2010; pp. 81–103. [CrossRef]
93. Bateni, C.; Ventura, M.; Tonon, G.; Pisanelli, A. Soil carbon stock in olive groves agroforestry systems under different management and soil characteristics. *Agroforest Syst.* **2021**, *95*, 951–961. [CrossRef]
94. Lacolla, G.; Fortunato, S.; Nigro, D.; De Pinto, M.C.; Mastro, M.A.; Caranfa, D.; Gadaleta, A.; Cucci, G. Effects of mineral and organic fertilization with the use of wet olive pomace on durum wheat performance. *Int. J. Recycl. Org. Waste Agric.* **2019**, *8* (Suppl. S1), 245–254. [CrossRef]
95. Mpanga, I.K.; Neumann, G.; Brown, J.K.; Blankinship, J.; Tronstad, R.; Idowu, O. Grape pomace's potential on semi-arid soil health enhances performance of maize, wheat, and grape crops. *JPNSS* **2023**, *186*, 276–285. [CrossRef]
96. Ozlu, E.; Kumar, S. Response of soil organic carbon, pH, electrical conductivity, and water stable aggregates to long-term annual manure and inorganic fertilizer. *Soil Sci. Soc. Am. J.* **2018**, *82*, 1243–1251. [CrossRef]
97. Ouyang, W.; Wu, Y.; Hao, Z.; Zhang, Q.; Bu, Q.; Gao, X. Combined impacts of land use and soil property changes on soil erosion in a mollisol area under long-term agricultural development. *Sci. Total Environ.* **2018**, *613*, 798–809. [CrossRef]
98. Sharma, A.; Tiwari, K.N.; Bhadoria, P.B.S. Effect of land use land cover change on soil erosion potential in an agricultural watershed. *Environ. Monit. Assess.* **2011**, *173*, 789–801. [CrossRef]
99. Li, Z.; Fang, H. Impacts of climate change on water erosion: A review. *Earth-Sci. Rev.* **2016**, *163*, 94–117. [CrossRef]
100. Bogunovic, I.; Pereira, P.; Kisic, I.; Sajko, K.; Sraka, M. Tillage management impacts on soil compaction, erosion and crop yield in Stagnosols (Croatia). *Catena* **2018**, *160*, 376–384. [CrossRef]
101. Wang, Y.; Zhang, J.H.; Zhang, Z.H.; Jia, L.Z. Impact of tillage erosion on water erosion in a hilly landscape. *Sci. Total Environ.* **2016**, *551*, 522–532. [CrossRef]
102. Shi, Z.H.; Fang, N.F.; Wu, F.Z.; Wang, L.; Yue, B.J.; Wu, G.L. Soil erosion processes and sediment sorting associated with transport mechanisms on steep slopes. *J. Hydrol.* **2012**, *454*, 123–130. [CrossRef]
103. Bajracharya, R.M.; Lal, R.; Kimble, J.M. Soil Organic Carbon Distribution in Aggregates and Primary Particle Fractions as Influenced by Erosion Phases and Landscape Positions. In *Soil Processes and the Carbon Cycle*; Lal, R., Kimble, J.M., Follet, R.F., Stewart, B.A., Eds.; CRC Press: Boca Raton, FL, USA, 1997; pp. 353–368.
104. Stocking, M. Soil erosion and land degradation. In *Environmental Science for Environmental Management*, 2nd ed.; O'Riordan, T., Ed.; Routledge: Oxfordshire, UK, 2000; pp. 287–321. [CrossRef]
105. Kadlec, V.; Procházková, E.; Urbanová, J.; Tippl, M.; Holubík, O. Soil Organic Carbon Dynamics and its Influence on the Soil Erodibility Factor. *Soil Water Res.* **2012**, *7*, 97–108. [CrossRef]
106. Arunrat, N.; Sereenonchai, S.; Kongsurakan, P.; Hatano, R. Soil organic carbon and soil erodibility response to various land-use changes in northern Thailand. *Catena* **2022**, *219*, 106595. [CrossRef]
107. Rose, N.L.; Yang, H.; Turner, S.D.; Simpson, G.L. An assessment of the mechanisms for the transfer of lead and mercury from atmospherically contaminated organic soils to lake sediments with particular reference to Scotland, UK. *Geochim. Cosmochim. Acta* **2012**, *82*, 113–135. [CrossRef]
108. Huang, B.; Yuan, Z.; Li, D.; Zheng, M.; Nie, X.; Liao, Y. Effects of soil particle size on the adsorption, distribution, and migration behaviors of heavy metal(loid)s in soil: A review. *Environ. Sci. Process. Impacts* **2020**, *22*, 1596–1615. [CrossRef]
109. Nearing, M.A.; Xie, Y.; Liu, B.; Ye, Y. Natural and anthropogenic rates of soil erosion. *Int. Soil Water Conserv. Res.* **2017**, *5*, 77–84. [CrossRef]
110. Lal, R. Soil conservation and ecosystem services. *Int. Soil Water Conserv. Res.* **2014**, *2*, 36–47. [CrossRef]
111. Xiong, M.; Sun, R.; Chen, L. Effects of soil conservation techniques on water erosion control: A global analysis. *Sci. Total Environ.* **2018**, *645*, 753–760. [CrossRef]
112. Haregeweyn, N.; Tsunekawa, A.; Nyssen, J.; Poesen, J.; Tsubo, M.; Tsegaye Meshesha, D.; Schütt, B.; Adgo, E.; Tegegne, F. Soil erosion and conservation in Ethiopia: A review. *Prog. Phys. Geogr.* **2015**, *39*, 750–774. [CrossRef]
113. Rickson, R.J. Can control of soil erosion mitigate water pollution by sediments? *Sci. Total Environ.* **2014**, *468*, 1187–1197. [CrossRef] [PubMed]
114. Prosdocimi, M.; Tarolli, P.; Cerdà, A. Mulching practices for reducing soil water erosion: A review. *Earth-Sci. Rev.* **2016**, *161*, 191–203. [CrossRef]

115. Chen, D.; Wei, W.; Chen, L. Effects of terracing practices on water erosion control in China: A meta-analysis. *Earth-Sci. Rev.* **2017**, *173*, 109–121. [CrossRef]
116. Baumhardt, R.L.; Stewart, B.A.; Sainju, U.M. North American soil degradation: Processes, practices, and mitigating strategies. *Sustainability* **2015**, *7*, 2936–2960. [CrossRef]
117. Peng, X.; Zhu, Q.H.; Xie, Z.B.; Darboux, F.; Holden, N.M. The impact of manure, straw and biochar amendments on aggregation and erosion in a hillslope Ultisol. *Catena* **2016**, *138*, 30–37. [CrossRef]
118. Shi, Y.; Zhang, Q.; Liu, X.; Jing, X.; Shi, C.; Zheng, L. Organic manure input and straw cover improved the community structure of nitrogen cycle function microorganism driven by water erosion. *Int. Soil Water Conserv. Res.* **2022**, *10*, 129–142. [CrossRef]
119. Li, Y.; Zhang, F.; Yang, M.; Zhang, J.; Xie, Y. Impacts of biochar application rates and particle sizes on runoff and soil loss in small cultivated loess plots under simulated rainfall. *Sci. Total Environ.* **2018**, *649*, 1403–1413. [CrossRef] [PubMed]
120. Alaoui, A.; Rogger, M.; Peth, S.; Blöschl, G. Does soil compaction increase floods? A review. *J. Hydrol.* **2018**, *557*, 631–642. [CrossRef]
121. Kim, H.; Anderson, S.H.; Motavalli, P.P.; Gantzer, C.J. Compaction effects on soil macropore geometry and related parameters for an arable field. *Geoderma* **2010**, *160*, 244–251. [CrossRef]
122. Nawaz, M.F.; Bourrie, G.; Trolard, F. Soil compaction impact and modelling. A review. *Agron. Sustain. Dev.* **2013**, *33*, 291–309. [CrossRef]
123. Celik, I.; Gunal, H.; Budak, M.; Akpinar, C. Effects of long-term organic and mineral fertilizers on bulk density and penetration resistance in semi-arid Mediterranean soil conditions. *Geoderma* **2010**, *160*, 236–243. [CrossRef]
124. Gao, W.; Watts, C.W.; Ren, T.; Whalley, W.R. The effects of compaction and soil drying on penetrometer resistance. *Soil Tillage Res.* **2012**, *125*, 14–22. [CrossRef]
125. Hargreaves, P.R.; Baker, K.L.; Graceson, A.; Bonnett, S.; Ball, B.C.; Cloy, J.M. Soil compaction effects on grassland silage yields and soil structure under different levels of compaction over three years. *Eur. J. Agron.* **2019**, *109*, 125916. [CrossRef]
126. Shah, A.N.; Tanveer, M.; Shahzad, B.; Yang, G.; Fahad, S.; Ali, S.; Bukhari, M.A.; Tung, S.A.; Hafeez, A.; Souliyanonh, B. Soil compaction effects on soil health and cropproductivity: An overview. *Environ. Sci. Pollut. Res.* **2017**, *24*, 10056–10067. [CrossRef] [PubMed]
127. Munkholm, L.J.; Heck, R.J.; Deen, B. Long-term rotation and tillage effects on soil structure and crop yield. *Soil Tillage Res.* **2013**, *127*, 85–91. [CrossRef]
128. Capello, G.; Biddoccu, M.; Ferraris, S.; Cavallo, E. Effects of tractor passes on hydrological and soil erosion processes in tilled and grassed vineyards. *Water* **2019**, *11*, 2118. [CrossRef]
129. Alaoui, A.; Lipiec, J.; Gerke, H.H. A review of the changes in the soil pore system due to soil deformation: A hydrodynamic perspective. *Soil Tillage Res.* **2011**, *115*, 1–15. [CrossRef]
130. Prats, S.A.; Malvar, M.C.; Coelho, C.O.A.; Wagenbrenner, J.W. Hydrologic and erosion responses to compaction and added surface cover in post-fire logged areas: Isolating splash, interrill and rill erosion. *J. Hydrol.* **2019**, *575*, 408–419. [CrossRef]
131. Ahmadi, I.; Ghaur, H. Effects of soil moisture content and tractor wheeling intensity on traffic-induced soil compaction. *J. Cent. Eur. Agric.* **2015**, *16*, 489–502. [CrossRef]
132. Botta, G.F.; Tolon-Becerra, A.; Tourn, M.; Lastra-Bravo, X.; Rivero, D. Agricultural traffic: Motion resistance and soil compaction in relation to tractor design and different soil conditions. *Soil Tillage Res.* **2012**, *120*, 92–98. [CrossRef]
133. Becerra, A.T.; Botta, G.F.; Bravo, X.L.; Tourn, M.; Melcon, F.B.; Vazquez, J.; Rivero, D.; Linares, P.; Nardon, G. Soil compaction distribution under tractor traffic in almond (*Prunus amigdalus* L.) orchard in Almería España. *Soil Tillage Res.* **2010**, *107*, 49–56. [CrossRef]
134. Elaoud, A.; Chehaibi, S. Soil compaction due to tractor traffic. *J. Fail. Anal. Prev.* **2011**, *11*, 539–545. [CrossRef]
135. Shaheb, M.R.; Venkatesh, R.; Shearer, S.A. A review on the effect of soil compaction and its management for sustainable crop production. *Biosyst. Eng.* **2021**, *46*, 417–439. [CrossRef]
136. Bandyopadhyay, K.K.; Misra, A.K.; Ghosh, P.K.; Hati, K.M. Effect of integrated use of farmyard manure and chemical fertilizers on soil physical properties and productivity of soybean. *Soil Tillage Res.* **2010**, *110*, 115–125. [CrossRef]
137. Walters, R.D.; White, J.G. Biochar in situ decreased bulk density and improved soil-water relations and indicators in Southeastern US Coastal Plain Ultisols. *Soil Sci.* **2018**, *183*, 99–111. [CrossRef]
138. Zhang, X.; Wang, K.; Sun, C.; Yang, K.; Zheng, J. Differences in soil physical properties caused by applying three organic amendments to loamy clay soil under field conditions. *J. Soils Sediments* **2022**, *22*, 43–55. [CrossRef]
139. Bogunovic, I.; Dugan, I.; Pereira, P.; Filipovic, V.; Filipovic, L.; Krevh, V.; Defteradovic, J.; Matisic, M.; Kisic, I. Effects of Biochar and Cattle Manure under Different Tillage Management on Soil Properties and Crop Growth in Croatia. *Agriculture* **2023**, *13*, 2128. [CrossRef]
140. Blanco-Canqui, H. Biochar and soil physical properties. *Soil Sci. Soc. Am. J.* **2017**, *81*, 687–711. [CrossRef]
141. Chamizo, S.; Cantón, Y.; Lázaro, R.; Solé-Benet, A.; Domingo, F. Crust composition and disturbance drive infiltration through biological soil crusts in semiarid ecosystems. *Ecosystems* **2012**, *15*, 148–161. [CrossRef]
142. Indoria, A.K.; Rao, C.S.; Sharma, K.L.; Reddy, K.S. Conservation agriculture–A panacea to improve soil physical health. *Curr. Sci.* **2017**, *112*, 52–61. Available online: https://www.jstor.org/stable/24911616 (accessed on 6 December 2023). [CrossRef]
143. Šimanský, V.; Jonczak, J. Aluminium and iron oxides affect the soil structure in a long-term mineral fertilised soil. *J. Soils Sediments* **2020**, *20*, 2008–2018. [CrossRef]

144. Fujii, K.; Funakawa, S.; Kosaki, T. Soil acidification: Natural processes and human impact. *Pedologist* **2012**, *55*, 415–425. [CrossRef]
145. En-Qing, H.O.U.; Xiang, H.M.; Jian-Li, L.I.; Jiong, L.I.; Da-Zhi, W.E.N. Soil acidification and heavy metals in urban parks as affected by reconstruction intensity in a humid subtropical environment. *Pedosphere* **2015**, *25*, 82–92. [CrossRef]
146. Msimbira, L.A.; Smith, D.L. The roles of plant growth promoting microbes in enhancing plant tolerance to acidity and alkalinity stresses. *Front. Sustain. Food Syst.* **2020**, *4*, 106. [CrossRef]
147. Fageria, N.K.; Nascente, A.S. Management of soil acidity of South American soils for sustainable crop production. *Adv. Agron.* **2014**, *128*, 221–275. [CrossRef]
148. Han, J.; Shi, J.; Zeng, L.; Xu, J.; Wu, L. Effects of nitrogen fertilization on the acidity and salinity of greenhouse soils. *Environ. Sci. Pollut. Res.* **2015**, *22*, 2976–2986. [CrossRef] [PubMed]
149. Bojórquez-Quintal, E.; Escalante-Magaña, C.; Echevarría-Machado, I.; Martínez-Estévez, M. Aluminum, a friend or foe of higher plants in acid soils. *Front. Plant Sci.* **2017**, *8*, 1767. [CrossRef] [PubMed]
150. Penn, C.J.; Camberato, J.J. A critical review on soil chemical processes that control how soil pH affects phosphorus availability to plants. *Agriculture* **2019**, *9*, 120. [CrossRef]
151. Ch'ng, H.Y.; Ahmed, O.S.; Majid, N.M.A. Assessment of soil carbon storage in a tropical rehabilitated forest. *Int. J. Phys. Sci.* **2011**, *6*, 6210–6219. [CrossRef]
152. Bolan, N.; Sarmah, A.K.; Bordoloi, S.; Bolan, S.; Padhye, L.P.; Van Zwieten, L.; Sooriyakumar, P.; Khan, B.A.; Ahmad, M.; Solaiman, Z.M.; et al. Soil acidification and the liming potential of biochar. *Environ. Pollut.* **2023**, *317*, 120632. [CrossRef]
153. Dai, Z.; Zhang, X.; Tang, C.; Muhammad, N.; Wu, J.; Brookes, P.C.; Xu, J. Potential role of biochars in decreasing soil acidification-a critical review. *Sci. Total Environ.* **2017**, *581*, 601–611. [CrossRef] [PubMed]
154. Goulding, K.W.T. Soil acidification and the importance of liming agricultural soils with particular reference to the United Kingdom. *Soil Use Manag.* **2016**, *32*, 390–399. [CrossRef]
155. Zhang, S.; Zhu, Q.; de Vries, W.; Ros, G.H.; Chen, X.; Muneer, M.A.; Zhang, F.; Wu, L. Effects of soil amendments on soil acidity and crop yields in acidic soils: A world-wide meta-analysis. *J. Environ. Manag.* **2023**, *345*, 118531. [CrossRef]
156. Cai, A.; Xu, M.; Wang, B.; Zhang, W.; Liang, G.; Hou, E.; Luo, Y. Manure acts as a better fertilizer for increasing crop yields than synthetic fertilizer does by improving soil fertility. *Soil Tillage Res.* **2019**, *189*, 168–175. [CrossRef]
157. Nest, T.V.; Ruysschaert, G.; Vandecasteele, B.; Houot, S.; Baken, S.; Smolders, E.; Cougnon, M.; Reheul, D.; Merckx, R. The long term use of farmyard manure and compost: Effects on P availability, orthophosphate sorption strength and P leaching. *Agric. Ecosyst. Environ.* **2016**, *216*, 23–33. [CrossRef]
158. Mockeviciene, I.; Repsiene, R.; Amaleviciute-Volunge, K.; Karcauskiene, D.; Slepetiene, A.; Lepane, V. Effect of long-term application of organic fertilizers on improving organic matter quality in acid soil. *Arch. Agron. Soil Sci.* **2022**, *68*, 1192–1204. [CrossRef]
159. Vašák, F.; Černý, J.; Buráňová, Š.; Kulhanek, M.; Balík, J. Soil pH changes in long-term field experiments with different fertilizing systems. *Soil Water Res.* **2015**, *10*, 19–23. [CrossRef]
160. Aziz, M.A.; Ahmad, H.R.; Corwin, D.L.; Sabir, M.; Hakeem, K.R.; Öztürk, M. Influence of farmyard manure on retention and availability of nickel, zinc and lead in metal-contaminated calcareous loam soils. *J. Environ. Eng. Landsc. Manag.* **2017**, *25*, 289–296. [CrossRef]
161. Hale, S.E.; Nurida, N.L.; Mulder, J.; Sørmo, E.; Silvani, L.; Abiven, S.; Joseph, S.; Taherymoosavi, S.; Cornelissen, G. The effect of biochar, lime and ash on maize yield in a long-term field trial in a Ultisol in the humid tropics. *Sci. Total Environ.* **2020**, *719*, 137455. [CrossRef]
162. Chintala, R.; Mollinedo, J.; Schumacher, T.E.; Malo, D.D.; Julson, J.L. Effect of biochar on chemical properties of acidic soil. *Arch. Agron. Soil Sci.* **2014**, *60*, 393–404. [CrossRef]
163. Peña, D.; Fernández, D.; Albarrán, A.; Gómez, S.; Martín, C.; Sánchez-Terrón, J.; Vicente, L.; López-Piñeiro, A. Using olive mill waste compost with sprinkler irrigation as a strategy to achieve sustainable rice cropping under Mediterranean conditions. *Agron. Sustain. Dev.* **2022**, *42*, 36. [CrossRef]
164. Aranda, V.; Macci, C.; Peruzzi, E.; Masciandaro, G. Biochemical activity and chemical-structural properties of soil organic matter after 17 years of amendments with olive-mill pomace co-compost. *J. Environ. Manag.* **2015**, *147*, 278–285. [CrossRef]
165. Perri, S.; Molini, A.; Hedin, L.O.; Porporato, A. Contrasting effects of aridity and seasonality on global salinization. *Nat. Geosci.* **2022**, *15*, 375–381. [CrossRef]
166. Bui, E.N. Soil salinity: A neglected factor in plant ecology and biogeography. *J. Arid Environ.* **2013**, *92*, 14–25. [CrossRef]
167. Machado, R.M.A.; Serralheiro, R.P. Soil salinity: Effect on vegetable crop growth. Management practices to prevent and mitigate soil salinization. *Horticulturae* **2017**, *3*, 30. [CrossRef]
168. Shrivastava, P.; Kumar, R. Soil salinity: A serious environmental issue and plant growth promoting bacteria as one of the tools for its alleviation. *Saudi J. Biol. Sci.* **2015**, *22*, 123–131. [CrossRef]
169. Abdi, M.R.; Askarian, A.; Safdari Seh Gonbad, M. Effects of sodium and calcium sulphates on volume stability and strength of lime-stabilized kaolinite. *Bull. Eng. Geol. Environ.* **2020**, *79*, 941–957. [CrossRef]
170. Shabtai, I.A.; Shenker, M.; Edeto, W.L.; Warburg, A.; Ben-Hur, M. Effects of land use on structure and hydraulic properties of Vertisols containing a sodic horizon in northern Ethiopia. *Soil Tillage Res.* **2014**, *136*, 19–27. [CrossRef]
171. Rengasamy, P. Soil processes affecting crop production in salt-affected soils. *Funct. Plant Biol.* **2010**, *37*, 613–620. [CrossRef]

172. de Oliveira, A.B.; Alencar, N.L.M.; Gomes-Filho, E. Comparison between the water and salt stress effects on plant growth and development. *Responses Org. Water Stress* **2013**, *4*, 67–94. [CrossRef]
173. Grieve, C.M.; Grattan, S.R.; Maas, E.V. Plant salt tolerance. In *ASCE Manual and Reports on Engineering Practice No. 71 Agricultural Salinity Assessment and Management*, 2nd ed.; Wallender, W.W., Tanji, K.K., Eds.; ASCE: Reston, VA, USA, 2012; pp. 405–459.
174. Duan, M.; Liu, G.; Zhou, B.; Chen, X.; Wang, Q.; Zhu, H.; Li, Z. Effects of modified biochar on water and salt distribution and water-stable macro-aggregates in saline-alkaline soil. *J. Soils Sediments* **2021**, *21*, 2192–2202. [CrossRef]
175. Mahmoodabadi, M.; Yazdanpanah, N.; Sinobas, L.R.; Pazira, E.; Neshat, A. Reclamation of calcareous saline sodic soil with different amendments (I): Redistribution of soluble cations within the soil profile. *Agric. Water Manag.* **2013**, *120*, 30–38. [CrossRef]
176. Alcívar, M.; Zurita-Silva, A.; Sandoval, M.; Muñoz, C.; Schoebitz, M. Reclamation of saline–sodic soils with combined amendments: Impact on quinoa performance and biological soil quality. *Sustainability* **2018**, *10*, 3083. [CrossRef]
177. Mahdy, A.M. Comparative effects of different soil amendments on amelioration of saline-sodic soils. *Soil Water Res.* **2011**, *6*, 205–216. [CrossRef]
178. Abdel-Fattah, M.K. Reclamation of Saline-Sodic Soils for Sustainable Agriculture in Egypt. In *Sustainability of Agricultural Environment in Egypt: Part II. The Handbook of Environmental Chemistry*, 1st ed.; Negm, A., Abu-hashim, M., Eds.; Springer: Cham, Switzerland, 2018; Volume 77, pp. 69–92. [CrossRef]
179. Stockmann, U.; Adams, M.A.; Crawford, J.W.; Field, D.J.; Henakaarchchi, N.; Jenkins, M.; Zimmermann, M. The knowns, known unknowns and unknowns of sequestration of soil organic carbon. *Agric. Ecosyst. Environ.* **2013**, *164*, 80–99. [CrossRef]
180. Chenu, C.; Angers, D.A.; Barré, P.; Derrien, D.; Arrouays, D.; Balesdent, J. Increasing organic stocks in agricultural soils: Knowledge gaps and potential innovations. *Soil Tillage Res.* **2019**, *188*, 41–52. [CrossRef]
181. McDaniel, M.D.; Tiemann, L.K.; Grandy, A.S. Does agricultural crop diversity enhance soil microbial biomass and organic matter dynamics? A meta-analysis. *Ecol. Appl.* **2014**, *24*, 560–570. [CrossRef]

agriculture

MDPI

Review

Relationship among Soil Biophysicochemical Properties, Agricultural Practices and Climate Factors Influencing Soil Phosphatase Activity in Agricultural Land

Patrícia Campdelacreu Rocabruna [1,2,*], Xavier Domene [1,2], Catherine Preece [3] and Josep Peñuelas [2,4]

1 Department of Animal Biology, Plant Biology and Ecology, Universitat Autònoma de Barcelona, 08193 Cerdanyola del Vallès, Spain; x.domene@creaf.uab.cat
2 Ecological and Forestry Applications Research Centre (CREAF), Campus de Bellaterra (UAB) Edifici C, 08193 Cerdanyola del Vallès, Spain; josep.penuelas@uab.cat
3 Institute of Agrifood Research and Technology (IRTA), Sustainability in Biosystems Programme, Torre Marimon, 08140 Caldes de Montbui, Spain; catherine.preece@irta.cat
4 Spanish National Research Council (CSIC), Global Ecology Unit CREAF-CSIC-UAB, 08193 Cerdanyola del Vallès, Spain
* Correspondence: p.campdelacreu@creaf.uab.cat

Abstract: Phosphorus (P) is a vital macronutrient crucial for crop productivity. Plants absorb P salts, mainly orthophosphate, from the soil, yet the primary P source resides in organic materials. Acid and alkaline phosphatases (the predominant forms of soil phosphomonoesterases (APases)) are crucial for alleviating P deficiency in plants and play a vital role in releasing P from organic materials via hydrolysis. Our aim was to summarize the direction of the relationship between a variety of influential factors on acid and alkaline phosphatase activity in agricultural lands and identify gaps in knowledge. Our findings indicate a strong linkage between both APases and soil pH, positively influenced by clay content, organic matter, microbial biomass carbon, and nitrogen. Adopting healthy soil practices like balanced organic fertilizer usage, optimal soil water levels, reduced tillage, crop rotation, and using beneficial plant microbes help boost both APase activity. However, the connection between APases and crop productivity remains uncertain due to insufficient research in this area. We identified gaps in knowledge in relation to meso-macrofauna, alongside essential plant nutrients such as potassium, nutrient ratios, and the synergistic effects of various factors on APase response. Understanding the rapid, efficient assimilation of P through APases in the plant-soil and/or plant-microbiota ecosystem it can be crucial for crop productivity and yields.

Keywords: phosphomonoesterases; physicochemical properties; biological properties; management; fertilization; pollution; climate; yield

1. Introduction

Phosphorus (P) is an essential element for cell development in all living organisms [1]. As a component of nucleic acids (DNA, RNA), P is indispensable for reproduction and protein synthesis. Additionally, it plays a crucial role in energy-storing molecules like adenosine triphosphate (ATP) or cytidine triphosphate (CTP), among others, supplying the energy needed for diverse cellular endergonic processes [2]. This is why P is an important limiting nutrient for crop and plant growth in a range of natural and managed ecosystems, given that only 0.1% of the P available in the soil is in the inorganic form that can be assimilated by plants [3–5]. Soil enzymes released by plant roots, soil mesofauna, and living or dead microbes [6–8] contribute to the decomposition of organic matter and allow nutrient recycling [9,10]. The mechanisms governing how the composition, timing, spatial location, and quantity of soil enzymes adapt to environmental changes have been studied elsewhere [11,12]. These studies underscore the crucial role of soil enzymes in biogeochemical cycles and ecosystem responses to drivers of global change.

In the P cycle, soil phosphatase enzymes release P contained in organic matter for reuse by living organisms [13]. This process involves the hydrolysis of various P esters (carbon-oxygen-phosphorus monoesters, carbon-oxygen-phosphorus-oxygen-carbon diesters, carbon-phosphorus phosphonates, phosphoric triester hydrolases, triphosphoric acid monoester hydrolases) into soluble phosphate ions. This process provides soil-accessible and assimilable P for plant uptake [14,15]. Extracellular phosphatase enzymes are secreted by soil microorganisms, fauna, and plant roots [16], while intracellular (endogenous) phosphatase enzymes are within the cytoplasm of proliferating microbial, animal and plant cells, restricted to the periplasmic space of gram-negative bacteria or within non-proliferating cells such as fungal spores, protozoan cysts, plant seeds, and bacterial endospores [17,18]. Extracellular monoester hydrolases (APases) are included in a wide group of phosphoric monoester hydrolases (or phosphomonoesterases) [19], and its predominant forms across a wide range of soil pH conditions are acid phosphatase (ACP; EC 3.1.3.2) and alkaline phosphatase (ALP; EC 3.1.3.1). ACP is produced by plants in the phloem, cortex, epidermis, and roots [20,21] and also by microorganisms [22] and is active in acid/neutral soils with pH ≤ 7. ALP is produced by microorganisms and animals and is active in basic soils with pH > 7 [8,23–26]. The most well-studied group of ALP are encoded by different genes (i.e., phoA, phoD, phoX) [27], and the phoD gene is the form that has the highest abundance in soils [28].

Agricultural and livestock production covers approximately 5 billion hectares (38%) of the Earth's land surface, with around 66% consisting of livestock-grazed grasslands and 33% being cropland [29]. While APase activity in managed soils has been reported to be lower compared to natural ecosystems [30], its activity is, in turn, influenced by a combination of natural environmental conditions and anthropogenic factors, together with strong seasonal variations [31]. APase activity in agricultural soils is significantly impacted by management practices, including tillage, the crop species or crop rotation [32–34], as well as fertilization methods [35–37], in combination with various soil biophysicochemical and environmental factors [38,39]. Several quantitative studies have investigated APase response to various factors such as climatic effects [40–43], soil properties [30], fertilization [44–48], and pollution [49,50] across different ecosystems. However, a comprehensive global analysis specifically centered on APases in agricultural lands is yet to be conducted. Therefore, a preliminary qualitative analysis is needed to assess the APase response in agriculture-managed soils. This should be augmented by incorporating findings from quantitative analyses published to date, thereby enhancing the comprehensiveness of this qualitative study. Such an analysis should encompass all potential factors that could either augment, diminish, or have no effect on APase activity to address the challenge of identifying patterns within agricultural systems. To achieve this goal, we (i) summarize the direction of the relationships between a variety of influential factors on APase activity in agricultural lands and (ii) identify gaps in knowledge. This will help to direct future quantitative studies toward specific areas, leveraging a broad and well-documented qualitative foundation.

2. Materials and Methods

Using the Web of Science and Scopus databases, a bibliographic search was carried out, including research papers, reviews, and meta-analyses published from 1977 to December 2022. We carried out a search using different combinations of terms: "phosphatase* AND soil AND agriculture", "phosphatase* AND soil AND agricultural", "phosphatase* AND soil AND crop", "phosphatase* AND soil AND arable", and "phosphatase* AND grassland" in the title, abstract or keywords. We only selected papers reporting field, glasshouse, and laboratory studies using arable land and managed grassland and where soil APase was experimentally assessed. APase must be evaluated alongside other parameters from bulk soil. Only studies that used para-nitrophenol as a substrate to measure APase activity were included [8,51,52], where ACP and potential ALP activity following this method is usually measured at pH 6.5 and pH 11.0, respectively [53]. The article search and selection process is detailed in Figure S1.

Among all the selected studies for analysis, the response of ACP and ALP activity have been categorized according to these factors: biophysicochemical parameters, including total microbe activity, microbe abundance, microbial biomass P content, microbial biomass carbon content, microbial biomass nitrogen content, microbe diversity, phoD gene abundance and richness, earthworm abundance, soil depth, soil moisture content, clay content, sand content, microaggregate content, pH, cation exchange capacity, electrical conductivity, chlorine anion content, carbonate content, iron content, exchangeable aluminum content, grade of salinity, soil organic carbon/matter, total organic carbon, dissolved organic carbon, nitrate nitrogen form, ammonium nitrogen form, total nitrogen, soil C:N ratio, labile inorganic P, available P, organic P, labile organic P, soil C:P ratio and available potassium. Regarding the agricultural management practices factor, we registered any land use change, crop rotation, and cover cropping, tillage practices, types of inorganic and organic fertilization and rates, weed and pest management practices, irrigation practices, and livestock, grazing, and mowing management. Pollution was included as soil contaminant content. Concerning climatic variables and climate-change treatments, mean annual temperature, mean annual precipitation, drought, soil water scarcity, soil water availability, seasonal variations, and the impact of carbon dioxide fertilization in these studies were annotated. When available, crop yield responses were also taken.

All analyses underwent a review process involving vote counting, categorizing the direction of the effect as either positive, negative, or non-existent (neutral). When the papers were meta-analyses and reviews, it was not possible to separate the results obtained by different analytical methods. Therefore, only those that had selected studies agreeing with our selection criteria were included (Supplementary Table S1). Consequently, our dataset comprised 675 papers, encompassing 267 individual observations of ACP activity, 218 individual observations of ALP activity, and 190 paired observations involving both ACP and ALP. Additionally, twelve meta-analyses and one review were also considered in this study, acknowledging that certain studies within these publications overlap with those selected in order to function as a qualitative complement to this analysis (Supplementary material Tables S2–S20).

3. Results and Discussion

3.1. Soil Biophysicochemical Properties

3.1.1. Soil Microbes and Fauna

There is a positive relationship between the activity of soil microbes and APases (Figure 1, Table S3). This is influenced by the structure of bacterial and fungal communities [54,55], highlighting the role of microorganisms in facilitating nutrient movement within the soil [56]. Accordingly, the availability of soil P for plants is closely associated with the abundance of microorganisms and the presence of exoenzymes like APases [57]. When the activity of ACP in soil is low, microorganisms may adjust the activity of ALP in response to the nutritional needs of plants and microbes [58]. The activity of soil microorganisms varies throughout crop development, increasing in tandem with APase activity as a response to crop growth, thereby reflecting the complex interactions between soil, plants, and the atmosphere [59]. The activities of ACP and ALP are positively linked with the biomass of fungi, bacteria in general, and specifically actinobacteria (Figure 1, Table S4). Additionally, ALP activity is positively associated with soil respiration [60], as well as with the activities of dehydrogenase and urease enzymes [61].

A positive relationship between microbial biomass carbon (MBC) and microbial biomass nitrogen (MBN) has been demonstrated [62], and both are indicative of microbial biomass [63]. MBC serves as a crucial nutrient pool for ecosystem nutrient cycling [64], and soil properties, such as soil organic matter (SOM), are usually positively associated with MBC [65]. Our findings provide evidence of positive associations between APase activity and MBC, but also with microbial biomass P (MBP) and MBN (Figure 1, Table S3).

Although ALP activity has been proposed as an early indicator of change in soil biological status [64], it does not show a strong association with specific soil bacterial community composition [66], suggesting that it may be less sensitive compared to other enzymatic activities such as urease or dehydrogenase [55]. Consequently, ALP activity may not be a reliable indicator of soil microbial abundance [67], plausibly due to the diverse sources of this enzyme originating from both microorganisms and microbial plant secretions [68].

Figure 1. A number of single studies reported the direction of responses of ACP and ALP activities to soil biological factors. Factors with fewer or equal than three entries in ACP or ALP response are excluded from the figure as they are considered unrepresentative (e.g., microbe diversity (Shannon diversity index), phoD gene abundance and richness, and earthworm abundance).

The relationship between soil bacterial diversity (measured by the Shannon diversity index), phoD gene abundance and richness, and earthworm abundance and biomass with ACP and ALP activity is inconclusive (Table S3). Microbial richness demonstrated a moderate but positive linkage with plant diversity [69], and the abundance of the bacterial phoD gene is generally positively interrelated with ALP activity. On the other hand, soil microbial activity, in turn, influenced by plant root exudates, plays a more substantial role in driving APase activity compared to soil type [70]. This positive association with APases contributes to P availability in soil, potentially benefiting plant development [71,72]. The incorporation of earthworms alongside crop residues has demonstrated an increase in ALP activity [73,74]. This effect has been linked to the mitigation of soil compaction caused by crop residues, thereby, microbial conditions through improved water and oxygen supply [75,76]. Although ACP activity might also elevate with earthworm addition, it's noteworthy that available studies combined earthworms with biochar, lacking independent analysis of the isolated effects of earthworms [77]. Moreover, soil management practices influence earthworm metabolism and dynamic processes since enzyme activities in the casts produced in compacted soils are less stimulated [75]. Unfortunately, there is currently no available information regarding the impact of soil mesofauna groups on APase activity, despite their pivotal role in regulating organic matter decomposition and soil ecosystem functioning.

3.1.2. Soil Depth, Moisture, Texture and Structure

Several studies consistently demonstrated a decrease in ACP and ALP activities with increasing soil depth (Figure 2, Table S4). This decline aligns with root density and a lower abundance of heterotrophic microorganisms (bacteria and fungi). Notably, soil moisture content also has a positive linkage with APase activity and the functional potential of soil microbial communities [78], reflecting its role in optimizing soil conditions for plant roots and microbial growth [79] (Figure 2, Table S4). Some studies have consistently shown a positive trend between APase activity and soil structure (microporosity) and higher clay content (Figure 2, Table S4), which agrees with the well-studied connection between those properties and soil microbial and biochemical properties [80,81]. More specifically, ACP activity has been positively correlated with fine soil particle fractions such as silt [82] and clay [83]. The increase in ACP and ALP activity with higher clay content is also consistent with a meta-analysis conducted by Aponte et al. [49] and is likely associated with the increase in enzyme longevity in soil caused by clay minerals while preserving their activity [67]. In contrast, sandy soils often exhibit a decrease in APase activity owing to several factors, including their diminished organic matter content, limited water-holding capacity, and reduced microbial biomass [54]. Nevertheless, some studies have suggested a positive relationship between APase activity and soil sand content, potentially attributed to increased bioaccessibility and bioavailability of nutrients such as nitrates or exchangeable cations [58,84]. Regarding soil structure, there are no conclusive results to assess whether microaggregates play a significant role in the transformation of soil P via APases, thus lower concentrations of phosphate monoesters and diesters [85]. Consequently, a probable inverse relationship exists between the abundance of microaggregates (particle size < 0.25 mm) and the activities of ALP and ACP enzymes.

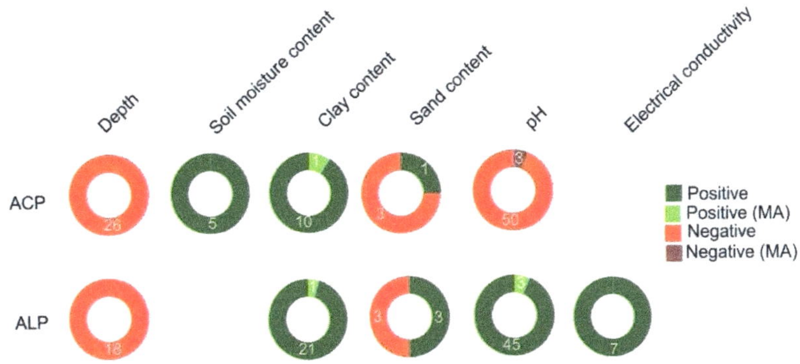

Figure 2. A number of single and meta-analysis studies reported the direction of ACP and ALP responses to soil depth, moisture, texture, and pH-related factors. Factors with fewer or equal to three entries in ACP or ALP response are excluded from the figure as they are considered unrepresentative (e.g., microaggregate content, cation exchange capacity, chlorine anion content, carbonate content, iron content, and exchangeable aluminum content). The number of meta-analyses (MA) has been counted in order to complement the qualitative analysis based on vote counting.

3.1.3. Soil pH and Associated Factors

Soil pH influences a variety of chemical and biochemical processes in soil [86]. In agricultural soil studies, the pH range typically spans from pH 5.5 to 7.5, and therefore, APase assessments are often focused on ACP due to the experimental buffer solutions that are typically adjusted to pH 6.5, followed by Tabatabai's method [8]. Consistently, the maximum ACP activity is observed in acidic to neutral soils, while the peak potential ALP activity is found in alkaline (calcareous) soils [54,87–90] (Figure 2, Table S5). This trend aligns with several meta-analyses [41,44,45]. Nevertheless, the measured activity of APases is potential activity, and it can be increased or reduced due to agricultural practices that modify soil pH. Factors such as high precipitation, acid rain, oxidative weathering, and crop management practices can lead to a decrease in soil pH, which promotes ACP activity. Conversely, weathering of silicates, aluminosilicates, or carbonate mineral compounds can increase soil pH, which promotes ALP activity. For instance, organic fertilizer application in maize cultivation [91] in acidic soils has demonstrated increased ALP activity due to its positive impact on soil pH. Conversely, practices like the use of rice straw biochar [92] or applying no-till management in maize and bean cropping [93,94] have resulted in decreased ACP activity by elevating soil pH.

Microelements and organic compounds in the soil, such as carbonates (CO_3^{2-}), iron (Fe), and aluminum (Al) oxides, influence the release of P from organic compounds, the size of P fractions, and P uptake, which in turn affect APase activity [95]. Specifically, soil CO_3^{2-} content could be negatively associated with ACP activity and positively associated with ALP activity, likely due to its neutralizing capacity, which shifts soil pH from neutral to alkaline [96]. The soil Fe content interacts positively with both ACP and ALP activity, as its availability increases with higher organic matter content [97]. Lastly, soil exchangeable aluminum (Al^{3+}) content has a negative connection with ACP activity due to pH increases after lime amendments, where calcium ions (Ca^{2+}) hydrolyze and react with soluble Al^{3+} to form insoluble Al hydroxide compounds [98] (Table S5).

The total cation exchange capacity (CEC) and electrical conductivity (EC) of the soil are partly related to soil pH [99,100], and available studies indicate a positive association between APase activity and CEC and EC (Figure 2, Table S5). Additionally, higher concentrations of chloride ions (Cl^-) in the soil can decrease ACP activity by inhibiting the growth of soil microflora, thus affecting enzymatic activity [101], but there are no conclusive results directly correlated with this ion. However, high salt content in soils is a growing issue exacerbated by climate change, and it poses significant challenges to agricultural production. Salinity and sodicity, the latter referring to high sodium (Na^+) content, have detrimental effects on crop growth and the biochemical processes essential for maintaining soil quality [102]. In relation to APase activity, although the results are not significant, it seems that salinity has a negative impact (Table S6) partly due to a decrease in the activity of soil microbes and associated microbial biomass with reductions in the release of enzymes [102] and partly due to the likely direct toxic effects of some ions, particularly Cl^-, on microbial growth [103] (Table S6).

3.1.4. Carbon

Soil organic carbon (SOC) is a crucial component of soil health and is derived from living and decomposing organic matter such as plant litter, root and microbial exudates, dead microorganisms and fauna, and fecal material [104]. Both single studies and meta-analyses have clearly demonstrated a positive linkage between indicators of soil organic matter, including SOM, SOC, total organic carbon, and dissolved organic carbon, and the two APases (ACP and ALP) (Figure 3, Table S7). This positive association is explained because the substrate for APases, soil organic P, is linked to SOC [105]. Quantifying soil organic matter (SOM) often does not provide detailed information about the underlying soil processes that contribute to its accumulation [84]. Certain agricultural practices, such as reduced tillage and cover cropping, have been shown to increase SOM levels [106,107] through higher levels of microbial biomass that stimulate decomposition processes and enhance the stabilization of organic compounds [108].

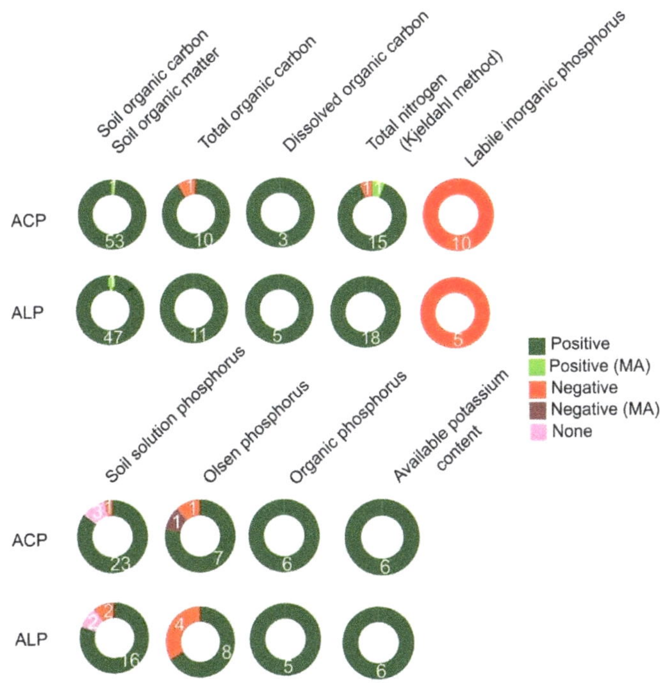

Figure 3. A number of single and meta-analysis studies reported the direction of ACP and ALP responses to carbon, nitrogen, phosphorus, and potassium. Factors with fewer than three entries in ACP and ALP response are excluded from the figure as they are considered unrepresentative (e.g., salt content, nitrate-N, ammonium-N, labile organic P, and soil C:P ratio). The number of meta-analyses (MA) has been counted in order to complement the qualitative analysis based on vote counting.

3.1.5. Nitrogen

Nitrogen is a crucial nutrient for plant growth and is considered an indicator of soil fertility and quality. Nitrate (NO_3^-) and ammonium (NH_4^+) are the primary forms of N available for plants, and their concentrations are often positively correlated with the activity of ACP and ALP. Higher concentrations of NO_3^- and NO. can have a positive impact on the formation and persistence of microbial biomass, which in turn can influence the activity of APases [39]. However, the fact that negative effects have sometimes also been found between both (NO_3^--N and NH_4^+-N) and APases indicates that there can be interactions among specific soil, environment, and management conditions leading to contrasting patterns (Table S8). For instance, when negative effects of NO_3^- on ALP activity have been reported, this has been attributed to the stabilization of ALP by soil colloids formed by organic matter and clay minerals [109] as well as the influence of SOC on the structure and composition phoD-harboring bacteria and ALP activity [66]. Both a meta-analysis and multiple studies have shown a positive association between APase activity and total soil N content, often determined using the Kjeldahl method (Figure 3). This relationship is likely due to the positive correlation between N and SOC content [41,110], suggesting that APase activity is induced by C and N mineralization and the availability of their decomposition products [111] (Figure 3, Table S8). The C:N ratio of soil organic matter also influences APase activity and a lower C:N ratio indicates rapid decomposition of organic matter, regardless of soil microbial biomass, and can result in increased APase activity. The positive connection between APases and the C:N ratio tends to be stronger than their connection with the C:P ratio [112].

3.1.6. Phosphorus

As expected and indicated by various studies, APase activity is closely associated with soil P content (Figure 3, Table S9). It is important for comprehending the dynamics of soil P and for effective P management in both natural and agricultural ecosystems [113]. The bulk of the soil P exists in three general groups of compounds, namely organic P, calcium-bound inorganic P, and iron or aluminum-bound inorganic P, where organic P is distributed among the biomass, labile or passive fractions of soil organic matter, inorganic P and calcium compounds predominate in most alkaline soils while the iron and aluminum forms are most important in acidic soils [114]. Since most of the P in each group is of very low solubility and not readily available for plant uptake, biotic processes controlled primarily by bacterial and fungal decomposition indirectly affect P availability for plants by influencing the form of soil minerals that chemically bind P [115]. For instance, in cropping systems with low levels of C and inorganic N, it becomes essential to supplement the soil with other mineral nutrients (e.g., P) and implement effective biological control strategies to ensure proper P cycling and availability for plants [116]. In terms of readily plant-available soil P content, studies considered different fractions, notably labile inorganic P (Pi), soil solution P, or other P fractions. The former comprises P fractions dissolved in the soil solution, directly accessible to plants, while the latter encompasses fractionation methods for inorganic P extraction. These extraction methods often involve sodium bicarbonate-P (commonly referred to as Olsen P, detailed separately) or P solubilization using reagents such as dilute acid-fluoride, dilute hydrochloric acid, sulfuric acid, or water, among other techniques [117]. Conversely, there are P fractions existing in organic forms, cited as organic P, that are not immediately available to plants, including labile organic P (Po). As previously mentioned, organic P denotes P bound within organic matter, while Po, like $NaHCO_3$-Po, represents P that can be relatively easily mineralized [4].

The activity of APases in soil is influenced by the P content, and its response is dynamic depending on the availability of P to plants. A priori, high levels of available soil P content can lead to a reduction in APase activity as plants and microbes adapt to the abundant P supply. Conversely, under P limitation, APase activity can increase to facilitate P uptake and meet or even surpass plant P demands [24] (Table S9). This trend is confirmed by Sun et al. [41], which showed a negative correlation between both APases and Olsen P and soil solution P. In this case, the negative association has been attributed to the hydrolysis of P compounds by other APase enzymes in the $NaHCO_3$-extractable fraction, leading to an increase in dissolved inorganic P in the soil solution. However, other studies showed a positive association between APase activity (both ACP and ALP) and Olsen P, soil solution P, and organic P (Figure 3), which means that the dynamics of P fractions, particularly Olsen P, are closely related to plant development and can be influenced by climate and intrinsic soil characteristics [118,119]. Additionally, the addition of organic P sources, which increase the soluble P content, can negatively impact APase activity, as they contribute to the pool of available inorganic P in the soil [120]. It seems that there is a relationship between these enzymes and the promotion of root growth and nutrient uptake by crops [121], which indicates that the positive relationship is directly associated with particular cases and that management is crucial to determine their correlation. When APase activity shows a negative linkage with the content of Po in the soil (Table S9), it suggests that APases are not the limiting factor in the utilization of organic P, but rather it is the availability of APase-hydrolysable P compounds that limits the process [24]. It is important to consider that when a wider group of phosphoric monoester hydrolase enzymes are assessed together, the high levels of inter-enzyme variation strengthen the relations of available P [122].

3.1.7. Potassium

Potassium (K) plays a crucial role in plant growth and soil fertility [123]. Therefore, the soil content of available K decreases more in cultivated soils than in natural ecosystems during plant growth due to erosion/runoff [61]. Studies indicate a positive linkage between the activity of ACP and ALP enzymes and the available K content in the soil (Figure 3,

Table S10). The studies do not delve deeply into K's relationship with other factors that may also affect APase activity. For this reason, further research is needed to fully understand the specific mechanisms and trade-offs associated with K and its impact on P acquisition in managed ecosystems [124].

3.2. Responses to Agro-Ecosystem Management

3.2.1. Conversion from Natural to Managed Ecosystems

Cropland soils experience more intensive human disturbance and receive lower inputs of plant residues, root exudates, and senescent leaves compared to soils in natural and semi-natural ecosystems [125]. This human activity negatively impacts the soil's biological and biochemical properties, leading to a decline in P and C cycling [126]. Non-managed ecosystems like native forests, on the other hand, exhibit higher microbial activity due to their abundant SOM and available P content [127], which facilitates the transformation of organic P into inorganic forms [128,129]. Cropland soils generally have lower SOC and MBC compared to non-managed soils [130], and the global activity of extracellular enzymes is diminished as a result [131]. Furthermore, the activity of APases is influenced by common management practices [132], with lower-intensity management systems generally exhibiting higher APase activity compared to higher-intensity management systems (Figure 4, Table S11).

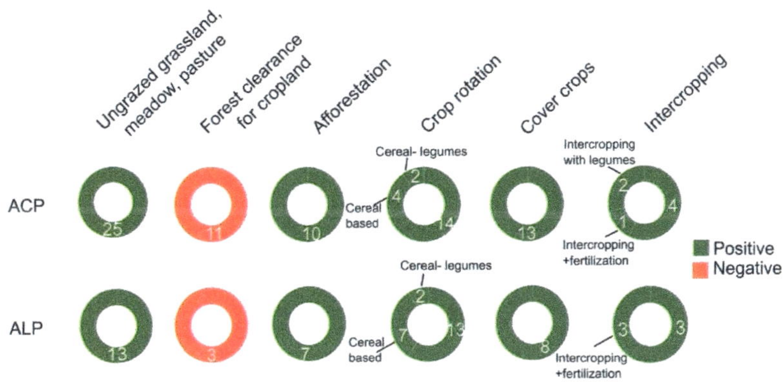

Figure 4. A number of single studies reported the direction of ACP and ALP responses to land use change, crop rotation, cover crops, and intercropping. Factors with fewer or equal to three entries in ACP or ALP response are excluded from the figure as they are considered unrepresentative (e.g., revegetation, plant invasion, and forest clearance for cropland).

On the other hand, the conversion of intensively managed agricultural land back to grassland and forest systems, using native plant species [127,133,134], improves the supply of organic matter, enhancing APase activity, especially ACP [135]. Furthermore, a meta-analysis made by Margalef et al. [40] has shown that invasive plant species can also increase ACP and ALP activity compared to native species, potentially due to differences in litter quality or quantity and related effects of changes in soil chemistry on microbial communities.

3.2.2. Crop Rotations and Species

APase activity in agricultural systems is influenced by crop rotation type, the crop species concerned, and also cover and intercropping practices (Figure 4, Table S12). Higher levels of ACP and ALP activity are observed in crop rotations in cereal-based rotations compared to cereal-legume rotations. This positive response has been attributed to increased ionic exchange capacity, SOC, MBC, and availability of essential nutrients such as P, K, and magnesium (Mg), as well as a greater presence of earthworms in rotation systems [62,136]. The inclusion of legumes and/or grasses in crop rotations, also as an intercrop, increases the synergism of microbial attributes (e.g., MBC, soil basal respiration, metabolic quotient, soil cultivable bacteria, fungi, actinobacteria and microorganisms with cellulolytic activity) [137] leading to higher productivity and economic profitability.

Different crop species influence soil N content, C sequestration, and P accumulation in long-term cropping systems [138], promoting efficient water, energy, and C use efficiency for crop production [139]. Maize monoculture, for example, exhibits higher soil APase activity compared to soybean, cowpea, or cotton, attributed to its deeper rooting system, and this links to its growth advantage in low P availability conditions [140,141]. Legume cultivation, especially lupine, which is the most well-studied, enhances soil nutrient availability in a broad sense [142] and results in higher ACP and ALP activity compared to grain crops like wheat and rice, as legumes offer benefits to soil microbial communities, ensuring stability in intensive production systems [143]. Additionally, genetically modified crops, such as transgenic cotton, have been found to enhance ACP and ALP activity, although the effect is crop-specific and may not apply uniformly (e.g., in rice, it is ACP that is enhanced). In horticultural crops like mango, kiwifruit, lettuce, potato, and tomato, the activity of ACP is higher compared to cereal crops, attributed partly to intensive fertilization and irrigation management [144] (Figure 5, Table S12).

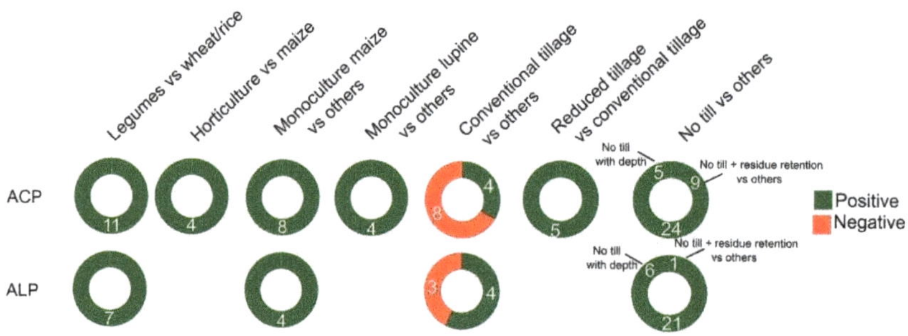

Figure 5. A number of single studies reported the direction of ACP and ALP responses to crop species and tillage. Factors with fewer or equal than three entries in ACP and ALP response are excluded from the figure as they are considered unrepresentative (e.g., comparison between wheat and maize/rice, barley vs. horticulture, monoculture sorghum vs. others, monoculture transgenic cotton vs. cotton, monoculture transgenic rice vs. rice, no-till with residue retention vs. others and no-till with depth vs. others).

The use of cover crops (i.e., specific crops planted primarily to improve soil health rather than for direct harvest) has a positive impact on soil and crop health by improving pest and disease control, increasing water availability, and enhancing the abundance and activity of soil microorganisms [145]. Cover crops have been shown to promote microbe-mediated processes that enhance ACP and ALP activities, likely through the increase of labile C and moisture in the soil, maintenance of high organic matter levels, and stabilization of soil temperature [146–148]. Intercropping, which involves cultivating two or more crop species within a single cropping season, results in greater ACP and ALP activities compared to monocropping. As mentioned before, the use of legumes leads to an increase in APase activity, as reflected in the study results (Figure 4, Table S12). This may be due to the differential secretion of root exudates by intercropped species, which might provide a higher diversity of labile C substrates with knock-on effects on soil microorganisms, thereby increasing enzyme activity [149]. Moreover, when intercropping is associated with fertilization (Section 3.2.4), APase activity is evidently enhanced.

3.2.3. Soil Tillage

Conventional soil tillage, which involves mechanical soil turning, aims to improve soil structure for sowing, seedling establishment, and weed control [150,151]. However, intensive tillage practices increase the risk of soil erosion and surface runoff, particularly following heavy rainfall, leading to the loss of SOM [94]. In contrast, reduced (conservation) tillage practices minimize soil disturbance, resulting in better conservation of SOM [152], increased MBC, MBN [153], and higher availability of K and Mg. Along with the improvements in soil physical properties, soil aggregation, and reduced decomposition, reduced tillage contributes to the promotion of APase activity [154] (Figure 5, Table S13).

No-till practices, which involve minimal soil disturbance and surface accumulation of crop residues, have distinct advantages in soil top layers even compared to reduced tillage. No-till practices lead to even greater reductions in the decomposition of labile organic matter, resulting in increased soil moisture, C, and N levels [63,155,156]. These practices also have positive effects on P fractions (e.g., inorganic, organic, and available P) [157]. The increased availability of substrates for enzymes in the presence of higher residue inputs enhances the activity of enzymes such as ACP and ALP [158] (Figure 5, Table S13).

3.2.4. Soil Fertilization

Fertilization of agricultural soils to increase crop yields tends to positively impact APase activity (Figure 6, Table S14), although concurrent factors such as fertilizer nutrient balance and type, crop species, and growth stage may determine its activity [159].

The application of combined (NPK) chemical (inorganic) fertilizer generally promotes APase activity [46]. Nitrogen fertilization, in particular, tends to enhance the activities of ACP [40,47,48] and ALP [40,48]. This suggests a connection between APases and the cycling of N. However, there are also reports indicating that ACP and ALP activity may decrease after mineral N fertilization, which suggests that substrate availability (i.e., specific organic N or P substrates in soil suspensions and soil filtrates) is more important than P deficiency [44,160]. Inorganic P fertilization alone tends to decrease the activity of both APases [40,48], although there is also a meta-analysis suggesting no significant effects [44].

The long-term application of organic fertilizer, derived from plant and animal material, is an important strategy for enhancing soil quality by increasing the abundance of soil microbes and the activity of extracellular enzymes such as ACP and ALP [46,161]. Organic fertilizers have a positive association with soil pH, especially in relation to ALP activity and P content [91], leading to improved availability of soil nutrients, including labile C, N, and P through mineralization, as well as enhanced microbial biomass and abundance [162,163]. Various soil amendments, such as vermicompost (i.e., organic material biodegraded by earthworms and microorganisms), biostimulants (e.g., humic substances, marine macro-algae, protein hydrolysates, microbial inoculants, and plant extracts), biowastes (i.e., optimal doses of organic compounds and metals), or sludge (i.e., rich in organic matter, NO_3^--N, copper (Cu), cadmium (Cd), and organic P), have also shown the ability to increase ACP and ALP activity [49] although do not report on their direct correlation over the very long term. The optimization of APase activity in soils without the addition of inorganic fertilizers can improve soil conservation, P release, and overall agricultural sustainability in ecosystems [164]. Finally, the co-application of inorganic and organic fertilizers in agricultural soils is a common practice due to their complementary composition and functions, resulting in increased ACP and ALP activity, thereby providing high levels of plant-available P [46,165].

Lime application to acid soils increases pH levels, improving plant access to essential nutrients for growth [166], and has positive effects on ACP and ALP activity. However, it should be noted that when Ca-based lime is applied, reductions in APase activity have been observed, indicating that Ca availability may not be a limiting factor for plant growth [167]

Combining fertilizers (organic or inorganic) with green manures (i.e., refer to non-crop plants, typically legumes, that are cultivated specifically to improve nutrient content in the soil) can have an impact on APase activity. While short-term trials combining green manure with fertilizers have not shown a significant effect on ACP activity, there is evidence for positive impacts on ACP and ALP activity in these trials when legume green manures are added with fertilizers [168,169]. This is attributed to the increase in SOM content and the contribution of N fixed by symbiotic legume root bacteria [148].

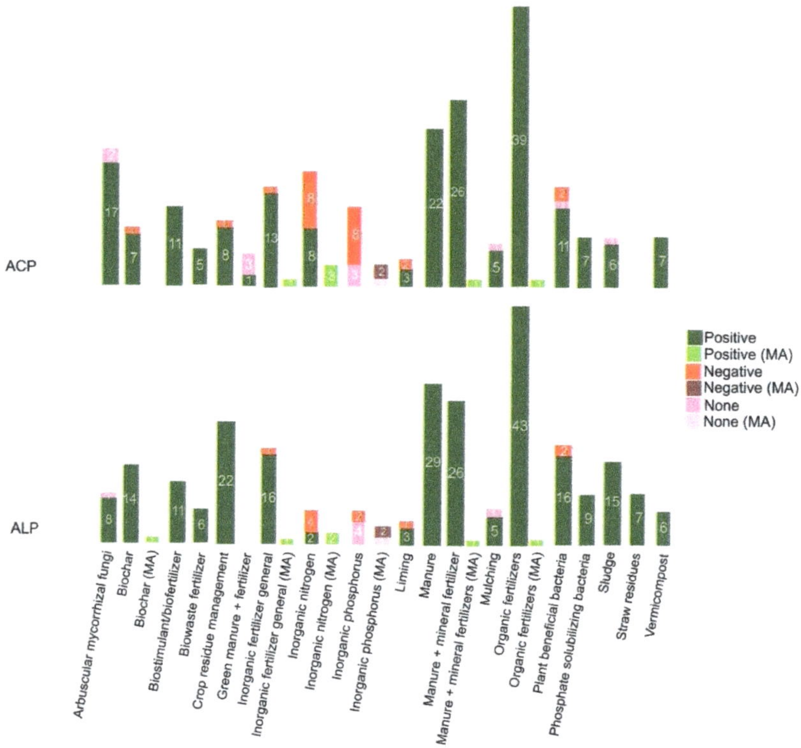

Figure 6. A number of single and meta-analysis studies reported the direction of ACP and ALP responses to agroecosystem fertilizer management practices. Factors with fewer or equal than three entries in ACP and ALP responses are excluded from the figure as they are considered unrepresentative (e.g., green manure alone). The number of meta-analyses (MA) has been counted in order to complement the qualitative analysis based on vote counting.

Crop residues, whether applied on the soil surface as mulch or incorporated into the soil, can have positive effects on P transformation rates and soil P plant available pool [170]. When crop residues, such as straw, are purposefully left on the soil surface, they gradually degrade over time, providing a greater and more sustained supply of substrate for soil [171]. Mulching also increases the supply of carbohydrates and available nutrients such as N, P, and K [172], having a positive impact on microbial communities [173]. This prolonged breakdown of residues contributes to an increase in SOC content [174], which in turn enhances ACP and ALP activity. The increased activity of APases resulting from crop residue mulching not only improves soil quality but also has the potential to reduce the need for chemical fertilizer inputs, leading to greater economic returns [175].

Generally, biochar amendments are known to have a positive effect on both ACP and ALP activity (Table S14). According to the findings of Pokharel et al. [45], the addition of biochar to soil increases the sensitivity of ALP to changes in pH. This heightened sensitivity results in an increased microbial demand for P and/or the potential limitation of P availability in the soil due to restricted microbial growth. However, despite these effects on ALP activity, the researchers did not observe significant impacts on ACP activity.

The practice of burning crop residues, on the other hand, releases environmental pollutants into the atmosphere (particulates carbon dioxide (CO_2) and carbon monoxide) and has negative impacts on ACP and ALP activity (Table S14). This is due to the changes it induces in soil chemical and biochemical processes, resulting in decreased soil nutrients, bacterial densities, and MBC [176,177].

Plant-beneficial microbes (PBMs) are increasingly used in biotechnology to reduce the reliance on agrochemicals with the aim of increasing soil nutrition, tolerance to stress, soil health, and crop yields [178,179]. Phosphate solubilizing bacteria (PSB) significantly contribute to the enhancement of APase activity and the availability of P to plants (Figure 6, Table S14). This is achieved through their possession of enzymes and metabolic mechanisms, enabling the conversion of insoluble forms of P into accessible forms for plant uptake [66]. They accomplish this through the mineralization of organic P and the solubilization of inorganic P minerals, leading to greater P uptake in plant biomass [180]. Incorporating PSB into the soil also results in faster humification of fresh organic matter and enhances mycorrhizal and endobacterial activities [181]. Likewise, soil inputs of bacteria, such as Bacillus, Pseudomonas, Aspergillus, Azospirillium, and Streptomyces, can increase both ACP and ALP activity, restore soil fertility, and promote plant productivity, taking into account addition parameters (e.g., EC, pH, and ionic concentration) to ensure proper nutritional management of the crop [182]. The input of arbuscular mycorrhizal fungi to soils assists the plants in absorbing nutrients by hydrolyzing organic P, similar to solubilizing bacteria, which enhances APase activity. Additionally, soil acidification caused by fungi increases the availability of organic P substrates for APases, particularly ACP [53].

3.2.5. Pest and Weed Management

Plant protection products, including herbicides, fungicides, and insecticides, are widely used in agriculture to mitigate the detrimental effects of competition, disease, and herbivory on crop yields. However, their application can lead to changes in soil function and health, affecting soil respiration, biomass, and APase activity (Figure 7, Table S15). The impact of fungicides on APases is a topic of debate, with one meta-analysis reporting an increase in ACP activity rather than ALP activity [50], possibly due to the predominance of ACP analysis in agricultural soils. Likewise, the effects of insecticides on APases do not exhibit a clear trend. The results found suggest decreases in ACP and ALP activity, followed by recovery in ALP activity within 7 to 30 days after insecticide application [50,183].

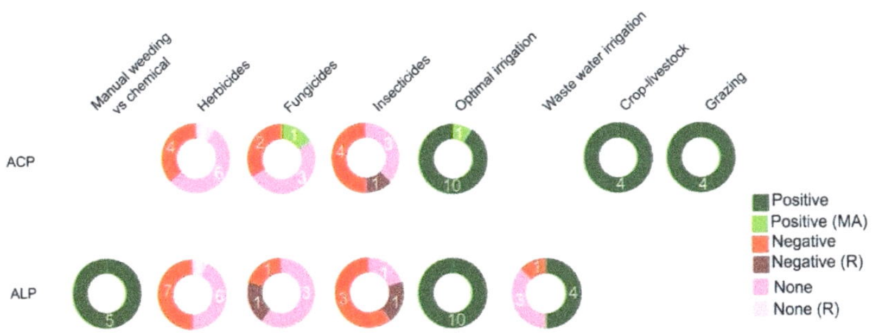

Figure 7. A number of single and meta-analysis/review studies reported the direction of ACP and ALP responses to pest and weed management, irrigation, crop-livestock, and grazing management. Factors with fewer or equal than three entries in ACP and ALP responses are excluded from the figure as they are considered unrepresentative (e.g., mowing). The number of meta-analyses (MA) and reviews (R) has been counted in order to complement the qualitative analysis based on vote counting.

Weed control plays a crucial role in reducing competition for resources by minimizing non-crop plant abundance. While manual weeding tends to increase APase activity (Figure 7, Table S15), the use of herbicides can result in either negative or negligible impacts on ACP and ALP activity [50]. Importantly, any adverse effects from herbicide use typically do not persist beyond 30 days after application [184]. However, cultivating crops in competition with weeds compared with weeds cultivated alone negatively impacts ACP activity, microbial activity, and inorganic P solubilization [185].

3.2.6. Irrigation

Crop irrigation is a practice that involves providing controlled amounts of fresh water or wastewater to sustain and enhance yields in water-scarce regions [186]. Optimal irrigation levels have been found to positively affect APase activity (Figure 7, Table S16). Irrigated soils have increased the availability of soil nutrients, leading to a higher demand for P by plants and microbes during plant growth [41]. Moreover, irrigation strategies can influence P availability, affecting P storage [187] and the abundance of bacteria, which may explain the observed impacts on APases. Research on wastewater irrigation has shown varying effects on ACP and ALP activity, as it affects soil microbial activity and the microbial community [188]. However, long-term use of wastewater may potentially reduce agricultural crop yield [189].

3.2.7. Livestock, Grazing and Mowing Management

Livestock can play a significant role in enhancing agroecosystem function [137], and it can be managed within a livestock-only system (pasture) or in combination with crop production (livestock integration). In both cases, the presence of livestock contributes to an increase in soil MBC content and ACP activity (Figure 7, Table S17). Grazing-based pasture management has been linked to various positive effects, including higher soil pH, increased water content, and elevated levels of NO_3^-, NH_4^+, organic matter, and C:N ratios [190]. These conditions promote greater APase activity, mostly ACP (Figure 7, Table S17). On the other hand, mowing encourages the growth of plant species with competitive strategies [191], while the contact between cut residues (substrates) and the soil reduces the activity of ALP [192] (Table S17).

3.3. Responses to Soil Pollutants

Soil pollution caused by heavy metals can disrupt biochemical, physiological, and metabolic processes. These pollutants alter nutrient stoichiometry and result in slower P cycling due to an imbalance between litter, soil organic matter, and the elemental composition of microbial biomass [49]. Heavy metals have an impact on APase activity (Figure 8, Table S18); negative responses of APase activity due to lead (Pb), chromium (Cr), nickel (Ni), zinc (Zn), cadmium (Cd), copper (Cu), manganese (Mn), arsenic (As), and mercury (Hg) have been observed, while positive responses are reported in one meta-analysis made by Aponte et al., [49] concerning Cu and Cd. The negative APase responses are attributed to the harmful effects of heavy metals on soil microorganisms [193], while the positive responses may indicate microbial metabolic stimulation resulting from increased levels of metals acting as micronutrients, such as Cu, Mn, cobalt (Co), Zn, and Cr [71]. Although heavy metals generally inhibit APase activity, the extent of the response depends on the initial metal composition in the soil, organic matter content, and the inhibition of microbial activity [80]. In soils with high organic matter content, heavy metal impact on APases is relatively lower compared to other enzymes due to the positive association between APase and soil C abundance [194].

Negative effects on APase activity have been observed following the use of sewage sludge compost with high concentrations of heavy metals such as Pb, As, Cr, Cd, Ba, and Ag [49]. Similarly, soil pollution caused by petroleum and nanomaterials (NMs) also negatively affects APase activity, also leading to a decrease in bacterial species richness and diversity [195]. The use of NMs as biocides and plant growth promoters influences soil properties and enzyme activity, and a meta-analysis made by Lin et al. [196] showed that C, Cu, and Ag NMs result in a decrease in ACP activity, whereas low soil concentrations of Fe NMs stimulate ACP activity (Table S18).

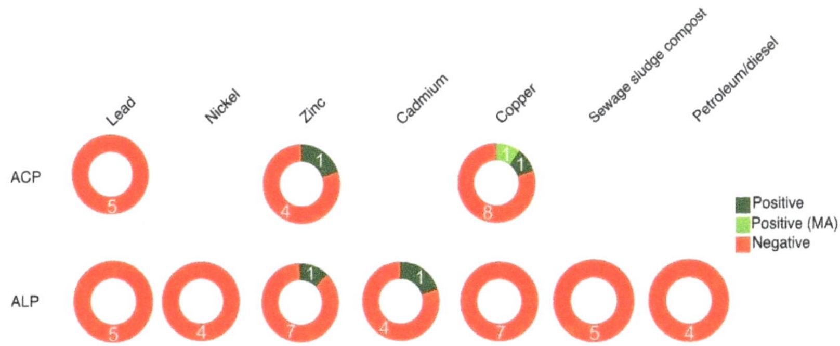

Figure 8. A number of single and meta-analysis studies reported the direction of ACP and ALP responses to soil pollutants. Factors with fewer or equal to three entries in ACP and ALP response are excluded from the figure as they are considered unrepresentative (e.g., chromium, manganese, arsenic, mercury, and nanomaterials). The number of meta-analyses (MA) has been counted in order to complement the qualitative analysis based on vote counting.

3.4. Impacts of Climate Change

The rapid global temperature increases, shifts in rainfall patterns, and rising atmospheric CO_2 concentrations that the planet is experiencing are significantly impacting plant stoichiometry and productivity, potentially affecting APase activity (Table S19). Existing meta-analyses have suggested that climate warming could increase ACP activity in agroecosystems and forests [41,42], primarily due to reduced soil P content (e.g., Olsen P and total soil P) resulting from accelerated plant growth and enhanced plant P acquisition [197]. However, another meta-analysis that encompassed grasslands and other natural ecosystems found no correlation between temperature and both APases [40].

The predicted increase in rainfall intensity in some areas under ongoing climate change is likely to lead to higher topsoil nutrient losses, as high soil water availability to plants can elevate groundwater chemistry, including the dissolved content of bicarbonate, sulfate, Cl^- anions, and Na^+, Ca^{2+}, Mg^{2+}, and K^+ cations [198]. Elevated mean annual precipitation (MAP) levels have been linked to increased ACP and ALP activity [41] compared to controls in models of humid grassland soils and irrigated soils [199,200]. Conversely, drier conditions are also expected to become more frequent in some regions under climate change, resulting in reduced demand for available P forms and associated enzyme activity [201]. APase activity tends to respond negatively to water scarcity and drought in agroecosystems (Figure 9, Table S19), particularly in grasslands and other natural ecosystems under Mediterranean climate conditions known for their seasonal aridity [41]. However, individual studies focused on temperate pasturelands have reported mixed responses, as changes in precipitation amounts may not significantly alter microbial biomass, allowing soil microbes to adapt to soil drying [202].

APase activity exhibits seasonal variations (Figure 9, Table S19), with higher activity recorded during periods of increased plant growth. In contrast, APase activity tends to be lower during drier cropping periods when human activities in agroecosystems are more pronounced [203].

The influence of anthropogenic CO_2 emissions on APases is not significant, but ongoing increases have been linked to enhanced ACP and ALP activities in grasslands and natural ecosystems [40], likely due to elevated microbial activity and increased soil P availability [204] (Table S19). However, this is not sufficient to determine the reason why this trend is the way it is.

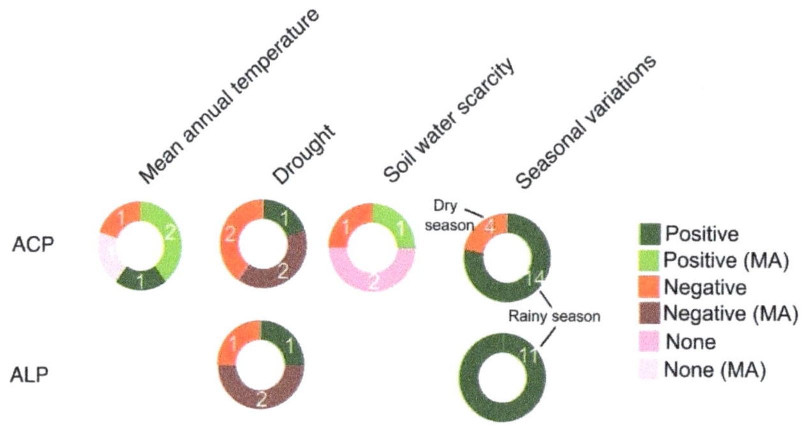

Figure 9. A number of single and meta-analysis studies reported the direction of ACP and ALP responses to climate change factors. Factors with fewer or equal than three entries in ACP and ALP response are excluded from the figure as they are considered unrepresentative (e.g., mean annual precipitation, soil water availability, CO_2 fertilization). The number of meta-analyses (MA) has been counted in order to complement the qualitative analysis based on vote counting.

3.5. Relationship between APases and Crop Yields

Investigating the potential effects of promoting APase activity in agricultural soils on crop yields is important for addressing global goals of increasing food security and crop productivity. Studies have primarily focused on cereals, although a few other crops have also been examined (Table S20). Positive connections have been observed between APase activity and yields of wheat, maize, barley, beet, fava bean, and lentil. However, the available literature does not show any association between APase activity and tree fruit yields (such as organic plum and orange). Interestingly, a negative relationship has been reported between rice yield and ALP activity, which can be attributed to variations in P availability from inorganic and organic sources, other P-regulating enzymes, and changes in soil pH [205]. Crop yield is influenced by various soil physicochemical parameters, including N, SOM, and high accumulation of dry matter [206,207]. Additionally, while crop yields are directly correlated with the amount of plant available P [208] and low soil available P directly affects APase release, there are limited studies that have directly associated APases with crop yield, suggesting that this link needs further research.

4. Conclusions

Due to the extensive number of studies evaluated and the results obtained, this systematic review, which is partly quantitative but predominantly qualitative, underscores the significance of APases in driving P uptake in agroecosystems and their role in the global P cycle. Observable changes in APase activity can be attributed to soil biophysicochemical properties, agricultural management practices, environmental pollutants, and climate change factors.

Firstly, microbial abundance, biomass, and activity demonstrate a positive relationship with both ACP and ALP. These enzymes are further correlated with pH levels, showing a positive association with soil texture—especially clay content—soil moisture, soil organic carbon, and available forms of N and P.

Secondly, the activity of ACP and ALP is generally enhanced by management practices promoting soil health. These practices include optimal irrigation, conservation or no-tillage techniques, crop rotation or intercropping, cover crops, and organic fertilization through the use of amendments such as organic manures, vermicompost, green manures, crop residue management, biochar, and biostimulants/biofertilizers containing beneficial bacteria and fungi (see Figure 10).

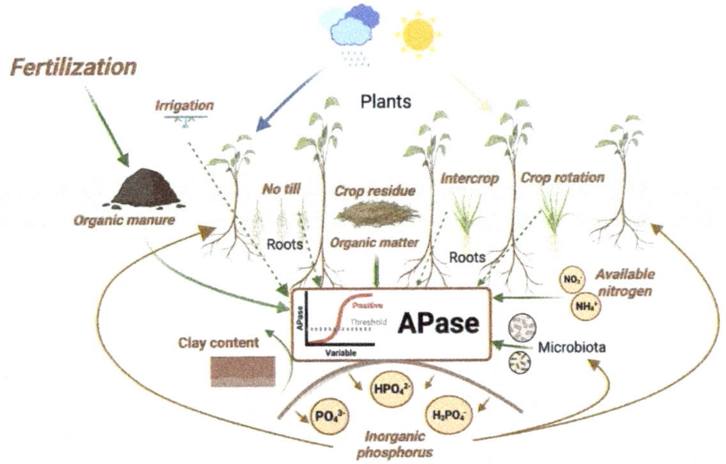

Figure 10. The factors influencing APase activity in belowground environments can be summarized through color-coded lines and names. Green lines represent the most positive influential factors. The solid lines refer to internal soil processes; the dashed lines correspond to crop management in the soil. When the concentration of inorganic P in the soil is low, plants, roots, and microbiota release APase. The brown lines represent the role of APase activity in providing assimilable P for plant and microbiota uptake. Physicochemical properties, such as soil organic matter, available N, clay content, and management practices like organic manure fertilization, no-till, crop residue utilization, intercropping, and crop rotation, are also depicted in brown as they enhance APase activity. Additionally, climate factors that increase APase activity, including optimal water levels, rainfall (indicated with a cloud), and temperature (indicated with a sun), are also shown.

On the other hand, factors such as soil depth, salinity, pesticide and sewage sludge use, and high concentrations of heavy metals or other pollutants in agricultural soils have a detrimental effect on APase activity. For this reason, the activity of APases is used as an indicator of soil quality in agricultural systems.

Perspectives on Knowledge Gaps

Several knowledge gaps have been identified in this review, such as the relationship between APases and crop productivity, which still remains unclear. However, there seems to be a direct relationship between cereal and legume production with the activity of APases that should be studied, especially when intercropping or crop rotations are used. Reviewing APase responses to crop management practices is problematic due to the diverse and complex nature of agronomic techniques. Thus, the interrelation between P availability, on one hand, and the production and activity of APase on the other hand, exhibits highly nuanced cause-and-effect dynamics. However, it is noteworthy that the adoption of conservative soil practices linked to non-intensive agricultural management holds promise for enhancing the response of APase activity.

The relationship between APases and P has been widely studied, but not the relationship with K, which is also important for plant growth and soil fertility. Plant uptake of P is influenced by the availability of K, which in turn depends on N and C levels. This extremely complex mechanism, involving microorganisms as well, should be experimentally studied, incorporating those strategies that increase enzymatic capacity investment and reduce competition and interference with other organisms.

Moreover, strategies to affect APase activity also involve other soil parameters altered by agricultural practices. For instance, increased CO_3^{2-}, which is carried by water and mobilized between soil horizons and is common in the pH range of agricultural soils, negatively affects the activity of ACP, which is directly linked to plants and consequently may affect their production. Moreover, assessing APase with respect to the availability of nutrients (P or N) in relation to C (e.g., C:P, C:N) would yield valuable information to designate it as a key soil quality variable. These ratios are crucial indicators of soil fertility, microbial activity, and plant nutrient uptake, influencing the overall health and productivity of the ecosystem. The repeated, excessive use of mineral fertilizers in agricultural soils for decades has substantially altered the microbial population adapting to this nutrient, excess which directly affects ALP activity mainly released by soil microorganisms. Studies evaluating the response of APases based on soil mesofauna, as well as macrofauna, which regulate soil organic matter transformations and significantly influence nutrient dynamics, are lacking. The activity of these organisms can notably change P availability in active soils and, in parallel, may enhance crop yield.

Ultimately, although the selected studies are too diverse to produce a meaningful summary estimate of the effect of more than two factors, the results demonstrate that there is sufficient data to focus on combined factors that clearly enhance APase activity. The information obtained will enable us to manage agricultural systems to promote the capabilities of plants and associated microorganisms to assimilate nutrients more effectively and rapidly and, at the same time, enhance our understanding of microbial-mediated processes and the dynamics of soil health. The results obtained could guide professional practice on one hand and future research on the other. This approach could achieve a cost-benefit ratio where APases, among other enzymes, would play a determining role.

Supplementary Materials: The following supporting information can be downloaded at: https://www.mdpi.com/article/10.3390/agriculture14020288/s1, Figure S1: Article search and selection process; Table S1: Comprehensive overview of meta-analyses and reviews investigating explanatory drivers for phosphatase activity (APase), including the total number of studies, enzyme analysis substrates, and ecosystem types. Table S2: Comprehensive overview of meta-analyses and reviews detailing factors influencing phosphatase activity (APase), encompassing number of observations, drivers, variables and acid and alkaline phosphatase (ACP and ALP, respectively) response. Table S3–S20: Summary and comprehensive tables inclusive of references detailing APase response relationships to biophysicochemical parameters, agricultural management practices, pollution, climatic variables and crop yield.

Author Contributions: Conceptualization, P.C.R., X.D., C.P. and J.P.; methodology, P.C.R.; data collection and formal analysis, P.C.R.; writing—original draft preparation, P.C.R.; writing—review and editing, P.C.R., X.D., C.P. and J.P.; funding acquisition, X.D. and J.P. All authors have read and agreed to the published version of the manuscript.

Funding: This research was funded by JP's research and was supported by the TED2021-132627B-I00 grant, funded by MCIN and the European Union NextGeneration EU/PRTR.

Institutional Review Board Statement: Not applicable.

Data Availability Statement: All data generated or analyzed during this study are included in this published article [and its supplementary information files].

Acknowledgments: Giovanni Peratoner at Laimburg Research Center (Italy) for help in the statistical understanding of field responses of APases.

Conflicts of Interest: The authors declare no conflicts of interest.

References

1. Wrage, N.; Chapuis-Lardy, L.; Isselstein, J. Phosphorus, Plant Biodiversity and Climate Change. In *Sociology, Organic Farming, Climate Change and Soil Science*; Lichtfouse, E., Ed.; Springer: Dordrecht, The Netherlands, 2010; Volume 3. [CrossRef]
2. Malhotra, H.; Vandana, R.; Sharma, S.; Pandey, R. Phosphorus Nutrition: Plant Growth in Response to Deficiency and Excess. In *Plant Nutrients and Abiotic Stress Tolerance*; Hasanuzzaman, M., Fujita, M., Oku, H., Nahar, K., Hawrylak-Nowak, B., Eds.; Springer: Singapore, 2018. [CrossRef]
3. Ghosh, P.; Rathinasabapathi, B.; Ma, L.Q. Phosphorus solubilization and plant growth enhancement by arsenic-resistant bacteria. *Chemosphere* **2015**, *134*, 1–6. [CrossRef] [PubMed]
4. Zhu, J.; Li, M.; Whelan, M. Phosphorus activators contribute to legacy phosphorus availability in agricultural soils: A review. *Sc. Total Environ.* **2018**, *612*, 522–537. [CrossRef] [PubMed]
5. Peñuelas, J.; Poulter, B.; Sardans, J.; Ciais, P.; van der Velde, M.; Bopp, L.; Boucher, O.; Godderis, Y.; Hinsinger, P.; Llusia, J.; et al. Human-Induced Nitrogen–Phosphorus Imbalances Alter Natural and Managed Ecosystems across the Globe. *Nat. Commun.* **2013**, *4*, 2934. [CrossRef] [PubMed]
6. Bandick, A.K.; Dick, R.P. Field management effects on enzyme activities. *Soil Biol. Biochem.* **1999**, *31*, 1471–1479. [CrossRef]
7. Dick, W.A. Influence of long-term tillage and crop rotation combinations on soil enzyme activities. *Soil Sci. Soc. Am. J.* **1984**, *48*, 569–574. [CrossRef]
8. Tabatabai, M.A. Methods of soil analysis, Part 2. Microbiological and Biochemical properties. In *Chapter 37: Soil Enzymes*; John Wiley & Sons: Madison, WI, USA, 1994; pp. 775–833.
9. Burns, R.G. *Soil Enzymes*; c.9b01784; Academic Press: London, UK, 1978.
10. Kiss, S.; Pasca, D.; Drăgan-Bularda, M. *Enzymology of Disturbed Soils*; Elsevier: Amsterdam, The Netherlands, 1998; Volume 26.
11. Allison, S.D.; Weintraub, M.N.; Gartner, T.B.; Waldrop, M.P. Evolutionary-economic principles as regulators of soil enzyme production and ecosystem function. In *Soil Enzymology*; Shukla, G., Varma, A., Eds.; Springer: Berlin, Germany, 2010; Volume 22. [CrossRef]
12. Zuccarini, P.; Sardans, J.; Asensio, L.; Peñuelas, J. Altered activities of extracellular soil enzymes by the interacting global environmental changes. *Glob. Chang. Biol.* **2023**, *29*, 2067–2091. [CrossRef]
13. Burns, R.G.; Dick, R.P. *Enzymes in the Environment: Activity, Ecology and Applications*; Marcel Dekker: New York, NY, USA, 2002.
14. Schmidt, G.; Laskowski, M., Sr. Phosphate ester cleavage (Survey). In *The Enzymes*, 2nd ed.; Boyer, P.D., Ed.; Academic Press: New York, NY, USA, 1961; pp. 3–35.
15. Acosta-Martínez, V.; Lascano, R.; Calderón, F.; Booker, J.D.; Zobeck, T.M.; Upchurch, D.R. Dryland Cropping Systems Influence the Microbial Biomass and Enzyme Activities in a Semiarid Sandy Soil. *Biol. Fertil. Soils* **2011**, *47*, 655–667. [CrossRef]
16. Cawley, G.C. Leave-one-out cross-validation based model selection criteria for weighted LS-SVMs. In Proceedings of the 2006 IEEE International Joint Conference on Neural Network Proceedings, Vancouver, BC, Canada, 16–21 July 2006; pp. 1661–1668. [CrossRef]
17. Burns, R.G. Enzyme activity in soil: Location and a possible role in microbial ecology. *Soil Biol. Biochem.* **1982**, *4*, 423–427. [CrossRef]
18. Joner, E.J.; van Aarle, I.M.; Vosatka, M. Phosphatase activity of extra-radical arbuscular mycorrhizal hyphae: A review. *Plant Soil* **2000**, *226*, 199–210. [CrossRef]
19. Park, Y.; Solhtalab, M.; Thongsomboon, W.; Aristilde, L. Strategies of organic phosphorus recycling by soil bacteria: Acquisition, metabolism, and regulation. *Environ. Micro. Rep.* **2022**, *14*, 3–24. [CrossRef]
20. McLean, J.; Gahan, P.B. The Distribution of Acid Phosphatases and Esterases in Differentiating Roots of Vicia Faba. *Histochemie* **1970**, *24*, 41–49. [CrossRef] [PubMed]
21. Juma, N.G.; Tabatabai, M.A. Phosphatase activity in corn and soybean roots: Conditions for assay and effects of metals. *Plant Soil* **1988**, *107*, 39–47. [CrossRef]
22. Carricondo-Martínez, I.; Falcone, D.; Berti, F.; Orsini, F.; Salas-Sanjuan, M.D.C. Use of Agro-Waste as a Source of Crop Nutrients in Intensive Horticulture System. *Agronomy* **2022**, *12*, 447. [CrossRef]
23. Alef, K.; Nannipieri, P. (Eds.) 7—Enzyme activities. In *Methods in Applied Soil Microbiology and Biochemistry*; Academic Press: London, UK, 1995; pp. 311–373. [CrossRef]
24. Tarafdar, J.C.; Claassen, N. Organic Phosphorus Compounds as a Phosphorus Source for Higher Plants through the Activity of Phosphatases Produced by Plant Roots and Microorganisms. *Biol. Fertil. Soils* **1988**, *5*, 308–312. [CrossRef]
25. Juma, N.G.; Tabatabai, M.A. Effects of trace elements on phosphatase activity in soils. *Soil Sci. Soc. Am. J.* **1977**, *41*, 343–346. [CrossRef]
26. Juma, N.G.; Tabatabai, M.A. Distribution of phosphomonoesterases in soils. *Soil Sci.* **1978**, *126*, 101–108. [CrossRef]
27. Neal, A.L.; Blackwell, M.; Akkari, E.; Guyomar, C.; Clark, I.; Hirsch, P.R. Phylogenetic distribution, biogeography and the effects of land management upon bacterial non-specific acid phosphatase gene diversity and abundance. *Plant Soil* **2018**, *427*, 175–189. [CrossRef] [PubMed]
28. Ragot, S.A.; Kertesz, M.A.; Mészaros, E.; Frossard, E.; Bünemann, E.K. Soil phoD and phoX alkaline phosphatase gene diversity responds to multiple environmental factors. *FEMS Microbiol. Ecol.* **2017**, *93*, fiw212. [CrossRef]
29. FAOSTAT. Comparar Datos (2021). Organización de las Naciones Unidas para la Alimentación y la Agricultura. Available online: http://www.fao.org/faostat/es/#compare (accessed on 17 September 2021).

30. Margalef, O.; Sardans, J.; Fernández-Martínez, M.; Molowny-Horas, R.; Janssens, A.; Ciais, P.; Goll, D.; Richter, A.; Obersteiner, M.; Asensio, D.; et al. Global patterns of phosphatase activity in natural soils. *Sci. Rep.* **2017**, *7*, 1337. [CrossRef]
31. Arora, R.; Sharma, V.; Sharma, S.; Maini, A.; Dhaliwal, S.S. Temporal Changes in Soil Biochemical Properties with Seasons under Rainfed Land Use Systems in Shiwalik Foothills of Northwest India. *Agrofor. Syst.* **2021**, *95*, 1479–1491. [CrossRef]
32. Choudhary, M.; Jat, H.S.; Datta, A.; Yadav, A.K.; Sapkota, T.B.; Mondal, S.; Meena, R.P.; Sharma, P.C.; Jat, M.L. Sustainable Intensification Influences Soil Quality, Biota, and Productivity in Cereal-Based Agroecosystems. *Appl. Soil Ecol.* **2018**, *126*, 189–198. [CrossRef]
33. Dick, R.P.; Rasmussen, P.E.; Kerle, E.A. Influence of Long-Term Residue Management on Soil Enzyme Activities in Relation to Soil Chemical Properties of a Wheat-Fallow System. *Biol. Fertil. Soils* **1988**, *6*, 159–164. [CrossRef]
34. Eichler-Löbermann, B.; Zicker, T.; Kavka, M.; Busch, S.; Brandt, C.; Stahn, P.; Miegel, K. Mixed Cropping of Maize or Sorghum with Legumes as Affected by Long-Term Phosphorus Management. *Field Crops Res.* **2021**, *265*, 108120. [CrossRef]
35. Chen, S.; Cade-Menun, B.J.; Bainard, L.K.; Luce MSt Hu, Y.; Chen, Q. The Influence of Long-Term N and P Fertilization on Soil P Forms and Cycling in a Wheat/Fallow Cropping System. *Geoderma* **2021**, *404*, 115274. [CrossRef]
36. Dutta, D.; Meena, A.L.; Chethan Kumar, G.; Mishra, R.P.; Ghasal, P.C.; Kumar, A.; Chaudhary, J.; Bhanu, C.; Kumar, V.; Kumar, A.; et al. Long Term Effect of Organic, Inorganic and Integrated Nutrient Management on Phosphorous Dynamics under Different Cropping Systems of Typic Ustochrept Soil of India. *Commun. Soil Sci. Plant Anal.* **2020**, *51*, 2746–2763. [CrossRef]
37. Singh, S.R.; Kundu, D.K.; Dey, P.; Singh, P.; Mahapatra, B.S. Effect of Balanced Fertilizers on Soil Quality and Lentil Yield in Gangetic Alluvial Soils of India. *J. Agric. Sci.* **2018**, *156*, 225–240. [CrossRef]
38. Grafe, M.; Kurth, J.K.; Panten, K.; Raj, A.D.; Baum, C.; Zimmer, D.; Leinweber, P.; Schloter, M.; Schulz, S. Effects of Different Innovative Bone Char Based P Fertilizers on Bacteria Catalyzing P Turnover in Agricultural Soils. *Agric. Ecos. Environ.* **2021**, *314*, 107419. [CrossRef]
39. Monkiedje, A.; Spiteller, M.; Fotio, D.; Sukul, P. The Effect of Land Use on Soil Health Indicators in Peri-Urban Agriculture in the Humid Forest Zone of Southern Cameroon. *J. Environ. Qual.* **2006**, *35*, 2402–2409. [CrossRef]
40. Margalef, O.; Sardans, J.; Maspons, J.; Molowny-Horas, R.; Fernández-Martínez, M.; Janssens, I.A.; Richter, A.; Ciais, P.; Obersteiner, M.; Peñuelas, J. The effect of global change on soil phosphatase activity. *Glob. Chang. Biol.* **2021**, *27*, 5989–6003. [CrossRef]
41. Sun, Y.; Goll, D.S.; Ciais, P.; Peng, S.; Margalef, O.; Asensio, D.; Sardans, J.; Peñuelas, J. Spatial Pattern and Environmental Drivers of Acid Phosphatase Activity in Europe. *Fron. Big Data* **2020**, *3*, 51. [CrossRef]
42. Meng, C.; Tian, D.; Zeng, H.; Li, Z.; Chen, H.Y.H.; Niu, S. Global Meta-Analysis on the Responses of Soil Extracellular Enzyme Activities to Warming. *Sci. Total Environ.* **2020**, *705*, 135992. [CrossRef] [PubMed]
43. Gao, D.; Bai, E.; Li, M.; Zhao, C.; Yu, K.; Hagedorn, F. Responses of Soil Nitrogen and Phosphorus Cycling to Drying and Rewetting Cycles: A Meta-Analysis. *Soil Biol. Biochem.* **2020**, *148*, 107896. [CrossRef]
44. Janes-Bassett, V.; Blackwell, M.S.A.; Blair, G.; Davies, J.; Haygarth, P.M.; Mezeli, M.M.; Stewart, G. A meta-analysis of phosphatase activity in agricultural settings in response to phosphorus deficiency. *Soil Biol. Biochem.* **2022**, *165*, 108537. [CrossRef]
45. Pokharel, P.; Ma, Z.; Chang, S.X. Biochar increases soil microbial biomass with changes in extra- and intracellular enzyme activities: A global meta-analysis. *Biochar* **2020**, *2*, 65–79. [CrossRef]
46. Miao, F.; Li, Y.; Cui, S.; Jagadamma, S.; Yang, G.; Zhang, Q. Soil extracellular enzyme activities under long-term fertilization management in the croplands of China: A meta-analysis. *Nut. Cycl. Agroecosyst.* **2019**, *114*, 125–138. [CrossRef]
47. Jian, S.; Li, J.; Chen, J.; Wang, G.; Mayes, M.A.; Dzantor, K.E.; Hui, D.; Luo, Y. Soil extracellular enzyme activities, soil carbon and nitrogen storage under nitrogen fertilization: A meta-analysis. *Soil Biol. Biochem.* **2016**, *101*, 32–43. [CrossRef]
48. Marklein, A.R.; Houlton, B.Z. Nitrogen inputs accelerate phosphorus cycling rates across a wide variety of terrestrial ecosystems. *New Phyt.* **2012**, *193*, 696–704. [CrossRef]
49. Aponte, H.; Meli, P.; Butler, B.; Paolini, J.; Matus, F.; Merino, C.; Cornejo, P.; Kuzyakov, Y. Meta-analysis of heavy metal effects on soil enzyme activities. *Sci. Total Environ.* **2020**, *737*, 139744. [CrossRef]
50. Riah, W.; Laval, K.; Laroche-Ajzenberg, E.; Mougin, C.; Latour, X.; Trinsoutrot-Gattin, I. Effects of pesticides on soil enzymes: A review. *Environ. Chem. Let.* **2014**, *12*, 257–273. [CrossRef]
51. Tabatabai, M.A.; Bremner, J.M. Use of p-nitrophenyl phosphate for assay of soil phosphatase activity. *Soil Biol. Biochem.* **1969**, *1*, 301–307. [CrossRef]
52. Eivazi, F.; Tabatabai, M.A. Phosphatases in Soils. *Soil Biol. Biochem.* **1977**, *9*, 167–172. [CrossRef]
53. Wang, F.; Jiang, R.; Kertesz, M.A.; Zhang, F.; Feng, G. Arbuscular mycorrhizal fungal hyphae mediating acidification can promote phytate mineralization in the hyphosphere of maize (*Zea mays* L.). *Soil Biol. Biochem.* **2013**, *65*, 69–74. [CrossRef]
54. Gesolmino, A.; Azzellino, A. Multivariate Analysis of Soils: Microbial Biomass, Metabolic Activity, and Bacterial-community Structure and Their Relationships with Soil Depth and Type. *J. Plant Nutr. Soil Sci.* **2011**, *174*, 381–394. [CrossRef]
55. Chowdhury, N.; Rasid, M.M. Evaluation of brick kiln operation impact on soil microbial biomass and enzyme activity. *Soil Sci. Annu.* **2021**, *72*, 132232. [CrossRef]
56. Sharma, P.; Singh, G.; Singh, R.P. Conservation Tillage and Optimal Water Supply Enhance Microbial Enzyme (Glucosidase, Urease and Phosphatase) Activities in Fields under Wheat Cultivation during Various Nitrogen Management Practices. *Arch. Agron. Soil Sci.* **2013**, *59*, 911–928. [CrossRef]

57. Scaramal da Silva, A.; Colozzi Filho, A.; Shigueyoshi Nakatani, A.; José Alves, S.; Souza de Andrade, D.; Guimarães, M.d.F. Atributos Microbiológicos Do Solo Em Sistema de Integração. *Rev. Bras. Ciência Solo* **2015**, *39*, 40–48. [CrossRef]
58. Woźniak, M.; Gałazką, A.; Siebielec, G.; Frąc, M. Can the Biological Activity of Abandoned Soils Be Changed by the Growth of *Paulownia elongata* × *Paulownia fortunei*?—Preliminary Study on a Young Tree Plantation. *Agriculture* **2022**, *12*, 128. [CrossRef]
59. Dubey, A.N.; Chattopadhyaya, N.; Mandal, N. Variation in Soil Microbial Population and Soil Enzymatic Activities Under Zincated Nanoclay Polymer Composites (ZNCPCs), Nano-ZnO and Zn Solubilizers in Rice Rhizosphere. *Agric. Res.* **2021**, *10*, 21–31. [CrossRef]
60. Antolín, M.C.; Pascual, I.; García, C.; Polo, A.; Sánchez-Díaz, M. Growth, Yield and Solute Content of Barley in Soils Treated with Sewage Sludge under Semiarid Mediterranean Conditions. *Field Crops Res.* **2005**, *94*, 224–237. [CrossRef]
61. Maini, A.; Sharma, V.; Sharma, S. Assessment of Soil Biochemical Properties and Soil Quality Index under Rainfed Land Use Systems in Submontane Punjab, India. *Indian J. Biochem. Biophys.* **2022**, *59*, 357–367. [CrossRef]
62. Borase, D.N.; Nath, C.P.; Hazra, K.K.; Senthilkumar, M.; Singh, S.S.; Praharaj, C.S.; Singh, U.; Kumar, N. Long-Term Impact of Diversified Crop Rotations and Nutrient Management Practices on Soil Microbial Functions and Soil Enzymes Activity. *Ecol. Ind.* **2020**, *114*, 106322. [CrossRef]
63. Hatti, V.; Ramachandrappa, B.K.; Mudalagiriyappa, S.A.; Thimmegowda, M.N. Soil properties and productivity of rainfed finger millet under conservation tillage and nutrient management in Eastern dry zone of Karnataka. *J. Environ. Biol.* **2018**, *19*, 612–624. [CrossRef]
64. Angers, D.A.; Bissonnette, N.; Legere, A.; Samson, N. Microbial and Biochemical Changes Induced by Rotation and Tillage in a Soil under Barley Production. *Can. J. Soil Sci.* **1993**, *73*, 39–50. [CrossRef]
65. Sepat, S.; Behera, U.K.; Sharma, A.R.; Das, T.K.; Bhattacharyya, R. Productivity, Organic Carbon and Residual Soil Fertility of Pigeonpea-Wheat Cropping System under Varying Tillage and Residue Management. *Proc. Natl. Acad. Sci. India Sect. B Biol. Sci.* **2014**, *84*, 561–571. [CrossRef]
66. Wang, M.; Wu, Y.; Zhao, J.; Liu, Y.; Chen, Z.; Tang, Z.; Tian, W.; Xi, Y.; Zhang, J. Long-Term Fertilization Lowers the Alkaline Phosphatase Activity by Impacting the PhoD-Harboring Bacterial Community in Rice-Winter Wheat Rotation System. *Sci. Total Environ.* **2022**, *821*, 153406. [CrossRef] [PubMed]
67. Banerjee, M.R.; Burton, D.L.; Grant, C.A. Influence of Urea Fertilization and Urease Inhibitor on the Size and Activity of the Soil Microbial Biomass under Conventional and Zero Tillage at Two Sites. *Can. J. Soil Sci.* **1999**, *79*, 255–263. [CrossRef]
68. Al-Taweel, J.L.S.; Al-Jubouri, G.A.A. Effect of Agricultural Exploitation on the Activity of Alkaline Phosphatase and Its Kinetic Properties in Some Soils. *Al-Qadisiyah J. Agric. Sci.* **2019**, *9*, 120–135. [CrossRef]
69. Liu, L.; Zhu, K.; Wurzburger, N.; Zhang, J. Relationships between plant diversity and soil microbial diversity vary across taxonomic groups and spatial scales. *Ecosphere* **2020**, *11*, e02999. [CrossRef]
70. Furtak, K.; Gawryjołek, K.; Gajda, A.M.; Gałązka, A. Effects of Maize and Winter Wheat Grown under Different Cultivation Techniques on Biological Activity of Soil. *Plant Soil Environ.* **2017**, *63*, 449–454. [CrossRef]
71. Mandal, N.; Datta, S.C.; Dwivedi, B.S.; Manjaiah, K.M.; Meena, M.C.; Bhowmik, A. Zincated Nanoclay Polymer Composite (ZNCPC): Effect on DTPA-Zn, Olsen-P and Soil Enzymatic Activities in Rice Rhizosphere. *Commun. Soil Sci. Plant Anal.* **2021**, *52*, 2032–2044. [CrossRef]
72. Wu, F.; Wan, J.H.C.; Wu, S.; Wong, M. Effects of Earthworms and Plant Growth-Promoting Rhizobacteria (PGPR) on Availability of Nitrogen, Phosphorus, and Potassium in Soil. *J. Plant Nut. Soil Sci.* **2012**, *175*, 423–433. [CrossRef]
73. Tao, J.; Griffiths, B.; Zhang, S.; Chen, X.; Liu, M.; Hu, F.; Li, H. Effects of Earthworms on Soil Enzyme Activity in an Organic Residue Amended Rice-Wheat Rotation Agro-Ecosystem. *Appl. Soil Ecol.* **2019**, *42*, 221–226. [CrossRef]
74. Balachandar, R.; Biruntha, M.; Yuvaraj, A.; Thangaraj, R.; Subbaiya, R.; Govarthanan, M.; Kumar, P.; Karmegam, N. Earthworm Intervened Nutrient Recovery and Greener Production of Vermicompost from Ipomoea Staphylina. An Invasive Weed with Emerging Environmental Challenges. *Chemosphere* **2021**, *263*, 128080. [CrossRef] [PubMed]
75. Buck, C.; Langmaack, M.; Schrader, S. Influence of Mulch and Soil Compaction on Earthworm Cast Properties. *Appl. Soil Ecol.* **2000**, *14*, 223–229. [CrossRef]
76. Soane, B.D.; van Ouwerkerk, C. Chapter 1—Soil Compaction Problems in World Agriculture. In *Developments in Agricultural Engineering*; Soane, B.D., van Ouwerkerk, C., Eds.; Elsevier: Amsterdam, The Netherlands, 1994; Volume 11, pp. 1–21. [CrossRef]
77. Noronha, F.R.; Manikandan, S.K.; Nair, V. Role of Coconut Shell Biochar and Earthworm (*Eudrilus euginea*) in Bioremediation and Palak Spinach (*Spinacia oleracea* L.) Growth in Cadmium-Contaminated Soil. *J. Environ. Manag.* **2022**, *302*, 114057. [CrossRef]
78. Brockett, B.F.T.; Prescott, C.E.; Grayston, S.J. Soil Moisture Is the Major Factor Influencing Microbial Community Structure and Enzyme Activities across Seven Biogeoclimatic Zones in Western Canada. *Soil Biol. Biochem.* **2012**, *44*, 9–20. [CrossRef]
79. Ojeda, G.; Patrício, J.; Navajas, H.; Comellas, L.; Alcañiz, J.M.; Ortiz, O.; Marks, E.; Natal-da Luz, T.; Sousa, J.P. Effects of Nonylphenols on Soil Microbial Activity and Water Retention. *Appl. Soil Ecol.* **2013**, *64*, 77–83. [CrossRef]
80. Calvarro, L.M.; de Santiago-Martín, A.; Quirós Gómez, J.; González-Huecas, C.; Quintana, J.R.; Vázquez, A.; Lafuente, A.L.; Rodríguez Fernández, T.M.; Ramírez Vera, R. Biological Activity in Metal-Contaminated Calcareous Agricultural Soils: The Role of the Organic Matter Composition and the Particle Size Distribution. *Environ. Sci. Pol. Res.* **2014**, *21*, 6176–6187. [CrossRef]
81. Gispert, M.; Emran, M.; Pardini, G.; Doni, S.; Ceccanti, B. The Impact of Land Management and Abandonment on Soil Enzymatic Activity, Glomalin Content and Aggregate Stability. *Geoderma* **2013**, *202–203*, 51–61. [CrossRef]

82. Garg, S.; Bahl, G.S. Phosphorus Availability to Maize as Influenced by Organic Manures and Fertilizer P Associated Phosphatase Activity in Soils. *Biores. Technol.* **2008**, *99*, 5773–5777. [CrossRef]
83. Nedyalkova, K.; Donkova, R.; Malinov, I. Acid Phosphatase Activity under the Impact of Erosion Level in Agricultural Soils of Different Type and Land Use. *Bulg. J. Agric. Sci.* **2020**, *26*, 1217–1222.
84. Bergstrom, D.W.; Monreal, C.M.; King, D.J. Sensitivity of Soil Enzyme Activities to Conservation Practices. *Soil Sci. Soc. Am. J.* **1998**, *62*, 1286–1295. [CrossRef]
85. Wei, K.; Chen, Z.; Zhu, A.; Zhang, J.; Chen, L. Application of 31P NMR Spectroscopy in Determining Phosphatase Activities and P Composition in Soil Aggregates Influenced by Tillage and Residue Management Practices. *Soil Tillage Res.* **2014**, *138*, 35–43. [CrossRef]
86. Odutola, O.S. Introductory Chapter: Relevance of Soil pH to Agriculture. In *Soil pH for Nutrient Availability and Crop Performance*; IntechOpen: London, UK, 2019.
87. Mandal, A.; Thakur, J.K.; Sahu, A.; Manna, M.C.; Rao, A.S.; Sarkar, B.; Patra, A.K. Effects of Bt-Cotton on Biological Properties of Vertisols in Central India. *Arch. Agron. Soil Sci.* **2018**, *65*, 670–685. [CrossRef]
88. Ortiz, J.; Faggioli, V.S.; Ghio, H.; Boccolini, M.F.; Ioele, J.P.; Tamburrini, P.; Garcia, F.O.; Gudelj, V. Long-Term Impact of Fertilization on the Structure and Functionality of Microbial Soil Community | Impacto a Largo Plazo de La Fertilización Sobre La Estructura y Funcionalidad de La Comunidad Microbiana Del Suelo. *Cien. Suelo* **2020**, *38*, 45–55.
89. Truu, M.; Truu, J.; Ivask, M. Soil Microbiological and Biochemical Properties for Assessing the Effect of Agricultural Management Practices in Estonian Cultivated Soils. *Eur. J. Soil Biol.* **2008**, *44*, 231–237. [CrossRef]
90. Laxminarayana, K. Effect of Mycorrhiza, Organic Sources, Lime, Secondary and Micro-Nutrients on Soil Microbial Activities and Yield Performance of Yam Bean (*Pachyrhizus erosus* L.) in Alfisols. *Commun. Soil Sci. Plant Anal.* **2017**, *48*, 186–200. [CrossRef]
91. Durrer, A.; Gumiere, T.; Rumenos Guidetti Zagatto, M.; Petry Feiler, H.; Miranda Silva, A.M.; Longaresi, R.H.; Homma, S.K.; Cardoso, E.J.B.N. Organic Farming Practices Change the Soil Bacteria Community, Improving Soil Quality and Maize Crop Yields. *PeerJ* **2021**, *9*, 1–24. [CrossRef]
92. Yang, L.; Zhao, F.; Chang, Q.; Li, T.; Li, F. Effects of Vermicomposts on Tomato Yield and Quality and Soil Fertility in Greenhouse under Different Soil Water Regimes. *Agric. Water Manag.* **2015**, *160*, 98–105. [CrossRef]
93. Roldán, A.; Salinas-García, J.R.; Alguacil, M.M.M.; Caravaca, F. Soil Sustainability Indicators Following Conservation Tillage Practices under Subtropical Maize and Bean Crops. *Soil Tillage Res.* **2007**, *93*, 273–282. [CrossRef]
94. Swedrzyńska, D.; Małecka, I.; Blecharczyk, A.; Swedrzyński, A.; Starzyk, J. Effects of various long-term tillage systems on some chemical and biological properties of soil. *Pol. J. Environ. Stud.* **2013**, *22*, 1835–1844.
95. Mahmood, M.; Xu, T.; Ahmed, W.; Yang, J.; Li, J.; Mehmood, S.; Liu, W.; Weng, J.; Li, W. Variability in Soil Parent Materials at Different Development Stages Controlled Phosphorus Fractions and Its Uptake by Maize Crop. *Sustainability* **2022**, *14*, 5048. [CrossRef]
96. Siddaramappa, R.; Wright, R.J.; Codling, E.E.; Gao, G.; McCarty, G.W. Evaluation of Coal Combustion Byproducts as Soil Liming Materials: Their Influence on Soil PH and Enzyme Activities. *Biol. Fertil. Soils* **1994**, *17*, 167–172. [CrossRef]
97. Yu, S.; He, Z.L.; Stoffella, P.J.; Calvert, D.V.; Yang, X.E.; Banks, D.J.; Baligar, V.C. Surface Runoff Phosphorus (P) Loss in Relation to Phosphatase Activity and Soil P Fractions in Florida Sandy Soils under Citrus Production. *Soil Biol. Biochem.* **2006**, *38*, 619–628. [CrossRef]
98. Meena, H.M.; Prakasha, H.C. The Impact of Biochar, Lime and Fertilizer on Soil Acidity and Microbiological Properties and Their Relationship with Yield of Rice and Cowpea in an Acidic Soil of Southern India. *J. Plant Nutr.* **2021**, *45*, 358–368. [CrossRef]
99. Purnamasari, L.; Rostaman, T.; Widowati LR Anggria, L. Comparison of Appropriate Cation Exchange Capacity (CEC) Extraction Methods for Soils from Several Regions of Indonesia. *IOP Conf. Ser. Earth Environ. Sci.* **2021**, *648*, 012209. [CrossRef]
100. Smith, J.L.; Doran, J.W. Measurement and Use of pH and Electrical Conductivity for Soil Quality Analysis. In *Methods for Assessing Soil Quality*; Wiley: Hoboken, NJ, USA, 1996; pp. 169–185. [CrossRef]
101. Dinesh, R.; Ramanathan, G.; Singh, H. Influence of Chloride and Sulphate Ions on Soil Enzymes. *J. Agron. Crop Sci.* **1995**, *175*, 129–133. [CrossRef]
102. Rietz, D.N.; Haynes, R.J. Effects of Irrigation-Induced Salinity and Sodicity on Soil Microbial Activity. *Soil Biol. Biochem.* **2003**, *35*, 845–854. [CrossRef]
103. Garcia, C.; Hernandez, T. Influence of Salinity on the Biological and Biochemical Activity of a Calciorthird Soil. *Plant and Soil* **1996**, *178*, 255–263. [CrossRef]
104. Turbé, A.; de Toni, A.; Benito, P.; Lavelle, P.; Lavelle, P.; Camacho, N.R.; van Der Putten, W.H.; Labouze, E.; Mudgal, S. Soil biodiversity: Functions, threats and tools for policy makers. 2010. Available online: https://hal-bioemco.ccsd.cnrs.fr/bioemco-00560420 (accessed on 17 September 2021).
105. Tipping, E.; Somerville, C.J.; Luster, J. The C:N:P:S stoichiometry of soil organic matter. *Biogeochemistry* **2016**, *130*, 117–131. [CrossRef]
106. Kooch, Y.; Ehsani, S.; Akbarinia, M. Stoichiometry of Microbial Indicators Shows Clearly More Soil Responses to Land Cover Changes than Absolute Microbial Activities. *Ecol. Eng.* **2019**, *131*, 99–106. [CrossRef]
107. Lungmuana Singh, S.B.B.; Choudhury, B.U.U.; Vanthawmliana Saha, S.; Hnamte, V. Transforming Jhum to Plantations: Effect on Soil Microbiological and Biochemical Properties in the Foot Hills of North Eastern Himalayas, India. *Catena* **2019**, *177*, 84–91. [CrossRef]

108. de Jesus Franco, A.; Valadares da Silva, A.P.; Silva Souza, A.B.; Loverde Oliveira, R.; Rodrigues Batista, E.; Damacena de Souza, E.; Oliveira Silva, A.; Carbone Carneiro, M.A. Plant diversity in integrated crop-livestock systems increases the soil enzymatic activity in the short term. *Pesq. Agropec. Trop. Goiânia* **2020**, *50*, e64026. [CrossRef]
109. Adrover, M.; Moyà, G.; Vadell, J. Seasonal and Depth Variation of Soil Chemical and Biological Properties in Alfalfa Crops Irrigated with Treated Wastewater and Saline Groundwater. *Geoderma* **2017**, *286*, 54–63. [CrossRef]
110. Palmer, J.; Thorburn, P.J.; Biggs, J.S.; Dominati, E.J.; Probert, M.E.; Meier, E.A.; Huth, I.N.; Dodd, M.; Snow, V.; Larsen, J.R.; et al. Nitrogen Cycling from Increased Soil Organic Carbon Contributes Both Positively and Negatively to Ecosystem Services in Wheat Agro-Ecosystems. *Front. Plant Sci.* **2017**, *8*, 731. [CrossRef] [PubMed]
111. Cattaneo, F.; Di Gennaro, P.; Barbanti, L.; Giovannini, C.; Labra, M.; Moreno, B.; Benitez, E.; Marzadori, C. Perennial Energy Cropping Systems Affect Soil Enzyme Activities and Bacterial Community Structure in a South European Agricultural Area. *Appl. Soil Ecol.* **2014**, *84*, 213–222. [CrossRef]
112. Singh, A.; Ghoshal, N. Impact of Herbicide and Various Soil Amendments on Soil Enzymes Activities in a Tropical Rainfed Agroecosystem. *Eur. J. Soil Biol.* **2013**, *54*, 56–62. [CrossRef]
113. Sigua, G.C.; Stone, K.C.; Bauer, P.J.; Szogi, A.A. Phosphorus Dynamics and Phosphatase Activity of Soils under Corn Production with Supplemental Irrigation in Humid Coastal Plain Region, USA. *Nutr. Cycl. Agroecosyst.* **2017**, *109*, 249–267. [CrossRef]
114. Weil, R.R.; Brady, N.C. *The Nature and Properties of Soils*, 15th ed.; Pearson: New York, NY, USA, 2017.
115. Cross, A.F.; Schlesinger, W.H. A literature review and evaluation of the Hedley fractionation: Applications to the biogeochemical cycle of soil phosphorus in natural ecosystems. *Geoderma* **1995**, *64*, 197–214. [CrossRef]
116. Zibilske, L.M.; Bradford, J.M. Tillage Effects on Phosphorus Mineralization and Microbial Activity. *Soil Sci.* **2003**, *168*, 677–685. [CrossRef]
117. Olsen, S.R.; Sommers, L.E. Phosphorus. In *Methods of Soil Analysis Part 2 Chemical and Microbiological Properties, American Society of Agronomy, Soil Science Society of America*; Agronomy Monographs, Wiley: Madison, WI, USA, 1982; pp. 403–430.
118. Koper, J.; Lemanowicz, J. Effect of Varied Mineral Nitrogen Fertilization on Changes in the Content of Phosphorus in Soil and in Plant and the Activity of Soil Phosphatases. *Ecol. Chem. Eng.* **2008**, *S 15*, 465–471.
119. Atoloye, I.A.; Jacobson, A.; Creech, E.; Reeve, J. Variable Impact of Compost on Phosphorus Dynamics in Organic Dryland Soils Following a One-Time Application. *Soil Sci. Soc. Am. J.* **2021**, *85*, 1122–1138. [CrossRef]
120. Madejón, E.; Burgos, P.; López, R.; Cabrera, F. Agricultural Use of Three Organic Residues: Effect on Orange Production and on Properties of a Soil of the "Comarca Costa de Huelva" (SW Spain). *Nutr. Cycl. Agroecosyst.* **2003**, *65*, 281–288. [CrossRef]
121. Li, Q.; Chen, J.; Wu, L.; Luo, X.; Li, N.; Arafat, Y.; Lin, S.; Lin, W. Belowground Interactions Impact the Soil Bacterial Community, Soil Fertility, and Crop Yield in Maize/Peanut Intercropping Systems. *Int. J. Mol. Sci.* **2018**, *19*, 622. [CrossRef]
122. Waldrop, M.P.; Balser, T.C.; Firestone, M.K. Linking Microbial Community Composition to Function in a Tropical Soil. *Soil Biol. Biochem.* **2000**, *32*, 1837–1846. [CrossRef]
123. Khan, S.A.; Mulvaney, R.L.; Ellsworth, T.R. The Potassium Paradox: Implications for Soil Fertility, Crop Production and Human Health. *Renew. Agric. Food Syst.* **2014**, *29*, 3–27. [CrossRef]
124. Honvault, N.; Houben, D.; Nobile, C.; Firmin, S.; Lambers, H.; Faucon, M.P. Tradeoffs among Phosphorus-Acquisition Root Traits of Crop Species for Agroecological Intensification. *Plant Soil* **2020**, *461*, 137–150. [CrossRef]
125. Riffaldi, R.; Saviozzi, A.; Levi-Minzi, R.; Cardelli, R. Biochemical Properties of a Mediterranean Soil as Affected by Long-Term Crop Management Systems. *Soil Tillage Res.* **2022**, *67*, 109–114. [CrossRef]
126. Cui, Y.; Fang, L.; Guo, X.; Wang, X.; Wang, Y.; Zhang, Y.; Zhang, X. Responses of Soil Bacterial Communities, Enzyme Activities, and Nutrients to Agricultural-to-Natural Ecosystem Conversion in the Loess Plateau, China. *J. Soils Sediments* **2019**, 19. [CrossRef]
127. Li, C.; Veum, K.S.; Goyne, K.W.; Nunes, M.R.; Acosta-Martinez, V.A. Chronosequence of soil health under tallgrass prairie reconstruction. *Appl. Soil Ecol.* **2021**, *164*, 103939. [CrossRef]
128. Da Cunha, J.R.; de Cassia de Freitas, R.; de Almeida Taveres Souza, D.J.; Santana Gualberto, A.V.; Antunes de Souza, H.; Fernando Carvalho Leite, L. Soil Biological Attributes in Monoculture and Integrated Systems in the Cerrado Region of Piaui State, Brazil. *Acta Sci.-Agron.* **2021**, *43*, e51814. [CrossRef]
129. Balota, E.L.; Machineski, O.; Truber, P.V. Soil Enzyme Activities under Pig Slurry Addition and Different Tillage Systems. *Agronomy* **2011**, *33*, 729–737. [CrossRef]
130. Barcelos Martins, L.N.; de Aguiar Santiago, F.L.; Montecchia, M.S.; Correa, O.S.; Saggin Junior, O.J.; Damacena de Souza, E.; Barbosa Paulino, H.; Carbone Carneiro, M.A. Biochemical and Biological Properties of Soil from Murundus Wetlands Converted into Agricultural Systems. *Rev. Bras. Ciência Solo* **2019**, *43*, e0180183. [CrossRef]
131. Carlos, F.S.; Schaffer, N.; Mariot, R.F.; Schmitt Fernandes, R.; Luiz Boechat, C.; Fernando Wurdig Roesch, L.; de Oliveira Camargo, F.A. Soybean Crop Incorporation in Irrigated Rice Cultivation Improves Nitrogen Availability, Soil Microbial Diversity and Activity, and Growth of Ryegrass. *Appl. Soil Ecol.* **2022**, *170*, 104313. [CrossRef]
132. Katsalirou, E.; Deng, S.; Gerakis, A.; Nofziger, D.L. Long-Term Management Effects on Soil P, Microbial Biomass P, and Phosphatase Activities in Prairie Soils. *Eur. J. Soil Biol.* **2016**, *76*, 61–69. [CrossRef]
133. Sciubba, L.; Mazzon, M.; Cavani, L.; Baldi, E.; Toselli, M.; Ciavatta, C.; Marzadori, C. Soil Response to Agricultural Land Abandonment: A Case Study of a Vineyard in Northern Italy. *Agronomy* **2021**, *11*, 1841. [CrossRef]
134. Garcia, C.; Roldan, A.; Hernandez, T. Changes in Microbial Activity after Abandonment of Cultivation in a Semiarid Mediterranean Environment. *J. Environ. Qual.* **1997**, *26*, 285–291. [CrossRef]

135. Paz-Ferreiro, J.; Trasar-Cepeda, C.; Leiros, M.C.; Seoane, S.; Gil-Sotres, F. Biochemical Properties in Managed Grassland Soils in a Temperate Humid Zone: Modifications of Soil Quality as a Consequence of Intensive Grassland Use. *Biol. Fertil. Soils* **2009**, *45*, 711–722. [CrossRef]
136. Woźniak, A.; Kawecka-Radomska, M. Crop Management Effect on Chemical and Biological Properties of Soil. *Inter. J. Plant Prod.* **2016**, *10*, 391–402.
137. Martins Sousa, H.; Ribeiro Correa, A.; de Motta Silva, B.; da Silva Oliveira, S.; da Silva Campos, D.T.; Wruck, F.J. Dynamics of soil microbiological attributes in integrated crop-livestock systems in the cerrado-amozonônia ecotone. *Rev. Catinga* **2020**, *33*, 9–20. [CrossRef]
138. Dou, F.; Wright, A.L.; Mylavaparu, R.S.; Jiang, X.; Matocha, J.E. Soil Enzyme Activities and Organic Matter Composition Affected by 26 Years of Continuous Cropping. *Pedosphere* **2016**, *26*, 618–625. [CrossRef]
139. Ansari, M.A.; Saha, S.; Das, A.; Lal, R.; Das, B.; Choudhury, B.U.; Roy, S.S.; Sharma, S.K.; Singh, I.M.; Meitei, C.B.; et al. Energy and Carbon Budgeting of Traditional Land Use Change with Groundnut Based Cropping System for Environmental Quality, Resilient Soil Health and Farmers Income in Eastern Indian Himalayas. *J. Environ. Manag.* **2021**, *293*, 112892. [CrossRef] [PubMed]
140. Gao, Y.; Zhou, P.; Mao, L.; Zhi, Y.; Zhang, C.; Shi, W. Effects of Plant Species Coexistence on Soil Enzyme Activities and Soil Microbial Community Structure under Cd and Pb Combined Pollution. *J. Environ. Sci.* **2010**, *22*, 1040–1048. [CrossRef] [PubMed]
141. Wang, X.; Deng, X.; Pu, T.; Song, C.; Yong, T.; Yang, F.; Sun, X.; Liu, W.; Yan, Y.; Du, J.; et al. Contribution of Interspecific Interactions and Phosphorus Application to Increasing Soil Phosphorus Availability in Relay Intercropping Systems. *Field Crops Res.* **2017**, *204*, 12–22. [CrossRef]
142. Saad, R.F.; Kobaissi, A.; Echevarria, G.; Kidd, P.; Calusinska, M.; Goux, X.; Benizri, E. Influence of New Agromining Cropping Systems on Soil Bacterial Diversity and the Physico-Chemical Characteristics of an Ultramafic Soil. *Sci. Total Environ.* **2018**, *645*, 380–392. [CrossRef] [PubMed]
143. Aparna, K.; Rao, D.L.N.; Balachandar, D. Microbial Populations, Activity and Gene Abundance in Tropical Vertisols Under Intensive Chemical Farming. *Pedosphere* **2016**, *26*, 725–732. [CrossRef]
144. Lago, M.d.C.F.; Gallego, P.P.; Briones, M.J.I. Intensive cultivation of kiwifruit alters the detrital foodweb and accelerates soil C and N losses. *Front. Microbiol.* **2019**, *10*, 686. [CrossRef]
145. Feng, H.; Sekaran, U.; Wang, T.; Kumar, S. On-Farm Assessment of Cover Cropping Effects on Soil C and N Pools, Enzyme Activities, and Microbial Community Structure. *J. Agric. Sci.* **2021**, *159*, 216–226. [CrossRef]
146. Adetunji, A.T.; Ncube, B.; Meyer, A.H.; Olatunji, O.S.; Mulidzi, R.; Lewu FBSoil, P.H. Nitrogen, Phosphatase and Urease Activities in Response to Cover Crop Species, Termination Stage and Termination Method. *Heliyon* **2021**, *7*, e05980. [CrossRef]
147. Yadav, D.; Shivay, Y.S.; Singh, Y.V.; Sharma, V.K.; Bhatia, A. Water Use and Soil Fertility under Rice–Wheat Cropping System in Response to Green Manuring and Zinc Nutrition. *Commun. Soil Sci. Plant Analys.* **2019**, *50*, 2836–2847. [CrossRef]
148. Balota, E.L.; Chaves, D.; César, J. Enzymatic activity and mineralization of carbon and nitrogen in soil cultivated with coffee and green manures. *Rev. Bras. Ciência Solo.* **2010**, *34*, 1573–1583. [CrossRef]
149. Wang, M.; Wu, C.; Cheng, Z.; Meng, H.; Zhang, M. Soil Chemical Property Changes in Eggplant/Garlic Relay Intercropping Systems under Continuous Cropping. *PLoS ONE* **2014**, *9*, e111040. [CrossRef] [PubMed]
150. Kroulík, M.; Kumhála, F.; Hůla, J.; Honzík, I. The Evaluation of Agricultural Machines Field Trafficking Intensity for Different Soil Tillage Technologies. *Soil Tillage Res.* **2009**, *105*, 171–175. [CrossRef]
151. Buhler, D.D. Weed management. In *Encyclopedia of Soils in the Environment*; Hillel, D., Ed.; Elsevier: Oxford, UK, 2005; pp. 323–328.
152. Melero, S.; López-Bellido, R.J.; López-Bellido, L.; Muñoz-Romero, V.; Moreno, F.; Murillo, J.M. Long-Term Effect of Tillage, Rotation and Nitrogen Fertiliser on Soil Quality in a Mediterranean Vertisol. *Soil Tillage Res.* **2011**, *114*, 97–107. [CrossRef]
153. Gajda, A.M.; Przewłoka, B. Soil Biological Activity as Affected by Tillage Intensity. *Inter. Agrophys.* **2012**, *26*, 15–23. [CrossRef]
154. Parihar, C.M.; Yadav, M.R.; Jat, S.L.; Singh, A.K.; Kumar, B.; Pradhan, S.; Chakraborty, D.; Jat, M.L.; Jat, R.K.; Saharawat, Y.S.; et al. Long Term Effect of Conservation Agriculture in Maize Rotations on Total Organic Carbon, Physical and Biological Properties of a Sandy Loam Soil in North-Western Indo-Gangetic Plains. *Soil Tillage Res.* **2016**, *161*, 116–128. [CrossRef]
155. Doran, J.W. Soil microbial and biochemical changes associated with reduced tillage. *Soil Sci. Soc. Am. J.* **1980**, *44*, 765–771. [CrossRef]
156. Redel, Y.D.; Rubio, R.; Rouanet, J.L.; Borie, F. Phosphorus Bioavailability Affected by Tillage and Crop Rotation on a Chilean Volcanic Derived Ultisol. *Geoderma* **2007**, *139*, 388–396. [CrossRef]
157. Yang, X.; Bao, X.; Yang, Y.; Zhao, Y.; Liang, C.; Xie, H. Comparison of Soil Phosphorus and Phosphatase Activity under Long-Term No-Tillage and Maize Residue Management. *Plant Soil Environ.* **2019**, *65*, 408–415. [CrossRef]
158. Ahmed, W.; Qaswar, M.; Jing, H.; Wenjun, D.; Geng, S.; Kailou, L.; Ying, M.; Ao, T.; Mei, S.; Chao, L.; et al. Tillage Practices Improve Rice Yield and Soil Phosphorus Fractions in Two Typical Paddy Soils. *J. Soils Sediments* **2019**, *20*, 850–861. [CrossRef]
159. Jiang, Y.; Arafat, Y.; Letuma, P.; Ali, L.; Tayyab, M.; Waqas, M.; Li, Y.; Lin, W.; Lin, S.; Lin, W. Restoration of Long-Term Monoculture Degraded Tea Orchard by Green and Goat Manures Applications System. *Sustainability* **2019**, *11*, 1011. [CrossRef]
160. Jarosch, K.A.; Kandeler, E.; Frossard, E.; Bünemann, E.K. Is the Enzymatic Hydrolysis of Soil Organic Phosphorus Compounds Limited by Enzyme or Substrate Availability? *Soil Biol. Biochem.* **2019**, *139*, 107628. [CrossRef]
161. Igalavithana, A.D.; Lee, S.S.; Niazi, N.K.; Lee, Y.H.; Kim, K.H.; Park, J.H.; Moon, D.H.; Ok, Y.S. Assessment of Soil Health in Urban Agriculture: Soil Enzymes and Microbial Properties. *Sustainability* **2017**, *9*, 310. [CrossRef]

162. Chatterjee, D.; Nayak, A.K.; Mishra, A.; Swain, C.K.; Kumar, U.; Bhaduri, D.; Panneerselvam, P.; Lal, B.; Gautam, P.; Pathak, H. Effect of Long-Term Organic Fertilization in Flooded Rice Soil on Phosphorus Transformation and Phosphate Solubilizing Microorganisms. *J. Soil Sci. Plant Nutr.* **2021**, *21*, 1368–1381. [CrossRef]
163. Dhanker, R.; Chaudhary, S.; Goyal, S.; Kumar, R. Soil Microbial Properties and Functional Diversity in Response to Sewage Sludge Amendments. *Arch. Agron. Soil Sci.* **2021**, *68*, 809–822. [CrossRef]
164. Adetunji, A.T.; Lewu, F.B.; Mulidzi, R.; Ncube, B. The biological activities of β-glucosidase, phosphatase and urease as soil quality indicators: A review. *J. Soil Sci. Plant Nutr.* **2017**, *7*, 794–807. [CrossRef]
165. Nobile, C.; Houben, D.; Michel, E.; Firmin, S.; Lambers, H.; Kandeler, E.; Faucon, M.-P. Phosphorus-Acquisition Strategies of Canola, Wheat and Barley in Soil Amended with Sewage Sludges. *Sci. Rep.* **2019**, *9*, 14878. [CrossRef] [PubMed]
166. Leirós, M.C.; Trasar-Cepeda, C.; García-Fernández, F.; Gil-Sotres, F. Defining the Validity of a Biochemical Index of Soil Quality. *Biol. Fertil. Soils* **1999**, *30*, 140–146. [CrossRef]
167. Makoi, J.H.J.R.; Bambara, S.; Ndakidemi, P.A. Rhizosphere Phosphatase Enzyme Activities and Secondary Metabolites in Plants as Affected by the Supply of Rhizobium, Lime and Molybdenum in *Phaseolus vulgaris* L. *Aust. J. Soil Sci.* **2010**, *4*, 590–597.
168. Bolton, H., Jr.; Elliott, L.F.; Papendick, R.I.; Bezdicek, D.F. Soil microbial biomass and selected soil enzyme activities: Effect of fertilization and cropping practices. *Soil Biol. Biochem.* **1985**, *17*, 297–302. [CrossRef]
169. Dhull, S.; Goyal, S.; Kapoor, K.; Mundra, M. Microbial Biomass Carbon and Microbial Activities of Soils Receiving Chemical Fertilizers and Organic Amendments. *Int. J. Phytoremed.* **2004**, *21*, 641–647. [CrossRef]
170. Singh, G.; Bhattacharyya, R.; Das, T.K.; Sharma, A.R.; Ghosh, A.; Das, S.; Jha, P. Crop Rotation and Residue Management Effects on Soil Enzyme Activities, Glomalin and Aggregate Stability under Zero Tillage in the Indo-Gangetic Plains. *Soil Tillage Res.* **2018**, *184*, 291–300. [CrossRef]
171. Chellappa, J.; Laxmisagara Sagar, K.; Sekaran, U.; Kumar, S.; Sharma, P. Soil Organic Carbon, Aggregate Stability and Biochemical Activity under Tilled and No-Tilled Agroecosystems. *J. Agric. Food Res.* **2021**, *4*, 100139. [CrossRef]
172. Tu, C.; Ristaino, J.B.; Hu, S. Soil Microbial Biomass and Activity in Organic Tomato Farming Systems: Effects of Organic Inputs and Straw Mulching. *Soil Biol. Biochem.* **2006**, *38*, 247–255. [CrossRef]
173. Wyszkowska, J. Effect of Soil Contamination with Treflan 480 EC on Biochemical Properties of Soil. *Pol. J. Environ. Stud.* **2002**, *11*, 71–77.
174. Sharma, S.; Dhaliwal, S.S. Effect of Sewage Sludge and Rice Straw Compost on Yield, Micronutrient Availability and Soil Quality under Rice–Wheat System. *Commun. Soil Sci. Plan. Anal.* **2019**, *50*(16), 1943–1954. [CrossRef]
175. Gaind, S.; Nain, L. Chemical and Biological Properties of Wheat Soil in Response to Paddy Straw Incorporation and Its Biodegradation by Fungal Inoculants. *Biodegradation* **2017**, *18*, 495–503. [CrossRef] [PubMed]
176. Hoyle, F.C.; Murphy, D.V. Seasonal Changes in Microbial Function and Diversity Associated with Stubble Retention versus Burning. *Aust. J. Soil Res.* **2006**, *44*, 407–423. [CrossRef]
177. Trujillo-Narcía, A.; Rivera-Cruz, M.C.; Magaña-Aquino, M.; Trujillo-Rivera, E.A. The Burning of Sugarcane Plantation in the Tropics Modifies the Microbial and Enzymatic Processes in Soil and Rhizosphere. *J. Soil Sci. Plant Nutr.* **2019**, *19*, 906–919. [CrossRef]
178. Emami, S.; Alikhani, H.A.; Pourbabaee, A.A.; Etesami, H.; Sarmadian, F.; Motesharezadeh, B.; Taghizadeh–Mehrjardi, R. Performance Evaluation of Phosphate-Solubilizing Fluorescent Pseudomonads in Minimizing Phosphorus Fertilizer Use and Improving Wheat Productivity: A Two-Year Field Study. *J. Soil Sci. Plant Nutr.* **2022**, *22*, 1224–1237. [CrossRef]
179. Parnell, J.J.; Berka, R.; Young, H.A.; Sturino, J.M.; Kang, Y.; Barnhart, D.M.; DiLeo, M.V. From the Lab to the Farm: An Industrial Perspective of Plant Beneficial Microorganisms. *Front. Plant Sci.* **2016**, *7*, 1110. [CrossRef] [PubMed]
180. Tian, J.; Ge, F.; Zhang, D.; Deng, S.; Liu, X. Roles of Phosphate Solubilizing Microorganisms from Managing Soil Phosphorus Deficiency to Mediating Biogeochemical P Cycle. *Biology* **2021**, *10*, 158. [CrossRef]
181. Valarini, P.J.; Alvarez, M.C.D.; Gasco, J.M.; Guerrero, F.; Tokeshi, H. Assessment of Soil Properties by Organic Matter and EM-Microorganism Incorporation. *Rev. Bras. Ciência Solo* **2003**, *27*, 519–525. [CrossRef]
182. Ruiz, J.L.; Salas, M.d.C. Evaluation of Organic Substrates and Microorganisms as Bio-Fertilisation Tool in Container Crop Production. *Agronomy* **2019**, *9*, 705. [CrossRef]
183. Mahapatra, B.; Adak, T.; Patil, N.K.B.; Pandi, G.G.P.; Gowda, G.B.; Jambhulkar, N.N.; Yadav, M.K.; Panneerselvam, P.; Kumar, U.; Munda, S.; et al. Imidacloprid Application Changes Microbial Dynamics and Enzymes in Rice Soil. *Ecotoxicol. Environ. Saf.* **2017**, *144*, 123–130. [CrossRef] [PubMed]
184. Meher, S.; Saha, S.; Tiwari, N.; Panneerselvam, P.; Munda, S.; Mahapatra, A.; Jangde, H.K. Herbicide-Mediated Effects on Soil Microbes, Enzymes and Yield in Direct Sown Rice. *Agric. Res.* **2021**, *10*, 592–600. [CrossRef]
185. Fialho, C.M.T.; Silva, A.A.; Melo, C.A.D.; Costa, M.D.; Souza, M.W.R.; Reis, L.A.C. Weed Interference in Soybean Crop Affects Soil Microbial Activity and Biomass. *Planta Daninha* **2020**, *38*, e020221853. [CrossRef]
186. Romero-Trigueros, C.; Díaz-López, M.; Vivaldi, G.A.; Camposeo, S.; Nicolás, E.; Bastida, F. Plant and Soil Microbial Community Responses to Different Water Management Strategies in an Almond Crop. *Sci. Total Environ.* **2021**, *778*, 146148. [CrossRef]
187. Zhang, Y.; Wang, X.; Xu, F.; Song, T.; Du, H.; Gui, Y.; Xu, M.; Cao, Y.; Dang, X.; Rensing, C.; et al. Combining Irrigation Scheme and Phosphorous Application Levels for Grain Yield and Their Impacts on Rhizosphere Microbial Communities of Two Rice Varieties in a Field Trial. *J. Agric. Food Chem.* **2019**, *67*, 10577–10586. [CrossRef]

188. García-Orenes, F.; Caravaca, F.; Morugán-Coronado, A.; Roldán, A. Prolonged Irrigation with Municipal Wastewater Promotes a Persistent and Active Soil Microbial Community in a Semiarid Agroecosystem. *Agric. Water Manag.* **2015**, *149*, 115–122. [CrossRef]
189. Kayikcioglu, H.H. Can Treated Wastewater Be Used as an Alternative Water Resource for Agricultural Irrigation? Changes in Soil and Plant Health after Three Years of Maize Cultivation in Western Anatolia, Turkey. *Appl. Ecol. Environ. Res.* **2018**, *16*, 8131–8161. [CrossRef]
190. Galindo, F.S.; Delate, K.; Heins, B.; Phillips, H.; Smith, A.; Pagliari, P.H. Cropping System and Rotational Grazing Effects on Soil Fertility and Enzymatic Activity in an Integrated Organic Crop-Livestock System. *Agronomy* **2020**, *10*, 803. [CrossRef]
191. Catorci, A.; Ottaviani, G.; Ballelli, S.; Cesaretti, S. Functional differentiation of central apennine grasslands under mowing and grazing disturbance regimes. *Pol. J. Ecol.* **2011**, *59*, 115–128.
192. Zibilske, L.M.; Makus, D.J. Black Oat Cover Crop Management Effects on Soil Temperature and Biological Properties on a Mollisol in Texas, USA. *Geoderma* **2009**, *149*, 379–385. [CrossRef]
193. Kunito, T.; Saeki, K.; Goto, S.; Hayashi, H.; Oyaizu, H.; Matsumoto, S. Copper and Zinc Fractions Affecting Microorganisms in Long-Term Sludge-Amended Soils. *Biores. Technol.* **2001**, *79*, 135–146. [CrossRef] [PubMed]
194. De Santiago-Martín, A.; Cheviron, N.; Quintana, J.R.; González, C.; Lafuente, A.L.; Mougin, C. Metal Contamination Disturbs Biochemical and Microbial Properties of Calcareous Agricultural Soils of the Mediterranean Area. *Arch. Environ. Contam. Toxicol.* **2013**, *64*, 388–398. [CrossRef]
195. Mitter, E.K.; Germida, J.J.; de Freitas, J.R. Impact of diesel and biodiesel contamination on soil microbial community activity and structure. *Sci. Rep.* **2021**, *11*, 10856. [CrossRef]
196. Lin, J.; Ma, K.; Chen, H.; Chen, Z.; Xing, B. Influence of Different Types of Nanomaterials on Soil Enzyme Activity: A Global Meta-Analysis. *Nano Today* **2021**, *42*, 101345. [CrossRef]
197. Sardans, J.; Peñuelas, J.; Estiarte, M. Warming and drought alter soil phosphatase activity and soil P availability in a Mediterranean shrubland. *Plant Soil* **2006**, *289*, 227–238. [CrossRef]
198. Yao, Y.; Dai, Q.; Gao, R.; Gan, Y.; Yi, X. Effects of rainfall intensity on runoff and nutrient loss of gently sloping farmland in a karst area of SW China. *PLoS ONE* **2021**, *16*, e0246505. [CrossRef]
199. Ghiloufi, W.; Chaieb, M. Environmental Factors Controlling Vegetation Attributes, Soil Nutrients and Hydrolases in South Mediterranean Arid Grasslands. *Ecol. Eng.* **2021**, *161*, 106155. [CrossRef]
200. Morugán-Coronado, A.; García-Orenes, F.; McMillan, M.; Pereg, L. The effect of moisture on soil microbial properties and nitrogen cyclers in Mediterranean sweet orange orchards under organic and inorganic fertilization. *Sci. Total Environ.* **2019**, *655*, 158–167. [CrossRef] [PubMed]
201. Sardans, J.; Peñuelas, J. Increasing drought decreases phosphorus availability in an evergreen Mediterranean forest. *Plant Soil* **2004**, *267*, 367–377. [CrossRef]
202. Landesman, W.J.; Dighton, J. Response of Soil Microbial Communities and the Production of Plant-Available Nitrogen to a Two-Year Rainfall Manipulation in the New Jersey Pinelands. *Soil Biol. Biochem.* **2010**, *42*, 1751–1758. [CrossRef]
203. Jaskulska, R. The Level of Luvisols Biochemical Activity in Midfield Shelterbelt and Winter Triticale (x*Triticosecale* Wittm. ex A. Camus) Cultivation. *Agronomy* **2020**, *10*, 1644. [CrossRef]
204. Dey, S.K.; Chakrabarti, B.; Purakayastha, T.J.; Prasanna, R.; Mittal, R.; Singh, S.D.; Pathak, H. Interplay of Phosphorus Doses, Cyanobacterial Inoculation, and Elevated Carbon Dioxide on Yield and Phosphorus Dynamics in Cowpea. *Environ. Monit. Assess.* **2019**, *191*, 223. [CrossRef]
205. Basak, N.; Mandal, B.; Datta, A.; Mitran, T.; Biswas, S.; Dhar, D.; Badole, S.; Saha, B.; Hazra, G.C. Impact of Long-Term Application of Organics, Biological, and Inorganic Fertilizers on Microbial Activities in Rice-Based Cropping System. *Commun. Soil Sci. Plant Anal.* **2017**, *48*, 2390–2401. [CrossRef]
206. de Castro Lopes, A.; Gomes de Sousa, D.M.; Chaer, G.M.; Bueno dos Reis Junior, F.; Goedert, W.J.; de Carvalho Mendes, I. Interpretation of Microbial Soil Indicators as a Function of Crop Yield and Organic Carbon. *Soil Sci. Soc. Am. J.* **2013**, *77*, 461. [CrossRef]
207. Tarafdar, J.C.; Rao, A.V. Contribution of Aspergillus Strains to Acquisition of Phosphorus by Wheat (*Triticum aestivum* L.) and Chick Pea (*Cicer arietinum* Linn.) Grown in a Loamy Sand Soil. *Appl. Soil Ecol.* **1996**, *3*, 109–114. [CrossRef]
208. Moharana, P.C.; Biswas, D.R. Phosphorus Delivery Potential in Soil Amended with Rock Phosphate Enriched Composts of Variable Crop Residues under Wheat–Green Gram Cropping Sequence. *Commun. Soil Sci. Plant Anal.* **2022**, *53*, 1000–1017. [CrossRef]

MDPI AG
Grosspeteranlage 5
4052 Basel
Switzerland
Tel.: +41 61 683 77 34

Agriculture Editorial Office
E-mail: agriculture@mdpi.com
www.mdpi.com/journal/agriculture

www.ingramcontent.com/pod-product-compliance
Lightning Source LLC
Chambersburg PA
CBHW061517210926
51850CB00008B/1952
9783725861798